CELL BIOLOGY RESEARCH PROGRESS

NEW DEVELOPMENTS ON SIGNAL TRANSDUCTION RESEARCH

CELL BIOLOGY RESEARCH PROGRESS

Additional books in this series can be found on Nova's website
under the Series tab.

Additional e-books in this series can be found on Nova's website
under the e-book tab.

CELL BIOLOGY RESEARCH PROGRESS

NEW DEVELOPMENTS ON SIGNAL TRANSDUCTION RESEARCH

MASAYOSHI YAMAGUCHI
EDITOR

NOVA BIOMEDICAL

New York

For permission to use material from this book please contact us:
Telephone 631-231-7269; Fax 631-231-8175
Web Site: http://www.novapublishers.com

Additional color graphics may be available in the e-book version of this book.

Library of Congress Cataloging-in-Publication Data

New developments on signal transduction research / editor, Masayoshi Yamaguchi (Division of Diabetes, Endocrinology and Metabolism, Department of Medicine, Baylor College of Medicine, Houston, TX, USA).

pages cm

Includes bibliographical references and index.

ISBN 978-1-62618-230-1 (hardcover)

1. Cellular signal transduction. 2. Cell interaction. I. Yamaguchi, Masayoshi, 1947-

QP517.C45N48 2013

571.7'4--dc23

2013006216

Published by Nova Science Publishers, Inc. † New York

CONTENTS

PREFACE

Signal transduction plays a pivotal role in cell regulation. Hormones, cytokines and neurotransmitters bind to their specific receptors in the plasma membranes of cells. The receptors of many hormones are coupled to G-protein coupled signaling pathways. Insulin and growth factors are coupled to receptor-linked kinase. The second messenger is generated in cells once the first messenger binds to the receptors of plasma membranes. Cyclic adenosine monophosphate (cyclic AMP), inositol 1,3,5-triphosphate (IP3) and calcium ion (Ca^{2+}) have been demonstrated to play a role as a second messenger in cells. The pivotal role of Ca^{2+} in cellular regulation was established with the finding of calmodulin and protein kinase C that modulates the effect of Ca^{2+} in the regulation of cellular functions. Many receptor classes and their downstream kinases activate Erk, Akt, Stat, Smad, NF-κB and other cascades to regulate gene expression and cellular function. This book provides recent research in the study of the new developments in signaling transduction in cells.

Chapter 1 - Calcium signal transduction is central to the growth and survival of all cells. As an intracellular messenger, calcium primarily works by binding to and regulating a diversity of calcium-binding proteins, of which calmodulin (CaM) is the most common. The central role of the calcium-sensor CaM has been well chronicled in a vast diversity of cellular functions in animals, plants and eukaryotic microbes. In response to increased levels of calcium, apo-CaM binds up to four Ca^{2+} undergoing a major conformational change as Ca^{2+}-CaM. Both apo-CaM and Ca^{2+}-CaM each bind to a specific complement of CaM-binding proteins (CaMBPs). While the roles of intracellular CaM and its CaMBPs have been well documented, the existence as well as the functions of these as extracellular proteins has been both controversial and enigmatic. This review summarizes the roles of extracellular CaM (extCaM) and extracellular CaMBPs (extCaMBPs) in animals, plants and eukaryotic microbes. The authors work with the amoebozoan *Dictyostelium*, which is the primary focus of this chapter, has shown that extCaM serves several functions in this organism. It regulates growth and chemotaxis. It also binds to and inhibits the proteolysis of at least one extCaMBP, the cysteine-rich matricellular protein CyrA. The presence of extracellular CaM with proven targets and functions in a eukaryotic microbe suggests that extracellular CaM is as ubiquitous and important evolutionarily as cytoplasmic CaM.

Chapter 2 - The loss of neurons in neurodegenerative disease leads to devastating loss of function. Because of their ability to give rise to new neurons, neural progenitors offer hope for functional restoration of neurons in neurodegenerative disease, either through pharmacologic intervention that induces adult neural stem cells to differentiate in situ, or by transplanting neurons differentiated in vitro. In order to advance these efforts, multiple groups are working to define the complex signaling pathways that govern neuronal differentiation. These pathways are activated by self-renewal and differentiation cues, typically in the form of growth factors or morphogens that bind to cell-surface receptors. Expression of growth factors and their receptors is dynamically regulated in the nervous system, establishing tightly controlled spatial and temporal gradients of activity. In the past decade, kinase-coupled receptors such as TGFβ, Lif, and FGF receptors have been clearly established as key regulators of stem cell multipotency and differentiation. These receptor classes and their downstream kinases activate Erk, Akt, Stat, and Smad cascades to regulate gene expression and cellular function. More recent studies highlight the overlooked importance of signaling through G protein coupled receptors (GPCRs), whose downstream effects overlap significantly with kinase-coupled receptors. Specifically, GPCRs have also been shown to regulate Smad, Stat, Akt and Erk pathways. The GPCR family is large and diverse, and specific GPCRs and their ligands have been shown to regulate stem cells from various sources, including neuronal stem cells. This chapter will highlight convergent kinase-coupled and G protein-coupled signaling pathways, and highlight examples of co-regulation of neural stem cell differentiation and multipotency by both receptor types.

Chapter 3 - For most moths, the ability of conspecific males to locate receptive females is governed by the detection of a blend of semiochemicals known as sex pheromones. Sex pheromone components are *de novo* synthesized in the female pheromone gland in response to pheromone biosynthesis activating neuropeptide (PBAN). The pheromonotropic effects of PBAN are mediated by a G protein-coupled receptor predominantly expressed in the pheromone gland, termed the PBAN receptor (PBANR). PBANR activation triggers the mobilization of extracellular Ca^{2+}, which activates a signal transduction cascade that activates the enzymatic processes involved in sex pheromone production. PBANR was initially identified in *Helicoverpa zea* and *Bombyx mori* based on sequence similarities with the mammalian neuromedin U family of receptors. Since then PBANRs have been identified in a number of other moth species. Functionally relevant structural variations suggested the presence of two PBANR isoforms, one expressed in moths using non-cAMP dependent biosynthetic pathways and the other in moths using cAMP dependent pathways. Recent studies, however, have shown that alternative splicing of the PBANR gene generates multiple variants, including the two initial isoforms that are concomitantly expressed in moth pheromone glands. Given the importance of sex pheromones to moth mating behaviors, PBANR has been suggested as a potential candidate for targeted disruption. This chapter provides a brief overview of PBANR identification and the role molecular dissection of the PBANR sequence has had in elucidating critical structure-function relationships. Recent developments in generating PBANR agonists/antagonists and the potential for RNAi interference-based disruption of PBANR functionality are also discussed.

Chapter 4 - Cyclin dependent kinase-5 (cdk5) belongs to the family of cyclin dependent kinases (cdks) based on its sequence homology, varying between 60-70 percent, with the other cdks. However, its activity is primarily restricted to neuronal cells. It is activated by neuron-specific molecules p35 and p39 with molecular weights of 35kD and 39kD, respectively, which have similar cyclin structures. Cdk5 has been shown to be a multifunctional protein kinase that regulates a large number of neuronal processes including synaptic activity, neural system development, and memory and learning. Cdk5 is not only necessary for nervous system development but is also essential for survival. The knockout of cdk5 is embryonically lethal. It has been shown that cdk5 not only phosphorylates its substrates such as NF, tau, and other cytoskeletal molecules, but also regulates the activity of kinases and phosphatases by kinase-phosphatase cross-talk. In this review, the cross-talk of cdk5 with different kinases and phosphatases and their role in the nervous system function and development is described.

Chapter 5 - A wide variety of environmental chemicals and drugs can interfere with normal embryonic development, leading to malformations and other anomalies. While the underlying cellular and molecular mechanisms are not always clear, there have been numerous studies implicating disruptions in signal transduction pathways in the induction of abnormalities by embryotoxic compounds. The types of intracellular signalling pathways influenced by teratogens are quite diverse and, in many cases, the pathway is activated by treatment. The Mitogen-Activated Protein Kinase family (MAPK) includes members that influence both cell proliferation and cell death. Stress-induced members of this family (c-Jun N-terminal kinase, (JNK) and p38) have been found to be induced by exposure of vertebrate embryos to toxic metals, cyclophosphamide, hydroxyurea, perfluorooctane sulfonate (PFOS) and staurosporine. The activation of these stress proteins is often linked to reactive oxygen species and oxidative stress, and typically culminates in apoptosis. Activation of Ras, which typically leads to cell growth and differentiation, is involved in phenytoin embryopathy. Rather than being activated, other pathways are inhibited. The Protein Kinase B/Akt pathway, which is linked to cell survival, has been found to be down-regulated after limb cells are exposed to the well-known teratogen, thalidomide. Developmentally important signalling pathways, such as Sonic Hedgehog (Shh), are disrupted by exposure to ethanol and retinoic acid. The primary consequences of the induction of many of these pathways include alterations in cell proliferation, differentiation, apoptosis and cell migration. These changes, in turn, disrupt the normal processes of embryonic development, leading to malformations or embryonic death.

This chapter will discuss the major disturbances in cell signalling by treatments that interfere with embryogenesis, with the aim of determining whether commonalities exist among the toxins, outcomes and pathways affected that would help to explain the mechanisms of toxin-induced abnormal development.

Chapter 6 - Regucalcin was discovered in 1978 as a novel calcium-binding protein that does not contain EF-hand motif of calcium-binding domain, which differs from calmodulin and other calcium-related proteins. The name of regucalcin was proposed for this calcium-binding protein, which regulates various Ca^{2+}- or Ca^{2+}/calmodulin-dependent enzyme activations. Regucalcin and its gene (RGN) are identified in over 15 species consisting of regucalcin family. The regucalcin gene is localized on the chromosome X, and the organization of the regucalcin gene consists of seven exons and six introns. AP-1, NF1-A1, and RGPR-p117, which is a transcription factor, bind to the promoter region of the regucalcin gene and enhance transcription activity of the regucalcin gene expression that is mediated through calcium and other signalings. Regucalcin mRNA expression and its protein content are pronounced in the liver, kidney cortex, and other tissues. Regucalcin has been demonstrated to play a multifunctional role in cell regulation: regulation of intracellular Ca^{2+} homeostasis, inhibition of Ca^{2+}-dependent and independent enzyme activations including protein kinases and protein phosphatases, suppression of nuclear signaling-related deoxyribonucleic acid (DNA) and ribonucleic acid (RNA) synthesis, and depression of apoptotic cell death and cell proliferation induced by various signaling factors in many cell types in vitro. Morever, regucalcin has been shown to have a suppressive effect on protein synthesis and an activating effect on protein degradation. Regucalcin suppresses the over-proliferation in regenerating liver cells with partial hepatoectomy and the development of carcinogenesis in liver tissues in vivo. Regucalcin is proposed to play a pivotal role in maintaining of cell homeostasis and function. The revelation of these effects is related to role of regucalcin as a suppressor in various cell-signaling systems. Regucalcin is the fisrt time finding as a suppressor protein in cell signaling. This chapter has been written to outline the recent advances that have been made concerning the role of regucalcin as a suppressor protein in cell signaling in liver, kidney, and other tissues.

Chapter 7 - By use of Paraquat (PQ) treatment during germination of pea seeds the authors obtained a high yielding mutants in the preceding works. The germinating seeds of *Oryza sativa* cv Koshihikari were treated in the presence of 4 μM of PQ, and the authors isolated PQ resistant mutants, R4-2-4 and R4-5-8 with high yielding. The two mutant lines, R4-2-4 and R4-5-8 showed significantly increased tiller numbers, ear numbers and ear weights and the leaves were deeper in green color than those of original wild type line. The mutant R4-2-4 and R4-5-8 lines showed resistant phenotype to PQ at the concentrations of 1-4 μM exhibiting good growth ability with green color of cotyledon. However, at the same concentration of PQ the seeds of original wild type showed germination with white cotyledon with no further growth ability. The leaves from both mutants were isolated and the crude extracts were prepared and these fractions were phosphorylated followed by the SDS-PAGE fractionation.

The increase in the phosphorylation bands in low molecular weights corresponding to the precursor of nucleoside diphospate kinase (NDPK) were observed in both mutant lines, and also increase in the phosphorylation bands in the higher molecular weights corresponding to high molecular weight histidine kinases could be detected in both of the mutant lines.

However, in the original wild type increased amounts of putative degradation product of phosphorylated precursor of NDPK could be detected. By the further fractionation of the soluble, chloroplast and mitochondrial fractions, the increase in the phosphorylation of protein bands corresponding to putative precursor of NDPK and higher molecular weight histidine kinases in the fractions of chloroplast could be observed. The nucleotide sequence analyses of the precursor of NDPK1, 2 and 3 showed no difference in the nucleotide sequence with those of original wild type line and mutant lines, R4-2-4 and R4-5-8. However, the authors detected that 78 amino acid sequence spanning in the transit peptide of Nipponbare NDPK2 was deleted in the Koshihikari transit peptide of NDPK2. The yield tests of original wild type and a mutant R4-2-4 with Kinuhikari as a reference line performed in Kanagawa Agricultural Technology Center demonstrated higher yielding of 1.11-, 1.12- and 1.35-fold than those of original line.

The authors estimate that higher molecular weight histidine kinase and the putative precursor of NDPKs in cytosol and on chloroplast may have capacity to function in the process of detoxification of reactive oxygen species including singlet oxygen ($^1O_2$).

In: New Developments on Signal Transduction Research
Editor: Masayoshi Yamaguchi
ISBN: 978-1-62618-230-1

Chapter 1

SIGNAL TRANSDUCTION MEDIATED VIA EXTRACELLULAR CALMODULIN AND EXTRACELLULAR CALMODULIN BINDING PROTEINS

Danton H. O'Day *

Department of Biology, University of Toronto Mississauga, Mississauga, Ontario, Canada
Department of Cell and Systems Biology, University of Toronto, Toronto, Ontario, Canada

ABSTRACT

Calcium signal transduction is central to the growth and survival of all cells. As an intracellular messenger, calcium primarily works by binding to and regulating a diversity of calcium-binding proteins, of which calmodulin (CaM) is the most common. The central role of the calcium-sensor CaM has been well chronicled in a vast diversity of cellular functions in animals, plants and eukaryotic microbes. In response to increased levels of calcium, apo-CaM binds up to four Ca^{2+} undergoing a major conformational change as Ca^{2+}-CaM. Both apo-CaM and Ca^{2+}-CaM each bind to a specific complement of CaM-binding proteins (CaMBPs). While the roles of intracellular CaM and its CaMBPs have been well documented, the existence as well as the functions of these as extracellular proteins has been both controversial and enigmatic. This review summarizes the roles of extracellular CaM (extCaM) and extracellular CaMBPs (extCaMBPs) in animals, plants and eukaryotic microbes. Our work with the amoebozoan *Dictyostelium*, which is the primary focus of this chapter, has shown that extCaM serves several functions in this organism. It regulates growth and chemotaxis. It also binds to and inhibits the proteolysis of at least one extCaMBP, the cysteine-rich matricellular protein CyrA. The presence of extracellular CaM with proven targets and functions in a eukaryotic microbe suggests that extracellular CaM is as ubiquitous and important evolutionarily as cytoplasmic CaM.

* Correspondence to: Dr. Danton H. O'Day; Email: danton.oday@utoronto.ca.

Keywords: extracellular calmodulin, calmodulin binding proteins, *Dictyostelium*, social amoebozoan, chemotaxis, matricellular

INTRODUCTION

Calmodulin (CaM) is a ubiquitous calcium-binding protein found in essentially all eukaryotes studied to date. Calcium-free or apo-CaM undergoes a dramatic conformational change in response to binding between one and four calcium ions as Ca^{2+}/CaM. CaM works by binding to and regulating calmodulin binding proteins (CaMBPs). Both apo-CaM and Ca^{2+}/CaM can each bind to a wide variety of calcium-independent and calcium-dependent CaMBPs respectively (Figure 1). The presence of intracellular CaM is well detailed and its diverse functions are widely known (Cohen and Klee, 1988; Vogel, 1994). In contrast, the existence of extracellular CaM (extCaM) has been studied in comparatively few different organisms.

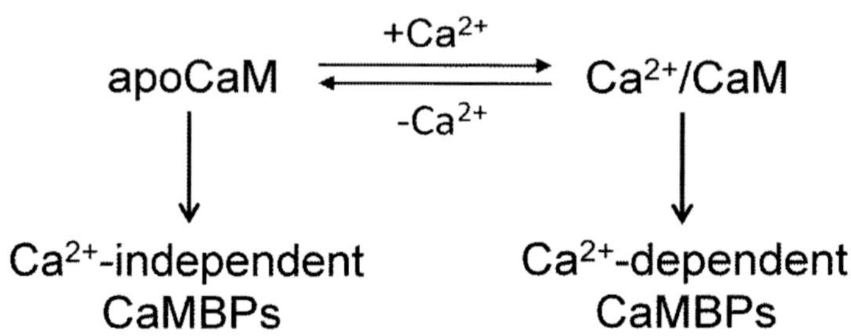

Figure 1. Calcium regulation of CaM and its effect on CaMBP binding.

The presence of extCaM has been conclusively demonstrated in several plant species where it has been shown to regulate cell proliferation (Sun et al, 1994), cell wall regeneration (Sun et al, 1995), gene activity (Zhou et al, 2001), germination (Sun et al, 2009) and stomatal closure (Li et al, 2009). In animals the literature is less focussed but the accumulated data argues strongly that extracellular CaM is a critical signaling component. For example, extCaM mediates nerve regeneration in frogs (Remgard et al, 1995). In hamsters, exogenous CaM amplifies peptide-mediated vasodilation (Ikezaki et al, 1999). Treatment with exogenous CaM stimulates DNA synthesis in rat liver cells (Boynton et al, 1980). In humans, extCaM appears to be required for cell division in the early preimplantation embryo (Woodward et al, 1993) and keratinocytes (Goberdhan et al, 1993). It also increases the proliferation of umbilical vein endothelial cells (Dawson and MacNeil, 1992).

Dictyostelium CaM (DdCaM) is encoded by a single gene (*calA*; DDB_G0279407). A single study indicates that DdCaM knockouts are lethal as they are in other organisms (Clarke, 1990). Intracellular DdCaM has been shown to regulate chemotaxis, development, spore germination as well as protein phosphorylation, among other events in *Dictyostelium* (O'Day, 2003). A large number of calcium-independent and calcium-dependent CaMBPs have been detected but only a handful of them have been identified and characterized (O'Day, 2003; Catalano and O'Day, 2008). As summarized in those reviews, CaMBPs have been detected in various cellular compartments: cytoplasm (calcineurin, DGAP1, DWWA, myosins, phosphoglycerate kinase, ribosomal protein L19, spectrin/fodrin, thymidine kinase), nucleus (histone H1, nucleomorphin), nucleolus (BAF60a/SMARCD1 homologue, nucleomorphin; Catalano and O'Day, 2012), and the contractile vacuole (VwkA).

CURRENT RESEARCH

Evidence for Extracellular Calmodulin in Dictyostelium

The presence of extracellular CaM (extCaM) in *Dictyostelium* has only recently been verified (O'Day et al, 2012). A proteomic analysis of the extracellular medium indicated the presence of DdCaM as well as the calcium binding protein cbp4a and some putative CaMBPs (Bakthavatsalam and Gomer, 2010). Subsequent work by our group revealed that CaM is secreted throughout growth and development where it regulates several different events. This review will focus on work done to date on extCaM and extCaMBPs in this social amoebozoan and will end with some of the questions, related to their role in signal transduction, that arise from this work.

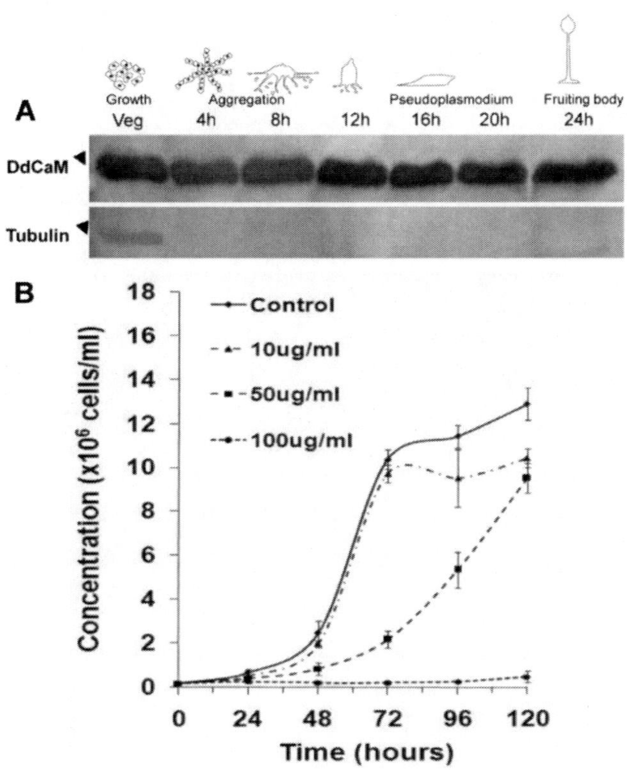

Figure 2. Extracellular CaM expression through growth and development (A) and the effects of exogenous bovine CaM on cell proliferation (B) in *Dictyostelium* (Suarez et al., 2011; O'Day et al, 2012).

Functions of Extracellular CaM during Growth

ExtCaM protein was detected throughout growth and development indicating that it might function at any and all stages (Figure 2, A). To examine the role of extCaM during the growth phase, proliferating cells were treated with different concentrations of bovine CaM (Figure 2,B). Bovine CaM has been show to share high sequence identity with and proven to be functionally identical to *Dictyostelium* CaM (Lydan and O'Day, 1994). The addition of bovine CaM dose-dependently inhibited cell proliferation of cells in axenic cultures with 100ug/ml being almost completely inhibitory. Depending on the organism (Boynton et al, 1980; Dawson and MacNeil, 1992; Goberdhan et al, 1993; Sun et al, 1994; Woodward et al, 1993) those and other results have led to the suggestion that extCaM acts as an autocrine growth factor.

EXTRACELLULAR CAM INCREASES CAMP-MEDIATED CHEMOTAXIS

The role of calcium and CaM in mediating chemotaxis has been known for many years mainly through results from pharmacological studies (Gauthier and O'Day, 2001). However, the importance of CaM was attributed to its intracellular locale since extracellular CaM was not known to exist. However, it is now clear that extracellular CaM dose-dependently enhances the rate of cAMP-mediated chemotaxis, providing another function for extCaM in *Dictyostelium* (Figure 3).

Secretion and Processing of CyrA, a CaM-Binding Protein

CyrA was first isolated using the CaM-binding overlay technique (CaMBOT) suggesting that it was a CaMBP (O'Day, 2003). Scanning of the sequence of CyrA further revealed the presence of a putative CaM-binding domain (CaMBD) between amino acids 249 to 275 (Figure 4A). Using CaM-agarose binding and CaM immunoprecipitation, it was subsequently shown that full-length CyrA and the predominant CyrA cleavage products all bind to CaM both intra- and extracellularly (Huber et al, 2012; Suarez et al., 2011).

Another defining aspect of CyrA is the presence of 4 tandem EGF-like (EGFL) repeats in its C-terminus. Detailed and careful studies by Robert Huber and colleagues verified that the first EGFL repeat increases both random cell motility and chemotaxis in *Dictyostelium* via binding to a surface receptor to activate specific downstream signaling events (Huber and O'Day, 2011, 2012a,b,c;).

In total, this work validated CyrA as the first matricellular calmodulin binding protein in a eukaryote (Huber and O'Day, 2012b,c). Developmental analyses showed that CyrA is secreted and processed to release two smaller EGFL repeat-containing polypeptides, CyrA-45 and CyrA-40 (Figure 4).

Bioinformatic analyses had indicated that CyrA was likely secreted since SignalP revealed the presence of a signal peptide in the N-terminus and TargetP revealed the protein is targeted to the secretory pathway (Figure 4).

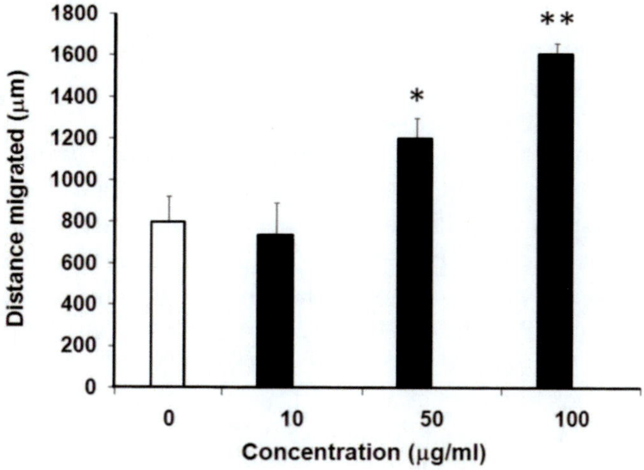

Figure 3. Extracellular CaM and its effect on chemotaxis. Exogenous CaM increases the rate of cAMP-mediated chemotaxis (O'Day et al, 2012). Data presented as the mean distance migrated ± s.e.m (n=4). Mann-Whitney two-sample rank U-test (*p=0.01 vs. 0 μg/ml; **p=0.005 vs. 0 μg/ml).

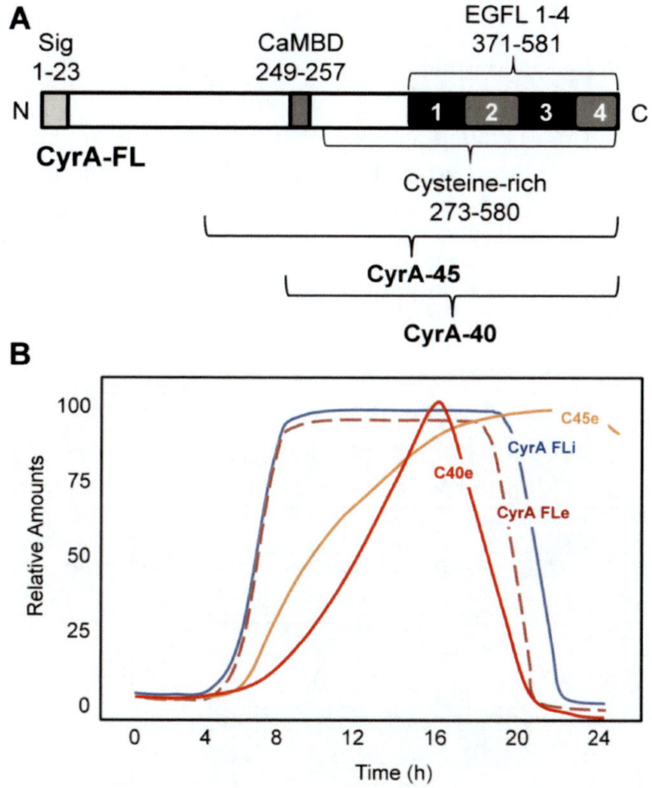

Figure 4. The structure of CyrA and its extracellular processing during development. (A) Full-length CyrA (CyrA-FL) possesses a signal sequence (Sig), a CaMBD and four tandem EGF-like (EGFL) repeats in its C-terminus. The approximate positions of the two major proteolytic products of CyrA of ~45KDa (CyrA-45) and ~40kDa (CyrA-40) are indicated. (B) Full length CyrA is present within the cells (CyrA-FLi) and is temporally secreted (CyrA-FLi). The two major proteolytic products of CyrA CyrA-45 (C45e) and ~40kDa (C40e) are produced during development in different temporal patterns.

Experimental support was provided with the localization of both endogenous CyrA and CyrA-GFP to the ER showing that it is secreted via a conventional secretory pathway (i.e., involving endoplasmic reticulum, Golgi, and vesicle release; Huber et al, 2012). Secretion of the protein itself was verified by western blotting with CyrA being secreted during both growth and development (Figure 4B). A pharmacological analysis further showed that CyrA secretion was dependent on intracellular Ca^{2+} release and active CaM (Huber et al, 2012).

CyrA is thus the first extracellular CaMBP to be experimentally verified in *Dictyostelium* or any other eukaryotic microbe. However, a proteomic analysis did reveal that several putative CaMBPs were present in the extracellular environment (Bakthavatsalam and Gomer, 2010). These include CaMBP15 (DDB0001413) and CaMBP46 (DDB0167501). In addition, puromycin-sensitive aminopeptidase-like protein (DDB0231240) was also found in the extracellular medium. Each of these proteins has been studied in our lab and has been verified as a true CaMBP via CaM-binding and other assays (Catalano and O'Day, 2008; Catalano et al, 2012; O'Day, unpublished).

Both CaM and CyrA Localize to the Extracellular Matrix during Development

During development, a motile multicellular structure known as a pseudoplasmodium is formed that migrates along the substratum in response to light and temperature. Isolation of the extracellular matrix (ECM; called a slime sheath) that surrounds this multicellular tissue, followed by western blotting revealed the presence of CaM, full-length CyrA (CyrA-FL) and CyrA cleavage produces (Figure 5, A; e.g., CyrA-C45, -C40; Huber et al, 2012; Suarez et al., 2011).

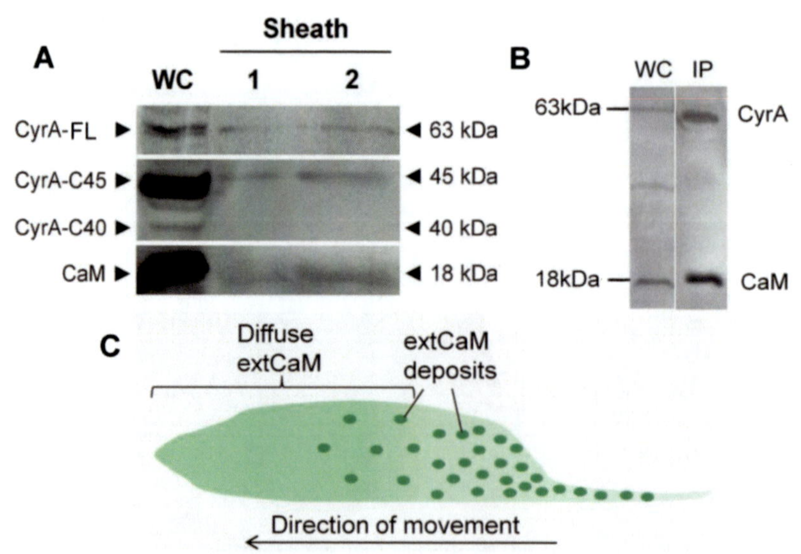

Figure 5. CaM and CyrA in the extracellular matrix. (A) Presence of CyrA and CaM in isolated slime sheath as shown via western blotting. (B) CyrA and CaM are both detected in CaM co-immunoprecipitates. (C) Diagrammatic summary of extCaM localization in the ECM of the pseudoplasmodium (O'Day et al, 2012).

Immunoprecipitation of extracellular CyrA pulled down both CyrA and CaM indicating that they are bound together in the ECM (Figure 5, B; Huber et al, 2012; Suarez et al, 2011). During development, extCaM is diffusely localized throughout the pseudoplasmodium in a gradient that is highest at the anterior end (Figure 5C). This apparently soluble extCaM becomes deposited in insoluble precipitates at the back of the pseudoplasmodium suggesting that the availability of extCaM in the ECM and its effects on CyrA processing and cell chemotaxis will vary along the length of this multicellular stage of development (Suarez et al., 2011; O'Day et al, 2012).

Extracellular CaM Inhibits CyrA Proteolysis

The ability of intracellular CaM to protect proteins from proteolysis has been described in several different organisms (Reddy et al., 2006; Suarez et al., 2011; Zakharov and Mosevitsky, 2007; Zhang et al., 2009). However, the ability of extCaM to similarly protect against proteolysis in the extracellular environment had not been studied. As seen in Figure 6, in the presence of CaM, CyrA proteolysis is inhibited such that the major products (i.e., CyrA-45, -40) are generated at much lower levels (Suarez et al, 2011). To our knowledge, CyrA represents the first extracellular CaMBP whose proteolytic processing is protected by CaM extracellularly.

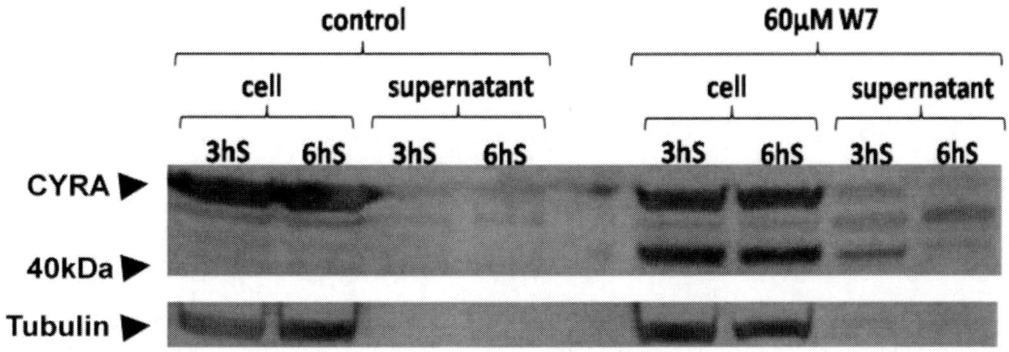

Figure 6. The inhibition of CyrA processing by CaM. Cells were either untreated (left side) or treated (right side) with 60mM of the CaM antagonist W7 after which cells and extracellular medium were isolated and subjected to western blotting (Suarez et al 2011).

Conclusion

As summarized in Figure 7, calmodulin serves multiple extracellular functions in *Dictyostelium*. In this chapter, the focus has been on the functions of extracellular calmodulin and its relationship with a secreted CaM-binding, matricellular protein, CyrA.

During growth, CaM is secreted and the extracellular CaM (extCaM) feeds back to inhibit growth. Secretion of CaM continues once cells are starved and embark on multicellular development. During early development extCaM enhances the chemotactic response of cells to cAMP. The mechanism by which CaM inhibits cell proliferation and stimulates chemotaxis in *Dictyostelium* remains unknown but likely involves its calcium-regulated interaction with specific CaMBPs. The CaMBP CyrA is also secreted as multicellular development progresses and this secretion is CaM-dependent. As development continues, extCaM and CyrA become incorporated into the extracellular matrix that surrounds the multicellular tissue called a pseudoplasmodium. At this time, CyrA is proteolytically processed to release EGFL-containing polypeptides that enhance cAMP-mediated chemotaxis. ExtCaM binds to CyrA within the ECM and this binding protects CyrA from proteolysis, slowing the release of these EGFL-containing fragments. In spite of this insight into the role of extCaM, many questions remain to be answered.

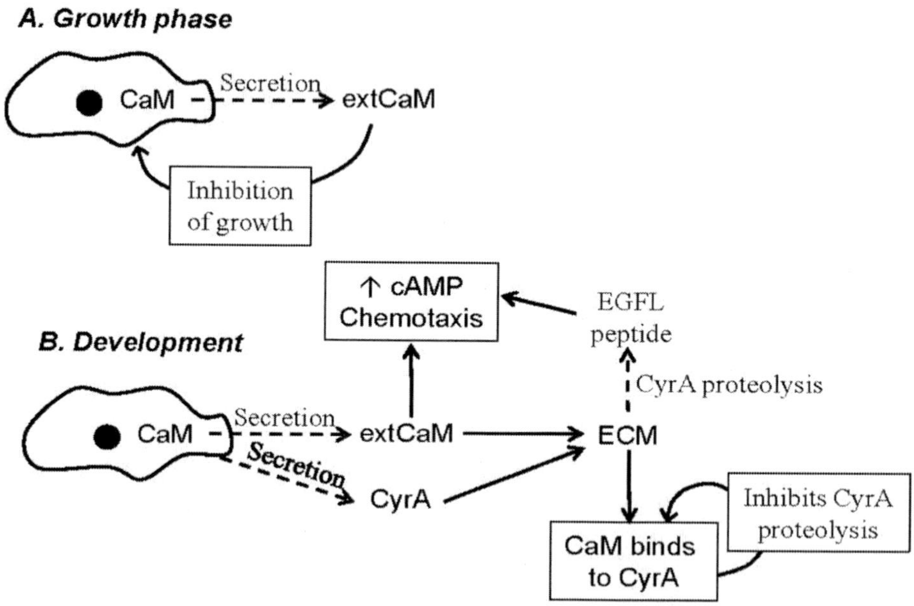

Figure 7. A summary of the extracellular functions of calmodulin and the calmodulin binding protein CyrA (O'Day et al, 2012).

How is CaM Translocated?

Typically proteins are targeted to different locales within cells and extracellularly by specific targeting sequences. Thus cytoplasmic proteins can be targeted to the endoplasmic reticulum via the KDEL sequence while mannose-6-phosphate residues target certain proteins to lysosomes. Targeting to the nucleus involves nuclear localisation signals (NLS) with targeting to the nucleolus being less clear and either utilizing related NLSs, other sequences of positive amino acids and as yet undefined mechanisms (O'Day and Catalano, 2013). Signal sequences commonly lead to protein secretion. CaM lacks any such sequences yet it has been verified in the cytoplasm, specific organelles, within the nucleus and in the extracellular environment (Figure 8). The presence of CaMBPs in the nucleolus suggests that CaM also translocates there if only temporarily. The question remains how does CaM enter all of these diverse locations to exert its role as a regulator of its target proteins? One possibility is that it translocates along with specific CaMBPs that do possess specific targeting sequences. In terms of secretion of extCaM, there is evidence a novel secretion mechanism is present in *Dictyostelium* that involves the contractile vacuole. This may be how extracellular CaM is released since CaM is associated with contractile vacuoles along with the CaMBP VwkA (see, Catalano and O'Day, 2008).

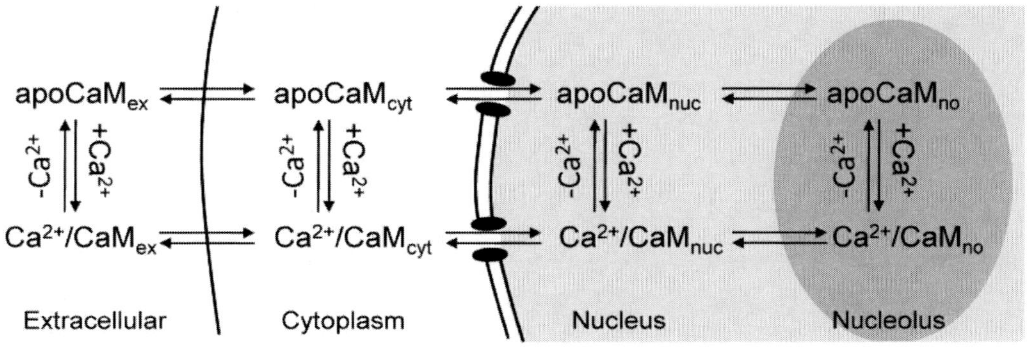

Figure 8. The potential translocations available to apoCaM and Ca^{2+}/CaM in *Dictyostelium*.

How is extCaM Signaling Regulated?

Another issue that arises is the signaling mechanisms that regulate extCaM function. Inside the cell calcium levels are precisely controlled with specific signaling events leading to pulses of localized calcium ions which in turn would bind to apo-CaM. This transition to calcium/CaM would cause a shift in CaMBP binding which in turn would generate the signal's response. Regulating the extracellular levels of calcium is a much more challenging task, especially in a eukaryotic microbe, and likely doesn't occur on a regular basis if at all. Thus the binding of extCaM to its targets and thus its functionality would be a more-or-less constant phenomenon dictated by calcium levels extracellularly.

While the afore-mentioned questions remain, this work reveals that extracellular CaM serves a number of different important functions during both growth and development and does this in conjunction with at least one extracellular CaMBP (CyrA). This work suggests that extCaM may be just as ubiquitous and significant a signaling molecule as it is within cells thus significantly widening understanding of the field of calcium-CaM signaling in eukaryotes.

ACKNOWLEDGMENTS

This work was supported by a Natural Sciences and Engineering Research Council of Canada grant (A6807).

REFERENCES

Bakthavatsalam, D., and R.H. Gomer, 2010. The secreted proteome profile of developing *Dictyostelium discoideum* cells. *Proteomics* 10: 2556-2559.

Boynton, A.L., J.F. Whitfield, and J.P. McManus, 1980. Calmodulin stimulated DNA synthesis by rat liver cells. *Biochem. Biophys. Res. Commun.* 95: 745-749.

Catalano, A., Y. Poloz and D. H. O'Day, 2011. *Dictyostelium* puromycin-sensitive aminopeptidase A is a nucleoplasmic nucleomorphin-binding protein that relocates to the cytoplasm during mitosis. *Histochem. Cell Biol.* 136: 677–688.

Catalano, A. and D.H. O'Day, 2012. Nucleoplasmic/nucleolar translocation and identification of a nuclear localization signal (NLS) in *Dictyostelium* BAF60a/SMARCD1 homologue Snf12. *Histochem. Cell Biol.* 138: 515-530.

Catalano, A. and D.H. O'Day, 2008. Calmodulin-binding proteins in the model organism *Dictyostelium*: a complete and critical review. *Cell. Signal.* 20: 277-291.

Clarke, M., 1990. Calmodulin structure, localization and expression in *Dictyostelium discoideum*, In: D.H. O'Day (Eds.), *Calcium as an Intracellular Messenger in Eukaryotic Microbes*, ASM Press, Washington, D.C., pp. 375-391.

Cohen, P. and C. B. Klee, *Calmodulin*, Elsevier, Amsterdam, the Netherlands, 1988.

Dawson, R.A. and S. MacNeil, 1992. Mitogenic role for extracellular calmodulin-like activity in normal human umbilical vein endothelial cells. *Brit. J. Hematol.* 82: 151-160.

Gauthier, M.L. and D.H.O'Day, 2001. Detection of calmodulin-binding proteins and calmodulin-dependent phosphorylation linked to calmodulin-dependent chemotaxis to folic and cAMP in *Dictyostelium. Cell Signal.* 13: 575-584.

Goberdhan, N.J., R.A. Dawson, E. Freedlander, and S. MacNeil, 1993. A calmodulin-like proteins as an extracellular mitogen for the keratinocyte. *Brit. J. Dermatol.* 129: 678-688.

Huber, R.J., and D.H. O'Day, D.H., 2011. EGF-like peptide-enhanced cell motility in *Dictyostelium* functions independently of the cAMP-mediated pathway and requires active Ca^{2+}/calmodulin signaling. *Cell. Signal.* 23, 731-738.

Huber, R.J., and D.H. O'Day, D.H., 2012a. EGF-like peptide-enhanced cell movement in *Dictyostelium* is mediated by protein kinases and the activity of several cytoskeletal proteins. *Cellular Signalling* 24: 1770-1580.

Huber, R.J. and D.H. O'Day, 2012b. A Matricellular protein and EGF-like repeat signalling in the social amoebozoan *Dictyostelium discoideum.* Cell. Mol. Life Sci. (in press; 19June2012) (DOI: 10.1007/s00018-012-1068-4)

Huber, R.J. A. Suarez, and D. H. O'Day, 2012c. CyrA, a matricellular protein that modulates cell motility in *Dictyostelium discoideum. Matrix Biol.* 3: 271-280.

Ikezaki, H., M. Patel, H. Onyuksel, S.R. Akhter, X.P. Gao, and I. Rubenstein, 1999. Exogenous calmodulin potentiates vasodilation elicited by phospholipip-associated VIP in vivo. *Am. J. Physiol. Regul. Integr. Comp. Physiol.* 276, 1359-1365.

Li, J-H., Y-Q Liu, P. Lü, et al, 2009. A signaling pathway linking nitric oxide production to heterotrimeric G protein and hydrogen peroxide regulates extracellular calmodulin induction of stomatal closure in *Arabidopsis. Plant Physiol.* 150: 114-124.

Lydan, M.A. and D.H. O'Day, 1994. ^{35}S labelling of recombinant calmodulin in E. coli. Chapter 36 In: Methods in Molecular Biology, Protocols for Gene Analysis 31: 389-396. J.M. Walker and A.J. Harwood, eds. Humana Press, Clifton N.J.

O'Day, D.H., 2003. CaMBOT: Profiling and characterizing calmodulin binding proteins. *Cell. Signal.* 15: 347-355.

O'Day, D.H. and A. Catalano, editors, 2013. *Proteins of the Nucleolus: Regulation, Translocation and Biomedical Functions*. Springer Publishing Company, UK. (in press)

O'Day, D.H., R.J. Huber and A. Suarez, 2012. Extracellular Calmodulin regulates growth and cAMP-mediated chemotaxis in *Dictyostelium discoideum. Biochem. Biophys. Res. Commun.* In press (online; dx.doi.org/10.1016/j.bbrc.2012.07.147)

Reddy, G.P., E.R. Barrack, Q.P. Dou, M. Menon, R. Pelley, F.H. Sarkar, and S. Shijie, 2006. Regulatory processes affecting androgen receptor expression, stability, and function: potential targets to treat hormone-refractory prostate cancer. *J. Cell. Biochem.* 98, 1408-1423.

Remgard, P., P.A.R. Ekstrom, P. Wiklund, and A. Edstrom, 1995. Calmodulin and in vitro regenerating frog sciatic nerves: Release and extracellular effects. *Eur. J. Neurosci.* 7: 1386-1392.

Suarez, A., R.J. Huber, M.A. Myre, and D.H. O'Day, 2011. An extracellular matrix, calmodulin-binding protein from *Dictyostelium* with EGF-like repeats that enhance cell motility. *Cell. Signal.* 23: 1197-1206.

Sun, D-Y., H.B. Li, and G. Cheng, 1994. Extracellular calmodulin accelerates the proliferation of suspension-cultures cells of *Angelica dahurica*. *Plant Sci.* 99: 1-8.

Sun, D-Y., Y.-Q. Bian, B-H. Zhao et al, 1995. The effects of extracellular calmodulin on cell wall regeneration of protoplasts and cell division. *Plant Cell Physiol.* 36: 133-138.

Sun, W., S. Li, J. Xu, T. Liu, and Z. Shang, 2009. H+-ATPase in the plasma membrane of *Arabidopsis* pollen cells is involved in extracellular calmodulin-promoted pollen germination. *Prog. Nat. Sci* 19: 1071-1078.

Vogel, H.J., 1994. Calmodulin: a versatile calcium mediator protein. *Biochem. Cell Biol.* 72: 357-376.

Woodward, B.J., E.A. Lenton, and S. MacNeil, 1993. Requirement of preimplantation human embryos for extracellular calmodulin for development. *Human Reprod.* 8: 272-276.

Zakharov, V.V., and M.I. Mosevitsky, 2007. M-calpain-mediated cleavage of GAP-43 near Ser41 is negatively regulated by protein kinase C, calmodulin and calpain-inhibiting fragment GAP-43-3. *J. Neurochem.* 101: 1539-1551.

Zhang, Z., V. Majava, A. Greffier, R.L. Hayes, P. Kursula, and K.K.W. Wang, 2009. Collapsin response mediator protein-2 is a calmodulin binding protein. *Cell. Mol. Life Sci.* 66, 526-536.

Zhou, J., L. Ma, S. Zhang, Y. Zhu, and D. Sun, 2001. Extracellular calmodulin stimulates light-independent *RbcS-GUS* expression in suspension-cultured cells of transgenic tobacco. *Plant Cell Physiol.* 42: 1049-1055.

Reviewed by Robert J. Huber, PhD

In: New Developments on Signal Transduction Research
Editor: Masayoshi Yamaguchi
ISBN: 978-1-62618-230-1

Chapter 2

KINASE COUPLED RECEPTORS AND G-PROTEIN COUPLED RECEPTORS REGULATE NEURAL STEM CELL DIFFERENTIATION VIA COMMON SIGNALING PATHWAYS: EMERGING EVIDENCE FOR CONVERGENT REGULATION

Shelley B. Hooks*

University of Georgia, Department of Pharmaceutical and Biomedical Sciences, Athens, Georgia, US

ABSTRACT

The loss of neurons in neurodegenerative disease leads to devastating loss of function. Because of their ability to give rise to new neurons, neural progenitors offer hope for functional restoration of neurons in neurodegenerative disease, either through pharmacologic intervention that induces adult neural stem cells to differentiate in situ, or by transplanting neurons differentiated in vitro. In order to advance these efforts, multiple groups are working to define the complex signaling pathways that govern neuronal differentiation. These pathways are activated by self-renewal and differentiation cues, typically in the form of growth factors or morphogens that bind to cell-surface receptors. Expression of growth factors and their receptors is dynamically regulated in the nervous system, establishing tightly controlled spatial and temporal gradients of activity. In the past decade, kinase-coupled receptors such as TGFβ, Lif, and FGF receptors have been clearly established as key regulators of stem cell multipotency and differentiation. These receptor classes and their downstream kinases activate Erk, Akt, Stat, and Smad cascades to regulate gene expression and cellular function. More recent studies highlight the overlooked importance of signaling through G protein coupled receptors (GPCRs), whose downstream effects overlap significantly with kinase-coupled receptors. Specifically, GPCRs have also been shown to regulate Smad, Stat, Akt and Erk pathways. The GPCR family is large and diverse, and specific GPCRs and their ligands have been shown to regulate stem cells from various sources, including neuronal stem cells. This chapter will highlight convergent kinase-coupled and G protein-coupled signaling pathways, and highlight examples of co-regulation of neural stem cell differentiation and multipotency by both receptor types.

1. Introduction: The Role of Receptor Signaling in Stem Cell Biology

Tissue homeostasis depends on stem cells balancing self-renewal, proliferation, and differentiation as instructed by cues from the microenvironment. Integration of multiple signaling networks is necessary to accurately balance self-renewal and lineage-specific differentiation of stem cell populations. Multiple receptor-mediated signaling cascades critically direct stem cell responses to autocrine or paracrine growth factors and morphogens present in the developing and adult nervous system. Canonical pathways that are established as critical regulators of cell proliferation, self-renewal, and differentiation are regulated by cell-surface kinase-coupled receptors that converge on Smad, Stat, Akt, GSK-3, and Erk signaling pathways to regulate gene expression and cellular identity. In many of the signaling pathways that control these steps, paradoxical effects of many signaling mediators have been observed. The same receptor agonist can cause seemingly opposite effects in different contexts, suggesting that the effect of a given receptor signaling pathway is dependent on which complimentary or opposing signaling pathways are co-activated or co-expressed. Given the therapeutic potential of stem cell biology, it is essential to understand how these cells will interact with complex mixtures of signaling mediators in the tissue microenvironment [1-3]. This chapter focuses on common signaling mechanisms shared between kinase-coupled receptors, which have been long established as pluripotency and differentiation regulators, and G-protein coupled receptors, which have emerged more recently as important regulators of stem cell biology.

2. Canonical Receptor Pathways

The major self-renewal and differentiation factors defined in the past decades function by binding cell surface receptors that relay activation signals to the nucleus via kinase signaling cascades. Several of the ligands that activate these pathways, namely BMP/activin, LIF, Wnt and FGF, have been identified as having critical roles in the regulation of the rise of neural progenitor populations and their terminal differentiation into neurons, oligodendrocytes, and astrocytes, both in vivo and in vitro.

2.1. Receptor Tyrosine Kinases: FGF

Fibroblast growth factor (FGF) ligands and receptors have broad and well-described roles in expansion and differentiation of the nervous system. FGF ligands are secreted proteins that act in an autocrine or paracrine manner by binding classic integral membrane receptor tyrosine kinases (RTK). Ligand binding causes receptor dimerization, activation of the kinase domain via auto-phosphorylation, and recruitment of adaptor proteins and/or substrates that trigger kinase cascades. The predominant target of FGF receptor activation is the Erk Map Kinase signaling cascade, resulting in transcriptional activation of known downstream Map kinase genes such as Fos, Jun, c-myc, CREB, and Elk [4-7]. The Erk activation pathway is commonly associated with the proliferative effects of FGF signaling. Additionally, FGFR activation can activate Phospholipase C (PLC) γ, leading to calcium signaling and protein kinase C (PKC) activation, associated with neurite outgrowth effects [8]. Basic FGF, or FGF-2, is particularly implicated in the regulation of neural differentiation. FGF-2 activation triggers activation of phosphoinositide-3 kinase (PI3K)/Akt signaling. This branch of FGF-2 signaling contributes to FGF-2 mediated pluripotency [6, 9] and provides potential crosstalk with cytokine and WNT signaling cascades (described below). Signaling by the related RTK ligand Epidermal Growth Factor (EGF) and its receptors is also implicated in the regulation of stem cell differentiation, and may function synergistically with FGF pathways [10].

2.2. TGF-Beta Receptors: Activin and BMP

Members of the Transforming growth factor (TGF) β family of receptors are serine/threonine kinase receptors that bind secreted protein ligands including TGFβ, activin A, and bone morphogenic protein (BMP). There are approximately 30 distinct TGFβ family ligands, all of which exert their effects by binding heterotetrameric receptors composed of type 1 and type II receptor subunits. Type II receptors bind ligand, while Type I receptors transduce signals into the cell via serine/threonine kinase activity. Type I receptors are activated upon ligand binding, and then phosphorylate their downstream targets, the R-smads, which are divided into two functional groups. The ligands TGFβ, Activin-A, and Nodal activate Smad2 and Smad3 R-smads, while BMP-like ligands activate Smad1,5,8 signaling [11]. Both classes of R-smads form complexes with Smad-4 and translocate to the nucleus to regulate transcription of target genes. Thus, Smad4 provides a common regulatory point for Activin and BMP-like TGF beta pathways. Although originally identified for their roles in bone development, it is now clear that BMPs are critical in neural development. BMP ligands are made and secreted in the neural plate and neural tube, and the gradient of BMP activity induces dorsal identity of the neural crest [12]. BMPs show pleiotropic effects, stimulating both proliferation and growth arrest under different conditions [13].

2.3. Cytokine Receptors: LIF

Leukemia inhibitory factor (LIF) is a member of the IL-6 family of cytokines, and regulates survival and differentiation of certain stem cell populations. The canonical activity of LIF is mediated through activation of a heteromeric receptor composed of gp130 and the LIF receptor [14] to activate the kinase Janus kinase (Jak) to phosphorylate its major target, signal transducer and activator of transcription (Stat). Once phosphorylated, Stat proteins dimerize and translocate to the nucleus, where they bind DNA and regulate transcription.

In 1998, LIF was identified as one of the major serum-derived molecules responsible for maintaining pluripotency in mouse ES cells [15, 16]. Since that time, it has become obvious that the effects of LIF are pleiotropic, reflecting activation of multiple pathways, and strikingly different between mouse and human stem cells. Activation of LIFR maintains transcription of pluripotency genes through Jak/Stat-3 activation and the downstream transcriptional activation of Sox-2 [17, 18]. LIF also activates the PI3K-Akt pathway which results in the transcriptional activation of Nanog, a key pluripotency gene [18]. Akt can further target GSK-3, providing a point of convergence with Wnt signaling. Finally, under certain conditions LIF also activates Map kinase cascades, which promote differentiation and inhibition of Nanog [18, 19].

2.4. Frizzled Receptors: WNTs

Signaling by WNT glycoproteins and their cognate frizzled (FZD) receptors affects pluripotency and differentiation in diverse and seemingly contradictory ways, and their overall function in regulating pluripotency remains unclear [9, 20].

WNT ligands are secreted glycoproteins that trigger signaling cascades by binding the extracellular domains of heteroreceptors comprised of LRP5/ LRP6 receptors and Frizzled receptors. Frizzled receptors are distantly related members of the seven-transmembrane G-protein coupled receptor superfamily, and mediate signaling by both G-protein independent and G-protein dependent mechanisms. The canonical Wnt mechanism is the GSK-3/β-catenin pathway, which regulates stem cell self-renewal [21] and neural progenitor proliferation [22]. In this pathway, WNT binding to the FZD receptor leads to phosphorylation of the protein Disheveled, which is then able to inhibit the activity of glycogen synthase kinase 3 (GSK-3).

This inhibition of GSK-3 decreases the activity of the 'destruction complex', which is responsible for the degradation of β-catenin, resulting in increased β-catenin levels for translocation to the nucleus [23]. In the nucleus, β-catenin regulates TCF/LEF transcription factor activity. The WNT-activated transcription factor TCF3 has been shown to bind to the promoters of pluripotency genes Sox-2, Oct-4 and Nanog; knock-down of TCF3 induced an increase in transcription of these genes, suggesting that WNT/TCF is a negative regulator of pluripotency in mESCs [24].

FGF-2 signaling converges with the WNT pathway through Akt mediated phosphorylation of GSK-3 and synergistic accumulation of β-catenin [9]. Wnt signaling can also lead to mTOR activation, leading to stem cell senescence and exhaustion [25]. Thus, WNT signaling has pleiotropic and potentially opposing effects.

2.5. Notch Receptors

Unlike the other receptors discussed which bind secreted ligands, Notch receptors bind cell-surface protein ligands expressed on adjacent cells. Once activated, an intracellular domain of Notch is cleaved and translocated to the nucleus, where it activates transcription of target genes such as Hes and Hey genes [26]. Notch maintains the neural stem cell pool during mammalian neurogenesis [27, 28] by maintaining progenitor proliferation and inhibiting neurogenesis. Notch signaling can also promote gliagenesis [29]. There is no evidence for convergent signaling between Notch and GPCR pathways, and thus Notch signaling is not discussed in detail in this review.

3. Kinase Coupled Receptor Regulation of Neural Stem Cell Differentiation

3.1. Stem Cell Pluripotency

Pluripotent stem cells can be maintained in culture in a self-renewal state by the action of multiple cell signaling mediators that suppress differentiation and promote cell proliferation. Recent advances have identified specific culturing conditions that can support stem cell survival in vitro and steer differentiation down specific lineages, and have revealed critical roles for TGF-β, RTK and cytokine receptors and their ligands. It is possible to maintain pluripotency of mouse ES cells (mESCs) without serum or feeder cells by culturing in the presence of BMP-4 and LIF [30]. In mESCs, BMP-4/Smad1,5,8 signaling promotes pluripotency via activation of the transcription factor Id [13, 30] and inhibition of p38 Map kinase [30, 31]. As described above, LIF is also a key pluripotency factor in maintaining mESCs, via activation of PI3K [15, 16]. In contrast, FGF-2 is not necessary for maintenance of pluripotency in mESCs, and FGF-2 signaling results in reduced transcription of pluripotency genes in mESCs [19, 32]. Ras/Raf/MapK (ERK) signaling stimulated by FGF opposes pluripotency, and inhibition of Erk signaling promotes pluripotency [33].

There are significant differences in the growth factor requirements for pluripotency in mouse and human ES cells (hESCs). Contrary to their prominent roles in mESCs, LIF and BMP-4 are not sufficient to maintain pluripotency in hESCs [5, 34, 35]. Instead, the activation of Smad1,5,8 signaling by BMP-4 induces differentiation in hESCs [36, 37]. Similarly, the addition of Noggin, an inhibitor of the pro-differentiation BMP-4/Smad1,5,8 pathway, is sufficient to promote self-renewal in the presence of FGF-2 [38]. Further, activation of Smad2,3 by the ligands Activin, TGFβ, or Nodal promotes pluripotency through direct transcriptional activation of the pluripotency gene Nanog [39, 40]. However, Activin can also promote mesoderm and endoderm differentiation via regulation of PI3K [41-43]. hESCs do require the addition of FGF-2 to maintain the pluripotent cell state, and in the presence of feeder cells and serum replacement, FGF-2 is sufficient to maintain pluripotency in hESCs [44]. Inhibition of FGF-2 promotes differentiation [45]. Recent studies indicate that the effect of FGF-2 on hESC pluripotency involves cooperativity with activin/nodal signaling [46]. Combining FGF-2 with Activin can support pluripotency in the absence of feeder cells or conditioned media in hESC [32, 46]. Further, hESCs do not require LIF for self-renewal. In hESC LIF may promote pluripotency [47, 48] or promote differentiation [9]. Thus, suppression of Smad1,5,8 and activation of the Smad2,3 signaling along with activation of Map kinase and PI3K by FGF-2 coordinately maintain pluripotency and self-renewal of hESCs [39, 45].

3.2. Neural Development *In Vivo*

Our understanding of early neural development and the roles of specific mediators is based largely on model organisms, and the ability of these models to predict human early development is unclear. In early embryonic development, the nervous system emerges from rapidly proliferating epithelial precursors in the neural tube. Differentiating neural cells delaminate from this cell layer, and overtime form the growing mantle layer. These changes result from complex autocrine, paracrine, and endocrine signals which direct cells to their terminal fate. A select few cells do not terminally differentiate, but instead maintain self-renewal genes that allow the cell to retain a multipotent potential into adulthood [49]. These nascent stem cells can be triggered to expand and differentiate through poorly understood mechanisms. At birth, the proliferative zone has reduced to very small populations of progenitor/stem cells in discreet regions of the brain that continues to generate new neurons into adulthood. Multiple signaling events precisely coordinate the process of neurogenesis.

The first step in the development of the nervous system is the irreversible process of neural induction, which specifies neuroepithelia from ectodermal cells. A critical mediator of neural induction is inhibition of BMP signaling, once thought to be the sole barrier to a default neural differentiation pathway [50]. Anti-BMP signaling via the receptor antagonist Noggin is critical in neural induction; the expression of Noggin prompts neural differentiation in mESCs, and promotes neural progenitor marker expression [51]. It is now apparent that positive neural induction signals also contribute, most notably via FGF [52]. FGF signaling functions partly by countering BMP pathways via inhibiting the BMP effector Smad1 and by inhibiting BMP expression. But FGF also acts to directly promote transcription of genes that promote neuroectoderm induction and oppose mesoderm differentiation [53, 54]. Thus, both BMP receptor inhibition and FGF receptor activation contribute to neural induction in model organisms.

Upon neuroectodermal specification, the neural tube forms from the neural plate and neural progenitors differentiate into various lineages of neurons and glia in a spatially and temporally coordinated manner. Cells along the neural tube receive positional cues based on their location along rostral-caudal and dorsal-ventral axes. Gradients of signaling mediators including FGF, BMP, and Wnt ligands establish a grid-like pattern that prompt regional differentiation of sub-populations of cells, eventually giving rise to the diversity of cell types and structure in the mature brain [55, 56]. The exposure to a unique balance of morphogens and growth factors at each coordinate gives rise to specific neuronal or glial lineages. For example, FGF2 and BMP cooperate to induce neural crest transition [57], and BMP promotes glial differentiation of neural precursors via Smad signaling and mTOR activation Stat3 [58]. BMPs also regulate axon pathfinding and dendritic arborization in the developing brain [59].

3.3. Neurogenesis

Small populations of progenitor cells remain in discreet regions of the brain into adulthood, where they continue to respond to intrinsic signals to promote neurogenesis as needed. Adult neural stem/progenitor cells maintain self-renewal capability and multi-potential capacity to generate neurons, astrocytes, and oligodendrocytes and maintain tissue homeostasis. In the adult brain, the neural progenitor cells are localized at the subgranular zone of the dentate gyrus, and the subventricular zone in the walls of the lateral ventricles. Neural epithelial progenitor cells are highly polarized, with their apical surfaces facing the ventricle and bathed in cerebral spinal fluid (CSF), while their basal surfaces are in contact with the basement membrane. During neural tube formation, the neuroepithelial cells surrounding the tube secretes factors including BMPs, WNTs, and FGFs into the amniotic fluid trapped in the tube, giving rise to CSF [60, 61]. Thus, these same pathways continue to regulate neurogenesis in the adult brain via paracrine and autocrine receptor activation.

FGFs play important roles in the survival and expansion of neural progenitors once they have been established. The early expansion of the progenitor population occurs via symmetric division of progenitors epithelial cells. As neurogenesis is initiated in vivo, neuroepithelial cells transform into radial glia cells, which then divide asymmetrically to give rise to post-mitotic neurons and daughter radial glia cells [62]. FGF ligands including FGF2 maintain the proliferative divisions of neuroepithelial cells. FGF2, unlike other FGF ligands, is required throughout the cortex for progenitor divisions during both early neurogenesis and the generation of projection neurons [63]. FGF2 promotes progenitor proliferation through adult neurogenesis [64]. The ERK Map kinase pathway is specifically implicated in the maintenance of adult neural progenitor proliferation, while PLCγ/Ca^{++} signaling is more important in inhibition of astroglial differentiation and maintenance of differentiation potential [65]. LIF is neuroprotective in rodent systems [66].

3.4. Neuronal Differentiation *In Vitro*

Neural progenitors (NPs) can be derived from embryonic stem cells with adherent culture conditions [67, 68] or through the formation of embryoid bodies (EBs) [69] followed by isolation of neural rosette-like structures. The addition of the morphogen retinoic acid induces neural induction in the embryoid body [70]. In vitro, FGF2 and EGF is used to expand neural stem cells in adherent and neurosphere cultures [10], and FGF2 induces responsiveness of neural progenitors to EGF, which may account for observed synergy between the two pathways. Neural precursor cells can be propagated over multiple passages in the presence of neural culture conditions containing FGF-2 [67, 69] or EGF and FGF-2 [68]. The removal of these growth factors induces differentiation into post mitotic cells types including oligodendrocytes, astrocytes and neurons [67-69, 71]. FGF2 is a critical growth and survival factor in early neural precursor cells, and can be used in vitro to promote proliferation of neural progenitors even in the absence of BMP antagonists [69]. Culturing ES cell aggregates in the presence of FGF2 and the absence of serum on an adhesive surface promotes selective survival of neural progenitors and selectively expands the neuroepithelial population. Activation of ERK is associated with neuronal differentiation in mouse stem cells [72], while Stat3 signaling is associated with proliferation in neural stem cells [73]. While it is not required for human ES cell survival or early neural induction, LIF promotes expansion of neural progenitors. LIF has contradictory effects on neural stem cell differentiation: either directly promoting astrocyte differentiation while reducing neuronal differentiation [74], promoting neural stem cell maintenance [75], or enhancing neuronal differentiation [76]. In embryonic neural stem cells, LIF increases clonogenicity and oligodendrocyte maturation [77]. LIF prevents apoptosis and enhances proliferation in human neural progenitor cells; LIF reduced caspase-mediated apoptosis in differentiating neural progenitors [78]. Further, neural progenitors differentiated in the presence of LIF exhibited higher ERK phosphorylation and Stat3 phosphorylation. LIF is also a strong inducer of neurite outgrowth [79].

4. G-Protein Coupled Receptors

G-protein coupled receptors (GPCRs) are the largest class of transmembrane receptors, and are responsible for mediating cellular responses to neurotransmitters, hormones, growth factors, and many widely used therapeutic and recreational drugs. GPCRs are the third largest gene family in the human genome, representing over 800 distinct genes and 2% of the human genome [80]. Subfamilies of GPCRs are activated by a diversity of ligands including biogenic amines, amino acids, ions, lipids, and peptides, as well as light, taste, and odor stimuli [81, 82]. The superfamily has traditionally been divided into three major families: Class A/ Rhodopsin-like receptors, Class B/Secretin-like receptors, and Class C/glutamate-like receptors. More recent bioinformatics analysis has updated the phylogenetic characterization to five distinct families: glutamate, rhodopsin, adhesion, frizzled and secretin [83]. GPCRs regulate the physiology of numerous systems including the cardiovascular, immune, nervous, endocrine, and respiratory systems. Additionally, these receptors are implicated in diseases, including numerous types of cancer and CNS disorders, and over 50% of therapeutic agents target GPCR mediated pathways [84]. GPCR activation has more recently been shown to affect pluripotency and differentiation via regulation of the pathways discussed above.

4.1. Canonical G-Protein Signaling

The GPCR superfamily is structurally defined by an extracellular N-terminus, seven transmembrane domains, and cytosolic C-terminus. Through intracellular loops and C-termini, the receptors couple to heterotrimeric guanine nucleotide binding proteins (G-proteins) consisting of α, β, and γ subunits. Ligand binding to extracellular and transmembrane domains of the GPCR induces the exchange of guanosine diphosphate (GDP) for guanosine triphosphate (GTP) on the Gα-subunit of the heterotrimeric G-protein. This exchange of GDP for GTP results in the dissociation of the Gα subunit from the receptor and the βγ dimer, leaving the Gα and Gβγ subunits available to interact with a diverse set of effectors. G-protein inactivation occurs through hydrolysis of GTP to GDP through intrinsic GTPase activity of the Gα subunit, which is accelerated by the Regulator of G-protein signaling (RGS) protein family of GTPase activating proteins (GAPs). G-proteins cycle between 'active' and 'inactive' states through this general mechanism. The Ras superfamily of monomeric GTPases is related to the Gα subunit of heterotrimeric G-proteins, and its members are similarly regulated by GDP/GTP exchange and GTP hydrolysis.

Individual receptors can couple to multiple Gα isoforms, and the Gα subunits define the potential signaling outcomes of receptor activation. Four families of Gα G-proteins and their canonical signaling pathways have been characterized in detail: Gαs, Gαi, Gαq, and Gα12 [85]. Gαi and Gαs G-proteins interact with adenylyl cyclase enzymes to inhibit and stimulate, respectively, the production of cyclic adenosine monophosphate (cAMP) and its regulatory target Protein Kinase A (PKA). Gαq G-proteins interact with PLC to stimulate the production of diacylglycerol (DAG) and inositol triphosphates (IP3) culminating in PKC activation. Gα12 G-proteins activate nucleotide exchange factor Rho family GTPases (RhoGEFs). Additionally, Gβγ dimers interact with numerous effectors including PI3K, PLCβ, adenylyl cyclase, and ion channels [86].

Finally, several GPCR families "transactivate" receptor tyrosine kinase pathways such as EGFR by stimulating release of their growth factor ligands, resulting in activation of Akt and Map kinase cascades [87]. GPCR transactivation of RTKs is typically mediated by Gi G-proteins and PKC activation of a matrix metalloproteinase, but the mechanisms in stem cells have not been established [88, 89]. Through these myriad signaling pathways, GPCRs regulate important cellular functions including proliferation, adhesion, migration, and differentiation.

4.2. Convergent Signaling

Growing evidence implicates heterotrimeric G-proteins in general and specific G-protein coupled receptors as critical regulators of stem cell maintenance and differentiation. The activity of Gi subfamily G proteins has been shown to affect the morphology and organization of human iPSC colonies, which may reflect contributions from multiple Gi-coupled receptors [90], and Gs signaling in self-renewing and differentiating mESCs promotes both proliferation and pluripotency [91]. Many of the G-protein mediated pathways downstream of GPCR activation directly regulate, or are synergistic with, the kinase-coupled receptor pathways described above that are critical in regulating stem cell pluripotency and differentiation. GPCR activation of Map kinase cascades, PI3K-AKT, and PLC-PKC activation are long-established and well documented [92-96]. These pathways provide a basis for cross-talk between GPCRs and FGF, LIF, and pathways.

More recent studies have also identified GPCR regulation of pluripotency and differentiation pathways not traditionally associated with GPCRs, namely the Smad and Stat signaling pathways, providing potential convergent signaling between GPCRs and TGFβ receptors, and further cross-talk with LIF receptors. Gαs and Gαq both activate Stat-3 signaling [97], and expression of constitutively active Gαs can induce Stat3 activation in hepatocellular carcinoma [98]. Further, several GPCRs have been shown to regulate Stat signaling pathways. CXCR5, a chemokine receptor homolog of Kaposi's sarcoma herpes virus, transforms cells when stimulated, and this transformation requires Stat3 activation. Stat3 is constitutively phosphorylated in cells expressing constitutively activated CXCR5 or KSHV, and transformation was blocked by inhibition of Jak2 [99]. Similarly, the Gq-coupled CCK2 receptor, expressed in pancreatic cancer, activates Jak2 and Stat3 via Gq, and this activity contributes to the proliferative effects of CCK2. Constitutively activated Gαq also activated Jak2 in a receptor-independent manner, suggesting that Gq is sufficient for this activity [100]. Finally, thrombin-mediated vascular smooth muscle cell motility is dependent on Stat-3 activation and Stat-3 dependent PLA2 activity [101]. Hematopoietic-specific Gq family members G16 and G14 activate Stats, and there appears to be significant G-protein selectivity, in that constitutively active mutants of Gα16 and Gα14 activate Stat3, but not Stat2, Stat4, or Stat5. In contrast, Gβγ dimers activate Stat3, but only dimers containing neuronal-specific gamma subunits Gγ2, Gγ4, Gγ7, or Gγ9 [97]. More recently, it has been shown that Thrombin-induced cyclin D expression in vascular smooth muscle cells requires Stat3 phosphorylation and acetylation [102]. Perhaps more relevant to neural cells, Stat5a is phosphorylated in response to μ opioid receptor (OR) activation [103], and Stat5b is associated with the δOR c-terminus, and is phosphorylated by and transcriptionally activated by receptor activation. This activation requires a complex including Src, Stat5b and G-proteins on the δOR c-terminus [104]. Further, opioid receptor-like 1 and formyl peptide receptor activate Stat3 phosphorylation via Gα16, which requires Jak and Raf1 [105].

There is also evidence for interaction between GPCR signaling and TGFβ receptor/Smad pathways. First, the Gi and Gs regulated second messenger cAMP activates the small GTPase Rap [106], which regulates Smad signaling [107]. Enhanced Gαs activity enhances adipogenesis and impaires osteoblast differentiation, and blocks BMP-induced osteogenesis [108], suggesting GPCR mediated roles in regulating BMP signaling. GRK-2, a kinase responsible for desensitization of multiple GPCRs, is induced by TGFβ. GRK2 associates with R-smads and inhibits their phosphorylation and translocation to nucleus. Thus, GRK2 inhibits TGFβ gene expression [109]. Further evidence for cross talk comes from studies of specific GPCRs. Endothelin-1 (ET1) is a vasoconstrictor and mitogenic GPCR agonist, and contributes to pulmonary hypertension. BMP-9 stimulates ET1 release and may contribute to pathological pulmonary function [110]. The GPCR chemokine receptor CXCR4 and its ligand, stromal cell-derived factor-1α (SDF-1α) are widely implicated in regulating differentiation pathways. SDF-1 regulates BMP2-stimulated osteogenic differentiation of mesenchymal progenitor cells. SDF-1 binding to its receptor CXCR4 is necessary for BMP-2 activation of osteogenesis, and this effect is dependent on SDF-1 activation of Smad and Erk [111]. Inhibition of SDF-1 blocks BMP-induced differentiation by blocking Smad and Erk activation, and exogenous SDF-1 enhances BMP-2 induced chondrogenic differentiation [112]. Additionally, activation of metabotropic, G-protein coupled glutamate receptor mGlu3 in neural stem cells impairs astrocyte differentiation and promotes progenitor cell proliferation via regulation of BMP pathways. Glutamate induces GFAP, a marker of astrocyte differentiation, via TGFβ receptors, which activate Smad2/3. Glutamate also induces Smad nuclear translocation via metabotropic glutamate receptors, and this effect was inhibited by blockade of MapK and PI3K pathways [113]. mGluR3 agonists block BMP4 mediated Smad 1/5/8 activation, and again the ability of mGluR3 to blunt BMP signaling is blocked by MapK inhibitors [114]. Thus, Glutamate appears to alter the balance between the two branches of Smad signaling.

Finally, there is evidence for coordinated regulation of TGFβ/Smad signaling pathways and the lysophospholipids Lysophosphatidic Acid (LPA) and Sphingosine 1-phosphate (S1P), through their GPCRs. S1P enhances BMP-2 induced differentiation of osteoblasts, and S1P agonists enhance BMP-2 induced Erk and Smad1/5/8 activation [115]. S1P has also been shown to activate Smad2/3 phosphorylation in multiple cell types [116]. S1P stimulates chemotaxis in immune cells dependent on Smad-3, and induces Smad activation and DNA binding in these cells [117]. Further, the ability of TGFβ to stimulate Erk phosphorylation, migration, and invasion in cancer cells is dependent on TGFβ-stimulated S1P activation of S1P2 receptors [118]. Thus, bidirectional regulation between S1P and TGFβ receptor signaling is observed.

Finally, we have recently shown that S1P activates Smad 2/3 phosphorylation in human neural progenitors, suggesting that S1P-Smad signaling interactions may regulate neural differentiation (discussed below) [119]. Less is known about the regulation of Smad signaling pathways by LPA, but it has been shown that LPA1 receptor expression is inhibited by TGFβ [120].

4.3. GPCR Regulation of Neural Stem Cell Differentiation

Given the ability of GPCRs to exert strong influence on stem and progenitor cell growth and differentiation, it is expected that expression of these receptors. We have shown that GPCR transcript expression is dynamically regulated during stem cell differentiation by comparing transcript levels of 350 GPCR genes in pluripotent hESCs, multipotent NPs, and differentiated neurons [67, 121].

Distinct genes were dramatically upregulated upon transition from hESCs to the intermediate NP stage and final differentiation from the progenitor to neurons, suggesting that these GPCRs may be either involved in the transitions to more highly differentiated cells, or may serve specific functions in the distinct cell populations [122].

Further, considerable functional evidence has accumulated for specific roles of distinct ligand-GPCR pairs. Several recent reviews have highlighted specific roles of GPCRs and their ligands in regulating stem cell pluripotency and differentiation [122-124]. In the following section, we briefly review selected examples of GPCR effects on neural stem cell populations in development and in culture.

4.3.1. Chemokines

CXCR4 activation by SDF-1 is reported to promote migration, differentiation, and quiescence in neural progenitors [125] [126, 127]. In primary rat cortical neural progenitors, SDF-1 mediated proliferation via both ERK Map kinase and PI3K pathways [126]. In bone marrow stromal stem cells differentiated to neural like cells, PI3K/Akt signaling upregulates CXCR4, thereby increasing migration of these cells toward SDF-1α [128, 129]. Finally, loss of CXCR4 signaling in oligodendrocyte progenitor cells decreased oligodendrocyte maturation and decreased myelination in a demyelinated mouse model [130].

4.3.2. Endothelins

In oligodendrocyte progenitor cells, endothelin-1 stimulation of ETa and ETb results in the phosphorylation of Erk and cAMP response element-binding protein, ultimately enhancing oligodendrocyte progenitor cells migration and blocking further differentiation [131]. Endothelin signaling also prevents adipocyte differentiation by increasing Erk phosphorylation and inhibiting Akt signaling [132].

4.3.3. Lysophospholipids

LPA and S1P mediate their extracellular effects by binding and activating multiple GPCRs, including five S1P receptors (S1P 1-5) and three LPA receptors (LPA 1-3) from the Edg gene family, and several additional LPA receptors that are related to purinergic receptors [133-135]. Lysophospholipids and their receptors have been studied in embryonic stem cell and neural progenitor cell populations, and their roles in these systems were recently reviewed [136, 137]. Both LPA and S1P are generated by neural cell types, and both are found in the developing and adult brain [138]. Further, LPA is generated in the cerebral spinal fluid at high concentrations [139]. In mouse ES cells, LPA and S1P increase proliferation [140], in the case of S1P through an S1P5-Gi-Erk1/2 dependent pathway [141]. S1P also mediates human ES cell survival [142, 143], and regulates the maintenance of hESCs in the presence of PDGF through Gi and Erk [144].

LPA and S1P receptors have also been implicated in the regulation of NP cell populations. In hESC-derived neurospheres, LPA inhibits neurosphere formation and neural cell differentiation through a Gi-independent mechanism requiring Rho/Rock and PI3K/Akt pathways [145]. LPA and S1P stimulate proliferation of hESC-derived neuroepithelial progenitor cell line through a Gi mediated transactivation of EGF receptors, and induce transient morphology changes through Rho/Rock [146]. LPA has been shown to both promote and inhibit the terminal differentiation of neural progenitor cells. LPA stimulates the differentiation of mouse cortical neuroblasts through a LPA1-Gi mediated pathway [147]. Further, LPA promotes the differentiation of rat neural stem cells towards cholinergic neurons [148]. However, in human neural progenitors, LPA blocks neurosphere formation, inhibits neuronal differentiation, and maintains astrocytic differentiation through PI3K- and Rho-dependent pathways [137, 145]. LPA and S1P both inhibit oligodendrocyte maturation [149, 150].

4.3.4. Cannabinoids

The cannabinoid receptors, CB1 and CB2 are related to the LPA and S1P receptors, and both CB1 and CB2 couple to Gi G proteins [82]. Ligands for these receptors include the endogenous cannabinoids anadamide and 2-arachidonyl glycerol, and the exogenous cannabinoids found in cannabis [151]. Both CB1 and CB2 are expressed in mES cells and neural progenitors, and promote the proliferation and differentiation of these cell types [152, 153]. Specifically, CB1 and Cb2 receptors increase proliferation of neural progenitor cells and stimulate the differentiation of these cells toward an astroglial lineage [154], inhibit the differentiation of progenitors through inhibition of Erk [155], and mediate the proliferation of adult hippocampal neural progenitors [156]. In mouse neural progenitor cells, the endogenous cannabinoid anandamide promotes glial and neuronal differentiation at distinct time points, but has no effect on survival or apoptosis [157]. CB1 or CB2 agonists increase neural stem cell proliferation, while CB antagonists inhibit proliferation [158]. Activation of CB1 and CB2 promotes neurosphere proliferation via PI3K/AKT pathways, and the endogenous cannabinoids also promotes neuronal survival though this pathway [159]. Finally, CB1 receptors promote astrogliogenesis, increasing differentiation of Oligo-2 positive cells [154].

4.3.5. Glutamate

Glutamate receptors that couple to G proteins are termed metabotropic glutamate (mGlu) receptors, and are distinct from the ionotropic ligand gated channel glutamate receptors. The mGlu receptor family is made up of 8 subtypes. mGluR1 and 5 couple to Gq G proteins, while mGluR 2,3,4,6, and 7 couple to Gi G proteins [160]. Metabotropic Glutamate receptors critically regulate the embryonic and adult brain, and have distinct effects on cell growth and differentiation in ES cells and NP cells. In mouse embryonic stem cells, activation of mGluR5 promotes self-renewal through interaction with a LIF signaling pathway [161, 162]. In this pathway, mGluR5 couples to Gq, which activates PKC thereby inhibiting GSK-3β and LIF activation of PI3K. This results in stabilization of c-Myc proteins leading to mESC proliferation. Inhibition of mGluR3 or mGluR5 blocks cell proliferation and survival of neural progenitors [163]. During differentiation of mESCs mGlu4 expression is upregulated and these receptors promote differentiation of GABAergic neurons [164, 165]. Gi-coupled mGluR3 receptor activation of Map kinase signaling inhibits the differentiation of neural stem cells through an inhibition of BMP signaling [114]. Inhibition of mGLUR1 inhibits hippocampal progenitor proliferation in slices [166], while mGluR2/3 inhibition reduces proliferation and mGluR5 activation promotes neurogenesis in cultured mouse neural progenitors [163]. Further, mGluR3 activation inhibits astrocyte differentiation of neural stem cells via a BMP-dependent pathway [114].

4.3.6. Monoamine Receptors

The biogenic amines epinephrine, norepinephrine, dopamine, and serotonin are major neurotransmitters that are released at virtually all brain regions, and function by activating GPCRs expressed widely in the central nervous system. Adrenergic receptors are activated by the sympathetic amine mediators epinephrine and norepinephrine, and regulate basic respiratory, cardiovascular, and neurological responses. The α adrenergic receptors (AR) are subdivided into Gq-coupled α_1AR and Gi-coupled α_2AR, while the β receptors are Gs-coupled. Norepinephrine contributes to early maturation of hippocampal neural progenitors [167], and endogenous adrenergic transmission enhances neurogenesis [168]. Activation of α-AR increases the proliferation of neuroepithelial progenitor cells [169] and stimulates the migration and survival of neural progenitors [170]. Further, α1-AR enhances neurogenesis and gliogenesis in adult neural differentiation via a PI3K mechanism that enhances apoptosis of astrocytes, but not other cell types [171]. In adult hippocampal progenitors, α2-AR agonists blunt proliferation [172], while chronic α2-AR blockade enhances progenitor cell survival [173]. In contrast, activation of β-AR with isoproterenol enhanced proliferation of neural precursors [174].

Serotonin, also known as 5-hydroxytryptamine (5-HT), is critical in adult neurogenesis [175], and loss of neurogenesis as a result of 5-HT depletion may contribute to clinical depressive disorder [176]. Serotonergic neurons widely innervate the brain, and serotonin acts on multiple receptor subtypes. Of the 15 serotonin receptor subtypes, all but the 5HT-3 receptor are GPCRs. 5-HT1A and 5-HT2A/C are strongly implicated in neurogenesis. 5-HT1A partial agonists enhance neurogenesis [177], and 5-HT2 agonists enhance cell proliferation in the subventricular zone [178]. Further, chronic 5-HT1A activation and 5-HT2A inhibition increases proliferation of adult hippocampal progenitors, and chronic 5-HT2 activation blunts proliferation while increasing survival and maturation [179]. Dopamine receptors are activated by the catecholamine dopamine, and are generally grouped into D1-like Gs-coupled receptors and D2-like Gi-coupled receptors. Dopamine generally enhances adult neurogenesis [180], in particular the D2-like D3 receptor. Pharmacological inhibition of D3 receptors blunts neurogenesis in the olfactory bulb, with no effect on stem cell maintenance [181].

CONCLUSION

The receptor pathways that dictate the proliferation and fate of neural stem cells are regulated by ligands that are generated and released in the nervous system, such that these pathways do not function in isolation. The common signal transduction mediators in multiple kinase cascades allow integration and crosstalk between these receptor-stimulated pathways. A cohesive view of GPCR regulation of neural stem cell maintenance and differentiation has not emerged, and will likely require consideration of input from multiple, overlapping receptor systems. Thus, future studies directed at decoding the synergy and antagonism among these pathways is required to predict and exploit stem cell responses to intrinsic signals in therapeutics.

REFERENCES

[1] McDonald, J.W., et al., Transplanted embryonic stem cells survive, differentiate and promote recovery in injured rat spinal cord. *Nat. Med.,* 1999. 5(12): p. 1410-2.

[2] Ben-Hur, T., et al., Transplantation of human embryonic stem cell-derived neural progenitors improves behavioral deficit in Parkinsonian rats. *Stem Cells,* 2004. 22(7): p. 1246-55.

[3] Keirstead, H.S., et al., Human embryonic stem cell-derived oligodendrocyte progenitor cell transplants remyelinate and restore locomotion after spinal cord injury. *J. Neurosci.,* 2005. 25(19): p. 4694-705.

[4] Kang, H.B., et al., Basic fibroblast growth factor activates ERK and induces c-fos in human embryonic stem cell line MizhES1. *Stem* Cells *Dev,* 2005. 14(4): p. 395-401.

[5] Cai, J., et al., Modeling co-expression across species for complex traits: insights to the difference of human and mouse embryonic stem cells. *PLoS Comput Biol.,* 2010. 6(3): p. e1000707.

[6] Dvorak, P., et al., Expression and potential role of fibroblast growth factor 2 and its receptors in human embryonic stem cells. *Stem Cells*, 2005. 23(8): p. 1200-11.

[7] Yordy, J.S. and R.C. Muise-Helmericks, Signal transduction and the Ets family of transcription factors. *Oncogene,* 2000. 19(55): p. 6503-13.

[8] Doherty, P. and F.S. Walsh, CAM-FGF Receptor Interactions: A Model for Axonal Growth. *Mol. Cell Neurosci*., 1996. 8(2/3): p. 99-111.

[9] Ding, V.M., et al., FGF-2 modulates Wnt signaling in undifferentiated hESC and iPS cells through activated PI3-K/GSK3beta signaling. *Journal Of Cellular Physiology,* 2010. 225(2): p. 417-28.

[10] Conti, L., et al., Niche-independent symmetrical self-renewal of a mammalian tissue stem cell. *PLoS Biol.,* 2005. 3(9): p. e283.

[11] Kitisin, K., et al., Tgf-Beta signaling in development. *Sci. STKE,* 2007. 2007(399): p. cm1.

[12] Liem, K.F., Jr., G. Tremml, and T.M. Jessell, A role for the roof plate and its resident TGFbeta-related proteins in neuronal patterning in the dorsal spinal cord. *Cell,* 1997. 91(1): p. 127-38.

[13] Panchision, D.M., et al., Sequential actions of BMP receptors control neural precursor cell production and fate. *Genes. Dev.,* 2001. 15(16): p. 2094-110.

[14] Ernst, M. and B.J. Jenkins, Acquiring signalling specificity from the cytokine receptor gp130. *Trends Genet*, 2004. 20(1): p. 23-32.

[15] Smith, A.G., et al., Inhibition of pluripotential embryonic stem cell differentiation by purified polypeptides. *Nature*, 1988. 336(6200): p. 688-90.

[16] Williams, R.L., et al., Myeloid leukaemia inhibitory factor maintains the developmental potential of embryonic stem cells. *Nature,* 1988. 336(6200): p. 684-7.

[17] Matsuda, T., et al., STAT3 activation is sufficient to maintain an undifferentiated state of mouse embryonic stem cells. *EMBO J.,* 1999. 18(15): p. 4261-9.
[18] Niwa, H., et al., A parallel circuit of LIF signalling pathways maintains pluripotency of mouse ES cells. *Nature,* 2009. 460(7251): p. 118-22.
[19] Ying, Q.L., et al., The ground state of embryonic stem cell self-renewal. *Nature,* 2008. 453(7194): p. 519-23.
[20] Kielman, M.F., et al., Apc modulates embryonic stem-cell differentiation by controlling the dosage of beta-catenin signaling. *Nat. Genet.*, 2002. 32(4): p. 594-605.
[21] Alonso, L. and E. Fuchs, Stem cells in the skin: waste not, Wnt not. *Genes. Dev.*, 2003. 17(10): p. 1189-200.
[22] Zechner, D., et al., beta-Catenin signals regulate cell growth and the balance between progenitor cell expansion and differentiation in the nervous system. Dev. Biol., 2003. 258(2): p. 406-18.
[23] Schulte, G. and V. Bryja, The Frizzled family of unconventional G-protein-coupled receptors. *Trends Pharmacol. Sci.*, 2007. 28(10): p. 518-25.
[24] Cole, M.F., et al., Tcf3 is an integral component of the core regulatory circuitry of embryonic stem cells. *Genes. Dev.*, 2008. 22(6): p. 746-55.
[25] Castilho, R.M., et al., mTOR mediates Wnt-induced epidermal stem cell exhaustion and aging. *Cell Stem Cell,* 2009. 5(3): p. 279-89.
[26] Iso, T., L. Kedes, and Y. Hamamori, HES and HERP families: multiple effectors of the Notch signaling pathway. *Journal Of Cellular Physiology,* 2003. 194(3): p. 237-55.
[27] Chenn, A. and S.K. McConnell, Cleavage orientation and the asymmetric inheritance of Notch1 immunoreactivity in mammalian neurogenesis. *Cell,* 1995. 82(4): p. 631-41.
[28] Mizutani, K., et al., Differential Notch signalling distinguishes neural stem cells from intermediate progenitors. *Nature*, 2007. 449(7160): p. 351-5.
[29] Louvi, A. and S. Artavanis-Tsakonas, Notch signalling in vertebrate neural development. *Nat. Rev. Neurosci.*, 2006. 7(2): p. 93-102.
[30] Ying, Q.L., et al., BMP induction of Id proteins suppresses differentiation and sustains embryonic stem cell self-renewal in collaboration with STAT3. *Cell*, 2003. 115(3): p. 281-92.
[31] Qi, X., et al., BMP4 supports self-renewal of embryonic stem cells by inhibiting mitogen-activated protein kinase pathways. Proceedings Of The National Academy Of Sciences Of The United States Of America, 2004. 101(16): p. 6027-32.
[32] James, D., et al., TGFbeta/activin/nodal signaling is necessary for the maintenance of pluripotency in human embryonic stem cells. *Development,* 2005. 132(6): p. 1273-82.
[33] Wray, J., T. Kalkan, and A.G. Smith, The ground state of pluripotency. *Biochemical Society Transactions,* 2010. 38(4): p. 1027-32.
[34] Ginis, I., et al., Differences between human and mouse embryonic stem cells. *Dev. Biol.,* 2004. 269(2): p. 360-80.
[35] Wei, C.L., et al., Transcriptome profiling of human and murine ESCs identifies divergent paths required to maintain the stem cell state. *Stem Cells,* 2005. 23(2): p. 166-85.
[36] Lee, L.H., et al., Micropatterning of human embryonic stem cells dissects the mesoderm and endoderm lineages. *Stem Cell Res.*, 2009. 2(2): p. 155-62.
[37] Zhang, P., et al., Short-term BMP-4 treatment initiates mesoderm induction in human embryonic stem cells. *Blood,* 2008. 111(4): p. 1933-41.

[38] Xu, R.H., et al., Basic FGF and suppression of BMP signaling sustain undifferentiated proliferation of human ES cells. *Nat. Methods,* 2005. 2(3): p. 185-90.

[39] Xu, R.H., et al., NANOG is a direct target of TGFbeta/activin-mediated SMAD signaling in human ESCs. *Cell Stem Cell,* 2008. 3(2): p. 196-206.

[40] Beattie, G.M., et al., Activin A maintains pluripotency of human embryonic stem cells in the absence of feeder layers. *Stem Cells* (Dayton, Ohio), 2005. 23(4): p. 489-95.

[41] Wardle, F.C. and J.C. Smith, Transcriptional regulation of mesendoderm formation in Xenopus. Semin. Cell Dev. Biol., 2006. 17(1): p. 99-109.

[42] McLean, A.B., et al., Activin a efficiently specifies definitive endoderm from human embryonic stem cells only when phosphatidylinositol 3-kinase signaling is suppressed. *Stem Cells* (Dayton, Ohio), 2007. 25(1): p. 29-38.

[43] Watanabe, S., et al., Activation of Akt signaling is sufficient to maintain pluripotency in mouse and primate embryonic stem cells. *Oncogene,* 2006. 25(19): p. 2697-707.

[44] Thomson, J.A., et al., Embryonic stem cell lines derived from human blastocysts. *Science,* 1998. 282(5391): p. 1145-7.

[45] Eiselleova, L., et al., A complex role for FGF-2 in self-renewal, survival, and adhesion of human embryonic stem cells. *Stem Cells,* 2009. 27(8): p. 1847-57.

[46] Vallier, L., M. Alexander, and R.A. Pedersen, Activin/Nodal and FGF pathways cooperate to maintain pluripotency of human embryonic stem cells. *J. Cell Sci.*, 2005. 118(Pt 19): p. 4495-509.

[47] Armstrong, L., et al., The role of PI3K/AKT, MAPK/ERK and NFkappabeta signalling in the maintenance of human embryonic stem cell pluripotency and viability highlighted by transcriptional profiling and functional analysis. *Human Molecular Genetics,* 2006. 15(11): p. 1894-913.

[48] Li, J., et al., MEK/ERK signaling contributes to the maintenance of human embryonic stem cell self-renewal. *Differentiation,* 2007. 75(4): p. 299-307.

[49] Morshead, C.M., et al., Neural stem cells in the adult mammalian forebrain: a relatively quiescent subpopulation of subependymal cells. *Neuron,* 1994. 13(5): p. 1071-82.

[50] Munoz-Sanjuan, I., et al., Gene profiling during neural induction in Xenopus laevis: regulation of BMP signaling by post-transcriptional mechanisms and TAB3, a novel TAK1-binding protein. *Development,* 2002. 129(23): p. 5529-40.

[51] Gratsch, T.E. and K.S. O'Shea, Noggin and chordin have distinct activities in promoting lineage commitment of mouse embryonic stem (ES) cells. *Dev. Biol.,* 2002. 245(1): p. 83-94.

[52] Wilson, S.I., et al., An early requirement for FGF signalling in the acquisition of neural cell fate in the chick embryo. *Curr. Biol.*, 2000. 10(8): p. 421-9.

[53] Albazerchi, A. and C.D. Stern, A role for the hypoblast (AVE) in the initiation of neural induction, independent of its ability to position the primitive streak. *Dev. Biol.,* 2007. 301(2): p. 489-503.

[54] Bertrand, V., et al., Neural tissue in ascidian embryos is induced by FGF9/16/20, acting via a combination of maternal GATA and Ets transcription factors. *Cell,* 2003. 115(5): p. 615-27.

[55] Osterfield, M., M.W. Kirschner, and J.G. Flanagan, Graded positional information: interpretation for both fate and guidance. *Cell,* 2003. 113(4): p. 425-8.

[56] Munoz-Sanjuan, I. and H.B. A, Early posterior/ventral fate specification in the vertebrate embryo. *Dev. Biol.,* 2001. 237(1): p. 1-17.

[57] Sailer, M.H., et al., BMP2 and FGF2 cooperate to induce neural-crest-like fates from fetal and adult CNS stem cells. *J. Cell Sci.*, 2005. 118(Pt 24): p. 5849-60.
[58] Rajan, P., et al., BMPs signal alternately through a SMAD or FRAP-STAT pathway to regulate fate choice in CNS stem cells. *J. Cell Biol.*, 2003. 161(5): p. 911-21.
[59] Charron, F. and M. Tessier-Lavigne, Novel brain wiring functions for classical morphogens: a role as graded positional cues in axon guidance. *Development,* 2005. 132(10): p. 2251-62.
[60] Currle, D.S., et al., Direct and indirect roles of CNS dorsal midline cells in choroid plexus epithelia formation. *Development,* 2005. 132(15): p. 3549-59.
[61] Fukuchi-Shimogori, T. and E.A. Grove, Neocortex patterning by the secreted signaling molecule FGF8. *Science,* 2001. 294(5544): p. 1071-4.
[62] Gotz, M. and W.B. Huttner, The cell biology of neurogenesis. *Nat. Rev. Mol. Cell Biol.,* 2005. 6(10): p. 777-88.
[63] Raballo, R., et al., Basic fibroblast growth factor (Fgf2) is necessary for cell proliferation and neurogenesis in the developing cerebral cortex. *J. Neurosci.,* 2000. 20(13): p. 5012-23.
[64] Zheng, W., R.S. Nowakowski, and F.M. Vaccarino, Fibroblast growth factor 2 is required for maintaining the neural stem cell pool in the mouse brain subventricular zone. *Dev. Neurosci.*, 2004. 26(2-4): p. 181-96.
[65] Ma, D.K., et al., Molecular genetic analysis of FGFR1 signalling reveals distinct roles of MAPK and PLCgamma1 activation for self-renewal of adult neural stem cells. *Mol. Brain,* 2009. 2: p. 16.
[66] Ikeda, K., et al., Neuroprotective effect of various cytokines on developing spinal motoneurons following axotomy. *J. Neurol. Sci.*, 1996. 135(2): p. 109-13.
[67] Shin, S., et al., Long-term proliferation of human embryonic stem cell-derived neuroepithelial cells using defined adherent culture conditions. *Stem Cells,* 2006. 24(1): p. 125-38.
[68] Reubinoff, B.E., et al., Neural progenitors from human embryonic stem cells. Nat. Biotechnol., 2001. 19(12): p. 1134-40.
[69] Zhang, S.C., et al., In vitro differentiation of transplantable neural precursors from human embryonic stem cells. *Nat. Biotechnol.*, 2001. 19(12): p. 1129-33.
[70] Edwards, M.K. and M.W. McBurney, The concentration of retinoic acid determines the differentiated cell types formed by a teratocarcinoma cell line. *Dev. Biol.,* 1983. 98(1): p. 187-91.
[71] Okabe, S., et al., Development of neuronal precursor cells and functional postmitotic neurons from embryonic stem cells in vitro. *Mech. Dev.*, 1996. 59(1): p. 89-102.
[72] Li, Z., M.H. Theus, and L. Wei, Role of ERK 1/2 signaling in neuronal differentiation of cultured embryonic stem cells. *Dev. Growth Differ.*, 2006. 48(8): p. 513-23.
[73] Kang, M.K. and S.K. Kang, Interleukin-6 induces proliferation in adult spinal cord-derived neural progenitors via the JAK2/STAT3 pathway with EGF-induced MAPK phosphorylation. *Cell Prolif.,* 2008. 41(3): p. 377-92.
[74] Bonni, A., et al., Regulation of gliogenesis in the central nervous system by the JAK-STAT signaling pathway. *Science,* 1997. 278(5337): p. 477-83.
[75] Hatta, T., et al., The Role of gp130 in cerebral cortical development: in vivo functional analysis in a mouse exo utero system. *J. Neurosci.,* 2002. 22(13): p. 5516-24.

[76] Richards, L.J., et al., Leukaemia inhibitory factor or related factors promote the differentiation of neuronal and astrocytic precursors within the developing murine spinal cord. *Eur. J. Neurosci.,* 1996. 8(2): p. 291-9.

[77] Pitman, M., et al., LIF receptor signaling modulates neural stem cell renewal. *Mol. Cell Neurosci.,* 2004. 27(3): p. 255-66.

[78] Majumder, A., et al., Neurotrophic effects of leukemia inhibitory factor on neural cells derived from human embryonic stem cells. *Stem Cells* (Dayton, Ohio), 2012. 30(11): p. 2387-99.

[79] Gillespie, L.N., et al., LIF is more potent than BDNF in promoting neurite outgrowth of mammalian auditory neurons in vitro. Neuroreport, 2001. 12(2): p. 275-9.

[80] Peroutka, S.J. and T.A. Howell, The molecular evolution of G protein-coupled receptors: focus on 5-hydroxytryptamine receptors. *Neuropharmacology,* 1994. 33(3-4): p. 319-24.

[81] Marinissen, M.J. and J.S. Gutkind, G-protein-coupled receptors and signaling networks: emerging paradigms. *Trends Pharmacol. Sci*., 2001. 22(7): p. 368-76.

[82] Gloriam, D.E., R. Fredriksson, and H.B. Schioth, The G protein-coupled receptor subset of the rat genome. *BMC Genomics*, 2007. 8: p. 338.

[83] Fredriksson, R., et al., The G-protein-coupled receptors in the human genome form five main families. Phylogenetic analysis, paralogon groups, and fingerprints. *Mol. Pharmacol.,* 2003. 63(6): p. 1256-72.

[84] Dorsam, R.T. and J.S. Gutkind, G-protein-coupled receptors and cancer. *Nat. Rev. Cancer*, 2007. 7(2): p. 79-94.

[85] Simon, M.I., M.P. Strathmann, and N. Gautam, Diversity of G proteins in signal transduction. Science, 1991. 252(5007): p. 802-8.

[86] Smrcka, A.V., G protein betagamma subunits: central mediators of G protein-coupled receptor signaling. *Cell Mol. Life Sci*., 2008. 65(14): p. 2191-214.

[87] Daub, H., et al., Signal characteristics of G protein-transactivated EGF receptor. *Embo J.,* 1997. 16(23): p. 7032-44.

[88] Prenzel, N., et al., EGF receptor transactivation by G-protein-coupled receptors requires metalloproteinase cleavage of proHB-EGF. *Nature*, 1999. 402(6764): p. 884-8.

[89] Shah, B.H., J.A. Olivares-Reyes, and K.J. Catt, The protein kinase C inhibitor Go6976 [12-(2-cyanoethyl)-6,7,12,13-tetrahydro-13-methyl-5-oxo-5H-indolo(2,3-a)pyrrolo(3,4-c)-carbazole] potentiates agonist-induced mitogen-activated protein kinase activation through tyrosine phosphorylation of the epidermal growth factor receptor. *Molecular Pharmacology*, 2005. 67(1): p. 184-94.

[90] Nakamura, K., et al., G(i)-coupled GPCR signaling controls the formation and organization of human pluripotent colonies. *PLoS One,* 2009. 4(11): p. e7780.

[91] Layden, B.T., et al., G protein coupled receptors in embryonic stem cells: a role for Gs-alpha signaling. PLoS One, 2010. 5(2): p. e9105.

[92] Katz, A., D. Wu, and M.I. Simon, Subunits beta gamma of heterotrimeric G protein activate beta 2 isoform of phospholipase C. *Nature*, 1992. 360(6405): p. 686-9.

[93] Maier, U., A. Babich, and B. Nurnberg, Roles of non-catalytic subunits in gbetagamma-induced activation of class I phosphoinositide 3-kinase isoforms beta and gamma. *The Journal Of Biological Chemistry*, 1999. 274(41): p. 29311-7.

[94] Luttrell, L.M., Composition and function of g protein-coupled receptor signalsomes controlling mitogen-activated protein kinase activity. *J. Mol. Neurosci.*, 2005. 26(2-3): p. 253-64.

[95] Rozengurt, E., Mitogenic signaling pathways induced by G protein-coupled receptors. *Journal Of Cellular Physiology*, 2007. 213(3): p. 589-602.

[96] Ye, X., et al., Lysophosphatidic acid as a novel cell survival/apoptotic factor. *Biochim. Biophys. Acta,* 2002. 1585(2-3): p. 108-13.

[97] Yuen, J.W., et al., Activation of STAT3 by specific Galpha subunits and multiple Gbetagamma dimers. *The International Journal Of Biochemistry and Cell Biology,* 2010. 42(6): p. 1052-9.

[98] Nault, J.C., et al., GNAS-activating mutations define a rare subgroup of inflammatory liver tumors characterized by STAT3 activation. *J. Hepatol,* 2012. 56(1): p. 184-91.

[99] Burger, M., et al., KSHV-GPCR and CXCR2 transforming capacity and angiogenic responses are mediated through a JAK2-STAT3-dependent pathway. *Oncogene*, 2005. 24(12): p. 2067-75.

[100] Ferrand, A., et al., A novel mechanism for JAK2 activation by a G protein-coupled receptor, the CCK2R: implication of this signaling pathway in pancreatic tumor models. *The Journal Of Biological Chemistry*, 2005. 280(11): p. 10710-5.

[101] Dronadula, N., et al., STAT-3-dependent cytosolic phospholipase A2 expression is required for thrombin-induced vascular smooth muscle cell motility. *The Journal Of Biological Chemistry,* 2005. 280(4): p. 3112-20.

[102] Kundumani-Sridharan, V., et al., Novel interactions between NFATc1 (Nuclear Factor of Activated T cells c1) and STAT-3 (Signal Transducer and Activator of Transcription-3) mediate G protein-coupled receptor agonist, thrombin-induced biphasic expression of cyclin D1, with first phase influencing cell migration and second phase directing cell proliferation. *The Journal Of Biological Chemistry*, 2012. 287(27): p. 22463-82.

[103] Mazarakou, G. and Z. Georgoussi, STAT5A interacts with and is phosphorylated upon activation of the mu-opioid receptor. *J. Neurochem.*, 2005. 93(4): p. 918-31.

[104] Georganta, E.M., A. Agalou, and Z. Georgoussi, Multi-component signaling complexes of the delta-opioid receptor with STAT5B and G proteins. *Neuropharmacology*, 2010. 59(3): p. 139-48.

[105] Wu, E.H., R.K. Lo, and Y.H. Wong, Regulation of STAT3 activity by G16-coupled receptors. *Biochemical And Biophysical Research Communications,* 2003. 303(3): p. 920-5.

[106] de Rooij, J., et al., Epac is a Rap1 guanine-nucleotide-exchange factor directly activated by cyclic AMP. Nature, 1998. 396(6710): p. 474-7.

[107] Choi, S.C., et al., Regulation of activin/nodal signaling by Rap2-directed receptor trafficking. *Dev. Cell*, 2008. 15(1): p. 49-61.

[108] Zhang, S., F.S. Kaplan, and E.M. Shore, Different roles of GNAS and cAMP signaling during early and late stages of osteogenic differentiation. *Horm. Metab. Res.,* 2012. 44(10): p. 724-31.

[109] Ho, J., et al., The G protein-coupled receptor kinase-2 is a TGFbeta-inducible antagonist of TGFbeta signal transduction. *The EMBO Journal*, 2005. 24(18): p. 3247-58.

[110] Star, G.P., M. Giovinazzo, and D. Langleben, Bone morphogenic protein-9 stimulates endothelin-1 release from human pulmonary microvascular endothelial cells: a potential mechanism for elevated ET-1 levels in pulmonary arterial hypertension. *Microvasc. Res.*, 2010. 80(3): p. 349-54.

[111] Hosogane, N., et al., Stromal derived factor-1 regulates bone morphogenetic protein 2-induced osteogenic differentiation of primary mesenchymal stem cells. *The International Journal Of Biochemistry and Cell Biology*, 2010. 42(7): p. 1132-41.

[112] Guang, L.G., A.L. Boskey, and W. Zhu, Regulatory role of stromal cell-derived factor-1 in bone morphogenetic protein-2-induced chondrogenic differentiation in vitro. *The International Journal Of Biochemistry and Cell Biology*, 2012. 44(11): p. 1825-33.

[113] Romao, L.F., et al., Glutamate activates GFAP gene promoter from cultured astrocytes through TGF-beta1 pathways. J. Neurochem., 2008. 106(2): p. 746-56.

[114] Ciceroni, C., et al., mGLU3 metabotropic glutamate receptors modulate the differentiation of SVZ-derived neural stem cells towards the astrocytic lineage. *Glia*, 2010. 58(7): p. 813-22.

[115] Sato, C., et al., Sphingosine 1-phosphate receptor activation enhances BMP-2-induced osteoblast differentiation. *Biochemical And Biophysical Research Communications*, 2012. 423(1): p. 200-5.

[116] Bu, S., et al., Opposite effects of dihydrosphingosine 1-phosphate and sphingosine 1-phosphate on transforming growth factor-beta/Smad signaling are mediated through the PTEN/PPM1A-dependent pathway. *The Journal Of Biological Chemistry*, 2008. 283(28): p. 19593-602.

[117] Radeke, H.H., et al., Overlapping signaling pathways of sphingosine 1-phosphate and TGF-beta in the murine Langerhans cell line XS52. *J. Immunol.*, 2005. 174(5): p. 2778-86.

[118] Miller, A.V., et al., Sphingosine kinases and sphingosine-1-phosphate are critical for transforming growth factor beta-induced extracellular signal-regulated kinase 1 and 2 activation and promotion of migration and invasion of esophageal cancer cells. *Mol. Cell Biol.*, 2008. 28(12): p. 4142-51.

[119] Callihan, P., et al., Regulation of stem cell pluripotency and differentiation by G protein coupled receptors. *Pharmacol. Ther.*, 2010. 129(3): p. 290-306.

[120] Wu, J., et al., Gene expression of the lysophosphatidic acid receptor 1 is a target of transforming growth factor beta. *Oncogene*, 2012.

[121] Dhara, S.K. and S.L. Stice, Neural differentiation of human embryonic stem cells. *J Cell Biochem*, 2008. 105(3): p. 633-40.

[122] Callihan, P., et al., Regulation of stem cell pluripotency and differentiation by G protein coupled receptors. *Pharmacol. Ther.*, 2011. 129(3): p. 290-306.

[123] Kobayashi, N.R., et al., G-protein coupled receptors in stem cell self-renewal and differentiation. *Stem Cell Rev.*, 2010. 6(3): p. 351-66.

[124] Doze, V.A. and D.M. Perez, G-protein-coupled receptors in adult neurogenesis. *Pharmacol. Rev.*, 2012. 64(3): p. 645-75.

[125] Krathwohl, M.D. and J.L. Kaiser, Chemokines promote quiescence and survival of human neural progenitor cells. *Stem Cells* (Dayton, Ohio), 2004. 22(1): p. 109-18.

[126] Gong, X., et al., Stromal cell derived factor-1 acutely promotes neural progenitor cell proliferation in vitro by a mechanism involving the ERK1/2 and PI-3K signal pathways. *Cell Biol. Int.*, 2006. 30(5): p. 466-71.

[127] Wu, Y., et al., CXCL12 increases human neural progenitor cell proliferation through Akt-1/FOXO3a signaling pathway. *J. Neurochem.*, 2009. 109(4): p. 1157-67.

[128] Li, S., et al., Upregulation of CXCR4 favoring neural-like cells migration via AKT activation. Neurosci. Res., 2010. 67(4): p. 293-9.

[129] Jujo, K., et al., CXCR4 blockade augments bone marrow progenitor cell recruitment to the neovasculature and reduces mortality after myocardial infarction. *Proceedings Of The National Academy Of Sciences Of The United States Of America,* 2010. 107(24): p. 11008-13.

[130] Patel, J.R., et al., CXCR4 promotes differentiation of oligodendrocyte progenitors and remyelination. *Proceedings Of The National Academy Of Sciences Of The United States Of America,* 2010. 107(24): p. 11062-7.

[131] Gadea, A., et al., Endothelin-1 regulates oligodendrocyte development. *J. Neurosci.,* 2009. 29(32): p. 10047-62.

[132] Bhattacharya, I. and A. Ullrich, Endothelin-1 inhibits adipogenesis: role of phosphorylation of Akt and ERK1/2. *FEBS Lett.,* 2006. 580(24): p. 5765-71.

[133] Pasternack, S.M., et al., G protein-coupled receptor P2Y5 and its ligand LPA are involved in maintenance of human hair growth. *Nature Genetics,* 2008. 40(3): p. 329-34.

[134] Murakami, M., et al., Identification of the orphan GPCR, P2Y(10) receptor as the sphingosine-1-phosphate and lysophosphatidic acid receptor. *Biochemical And Biophysical Research Communications,* 2008. 371(4): p. 707-12.

[135] Tabata, K., et al., The orphan GPCR GPR87 was deorphanized and shown to be a lysophosphatidic acid receptor. Biochemical And *Biophysical Research Communications,* 2007. 363(3): p. 861-6.

[136] Pebay, A., C.S. Bonder, and S.M. Pitson, Stem cell regulation by lysophospholipids. *Prostaglandins Other Lipid Mediat.*, 2007. 84(3-4): p. 83-97.

[137] Pitson, S.M. and A. Pebay, Regulation of stem cell pluripotency and neural differentiation by lysophospholipids. *Neurosignals,* 2009. 17(4): p. 242-54.

[138] Aoki, J., A. Inoue, and S. Okudaira, Two pathways for lysophosphatidic acid production. *Biochimica Et Biophysica Acta,* 2008. 1781(9): p. 513-8.

[139] Sato, K., et al., Identification of autotaxin as a neurite retraction-inducing factor of PC12 cells in cerebrospinal fluid and its possible sources. *J. Neurochem.,* 2005. 92(4): p. 904-14.

[140] Todorova, M.G., et al., Lysophosphatidic acid induces Ca2+ mobilization and c-Myc expression in mouse embryonic stem cells via the phospholipase C pathway. *Cell Signal,* 2009. 21(4): p. 523-8.

[141] Rodgers, A., et al., Sphingosine 1-phosphate regulation of extracellular signal-regulated kinase-1/2 in embryonic stem cells. *Stem Cells Dev,* 2009. 18(9): p. 1319-30.

[142] Inniss, K. and H. Moore, Mediation of apoptosis and proliferation of human embryonic stem cells by sphingosine-1-phosphate. *Stem Cells* Dev, 2006. 15(6): p. 789-96.

[143] Wong, R.C., et al., Anti-apoptotic effect of sphingosine-1-phosphate and platelet-derived growth factor in human embryonic stem cells. *Stem Cells Dev.*, 2007. 16(6): p. 989-1001.

[144] Pebay, A., et al., Essential roles of sphingosine-1-phosphate and platelet-derived growth factor in the maintenance of human embryonic stem cells. *Stem Cells,* 2005. 23(10): p. 1541-8.

[145] Dottori, M., et al., Lysophosphatidic Acid Inhibits Neuronal Differentiation of Neural Stem/progenitor Cells Derived from Human Embryonic *Stem Cells. Stem Cells* (Dayton, Ohio), 2008.

[146] Hurst, J.H., et al., Human neural progenitors express functional lysophospholipid receptors that regulate cell growth and morphology. *BMC Neuroscience,* 2008. 9: p. 118.

[147] Fukushima, N., et al., Lysophosphatidic acid stimulates neuronal differentiation of cortical neuroblasts through the LPA1-G(i/o) pathway. *Neurochem Int,* 2007. 50(2): p. 302-7.

[148] Cui, H.L. and J.T. Qiao, Promotive action of lysophosphatidic acid on proliferation of rat embryonic neural stem cells and their differentiation to cholinergic neurons in vitro. *Sheng Li Xue Bao*, 2006. 58(6): p. 547-55.

[149] Jaillard, C., et al., Edg8/S1P5: an oligodendroglial receptor with dual function on process retraction and cell survival. *J. Neurosci.*, 2005. 25(6): p. 1459-69.

[150] Dawson, J., et al., Lysophosphatidic acid induces process retraction in CG-4 line oligodendrocytes and oligodendrocyte precursor cells but not in differentiated oligodendrocytes. *J. Neurochem.*, 2003. 87(4): p. 947-57.

[151] Mackie, K., Cannabinoid receptors: where they are and what they do. *J. Neuroendocrinol.,* 2008. 20 Suppl 1: p. 10-4.

[152] Bari, M., et al., Characterization of the endocannabinoid system in mouse embryonic stem cells. *Stem Cells Dev.*, 2010.

[153] Aguado, T., et al., The endocannabinoid system drives neural progenitor proliferation. FASEB J., 2005. 19(12): p. 1704-6.

[154] Aguado, T., et al., The endocannabinoid system promotes astroglial differentiation by acting on neural progenitor cells. *J. Neurosci.,* 2006. 26(5): p. 1551-61.

[155] Rueda, D., et al., The endocannabinoid anandamide inhibits neuronal progenitor cell differentiation through attenuation of the Rap1/B-Raf/ERK pathway. *J. Biol. Chem.,* 2002. 277(48): p. 46645-50.

[156] Aguado, T., et al., The CB1 cannabinoid receptor mediates excitotoxicity-induced neural progenitor proliferation and neurogenesis. *J. Biol. Chem.*, 2007. 282(33): p. 23892-8.

[157] Soltys, J., M. Yushak, and Y. Mao-Draayer, Regulation of neural progenitor cell fate by anandamide. *Biochemical And Biophysical Research Communications,* 2010. 400(1): p. 21-6.

[158] Jiang, W., et al., Cannabinoids promote embryonic and adult hippocampus neurogenesis and produce anxiolytic- and antidepressant-like effects. *J. Clin. Invest.*, 2005. 115(11): p. 3104-16.

[159] Galve-Roperh, I., et al., Mechanisms of control of neuron survival by the endocannabinoid system. *Curr. Pharm. Des.*, 2008. 14(23): p. 2279-88.

[160] Melchiorri, D., et al., Metabotropic glutamate receptors in stem/progenitor cells. Neuropharmacology, 2007. 53(4): p. 473-80.

[161] Cappuccio, I., et al., Endogenous activation of mGlu5 metabotropic glutamate receptors supports self-renewal of cultured mouse embryonic stem cells. *Neuropharmacology,* 2005. 49 Suppl 1: p. 196-205.

[162] Spinsanti, P., et al., Endogenously activated mGlu5 metabotropic glutamate receptors sustain the increase in c-Myc expression induced by leukaemia inhibitory factor in cultured mouse embryonic stem cells. *J. Neurochem.*, 2006. 99(1): p. 299-307.

[163] Di Giorgi-Gerevini, V., et al., Endogenous activation of metabotropic glutamate receptors supports the proliferation and survival of neural progenitor cells. *Cell Death Differ.,* 2005. 12(8): p. 1124-33.

[164] Cappuccio, I., et al., Context-dependent regulation of embryonic stem cell differentiation by mGlu4 metabotropic glutamate receptors. *Neuropharmacology*, 2006. 51(3): p. 606-11.

[165] Sarichelou, I., et al., Metabotropic glutamate receptors regulate differentiation of embryonic stem cells into GABAergic neurons. *Cell Death Differ.*, 2008. 15(4): p. 700-7.

[166] Baskys, A., et al., Group I metabotropic glutamate receptors reduce excitotoxic injury and may facilitate neurogenesis. *Neuropharmacology,* 2005. 49 Suppl 1: p. 146-56.

[167] Karkkainen, V., et al., Neurotransmitter responsiveness during early maturation of neural progenitor cells. *Differentiation,* 2009. 77(2): p. 188-98.

[168] Malberg, J.E., et al., Chronic antidepressant treatment increases neurogenesis in adult rat hippocampus. *J. Neurosci.*, 2000. 20(24): p. 9104-10.

[169] Popovik, E. and L.W. Haynes, Survival and mitogenesis of neuroepithelial cells are influenced by noradrenergic but not cholinergic innervation in cultured embryonic rat neopallium. *Brain Res.,* 2000. 853(2): p. 227-35.

[170] Ohashi, H., et al., Alpha 1-adrenoceptor agonists protect against stress-induced death of neural progenitor cells. *European Journal Of Pharmacology,* 2007. 573(1-3): p. 20-8.

[171] Gupta, M.K., et al., alpha1-Adrenergic receptors regulate neurogenesis and gliogenesis. *Molecular Pharmacology*, 2009. 76(2): p. 314-26.

[172] Yanpallewar, S.U., et al., Alpha2-adrenoceptor blockade accelerates the neurogenic, neurotrophic, and behavioral effects of chronic antidepressant treatment. *J. Neurosci.,* 2010. 30(3): p. 1096-109.

[173] Rizk, P., et al., The alpha2-adrenoceptor antagonist dexefaroxan enhances hippocampal neurogenesis by increasing the survival and differentiation of new granule cells. *Neuropsychopharmacology*, 2006. 31(6): p. 1146-57.

[174] Jhaveri, D.J., et al., Norepinephrine directly activates adult hippocampal precursors via beta3-adrenergic receptors. *J. Neurosci.,* 2010. 30(7): p. 2795-806.

[175] Warner-Schmidt, J.L. and R.S. Duman, Hippocampal neurogenesis: opposing effects of stress and antidepressant treatment. *Hippocampus,* 2006. 16(3): p. 239-49.

[176] Kempermann, G., Regulation of adult hippocampal neurogenesis - implications for novel theories of major depression. *Bipolar. Disord.,* 2002. 4(1): p. 17-33.

[177] Grabiec, M., K. Turlejski, and R.L. Djavadian, The partial 5-HT1A receptor agonist buspirone enhances neurogenesis in the opossum (Monodelphis domestica). *Eur. Neuropsychopharmacol.*, 2009. 19(6): p. 431-9.

[178] Banasr, M., et al., Serotonin-induced increases in adult cell proliferation and neurogenesis are mediated through different and common 5-HT receptor subtypes in the dentate gyrus and the subventricular zone. *Neuropsychopharmacology,* 2004. 29(3): p. 450-60.

[179] Klempin, F., et al., Oppositional effects of serotonin receptors 5-HT1a, 2, and 2c in the regulation of adult hippocampal neurogenesis. *Front Mol. Neurosci,* 2010. 3.

[180] Winner, B., et al., Dopamine receptor activation promotes adult neurogenesis in an acute Parkinson model. *Exp. Neurol.*, 2009. 219(2): p. 543-52.

[181] Kim, Y., et al., Dopamine stimulation of postnatal murine subventricular zone neurogenesis via the D3 receptor. *J. Neurochem.,* 2010. 114(3): p. 750-60.

In: New Developments on Signal Transduction Research
Editor: Masayoshi Yamaguchi
ISBN: 978-1-62618-230-1

Chapter 3

PHEROMONE BIOSYNTHESIS ACTIVATING NEUROPEPTIDE RECEPTORS (PBANRS) IN MOTHS: NEW DEVELOPMENTS REGARDING ALTERNATIVE SPLICE VARIANTS AND THE POTENTIAL FOR TARGETED DISRUPTION OF PBANR IN PEST CONTROL

***J. Joe Hull*[1] *and Shogo Matsumoto*[2*]**

[1]USDA-ARS, US Arid Land Agricultural Research Center, Maricopa, AZ, US

[2]Molecular Entomology Laboratory, RIKEN Advanced Science Institute, Wako, Japan

ABSTRACT

For most moths, the ability of conspecific males to locate receptive females is governed by the detection of a blend of semiochemicals known as sex pheromones. Sex pheromone components are *de novo* synthesized in the female pheromone gland in response to pheromone biosynthesis activating neuropeptide (PBAN). The pheromonotropic effects of PBAN are mediated by a G protein-coupled receptor predominantly expressed in the pheromone gland, termed the PBAN receptor (PBANR). PBANR activation triggers the mobilization of extracellular Ca^{2+}, which activates a signal transduction cascade that activates the enzymatic processes involved in sex pheromone production. PBANR was initially identified in *Helicoverpa zea* and *Bombyx mori* based on sequence similarities with the mammalian neuromedin U family of receptors. Since then PBANRs have been identified in a number of other moth species. Functionally relevant structural variations suggested the presence of two PBANR isoforms, one expressed in moths using non-cAMP dependent biosynthetic pathways and the other in moths using cAMP dependent pathways. Recent studies, however, have shown that alternative splicing of the PBANR gene generates multiple variants, including

---

* Retired.

the two initial isoforms, that are concomitantly expressed in moth pheromone glands. Given the importance of sex pheromones to moth mating behaviors, PBANR has been suggested as a potential candidate for targeted disruption. This chapter provides a brief overview of PBANR identification and the role molecular dissection of the PBANR sequence has had in elucidating critical structure-function relationships. Recent developments in generating PBANR agonists/antagonists and the potential for RNAi interference-based disruption of PBANR functionality are also discussed.

INTRODUCTION

For most species of moths, female sexual receptivity is communicated to conspecific males through small molecular weight volatile compounds known as sex pheromones (Figure 1). Sex pheromones are typically species-specific blends of straight chain (C10-C18) acyclic aliphatic compounds with varying degrees of desaturation and an oxygenated functional group consisting of an alcohol, acetate ester, or aldehyde [1]. They are *de novo* synthesized from acetyl-CoA using modified fatty acid biosynthetic pathways in the pheromone gland (PG), a specialized tissue composed of modified epidermal cells comprising the intersegmental membrane between the eighth and ninth abdominal segments [1,2].

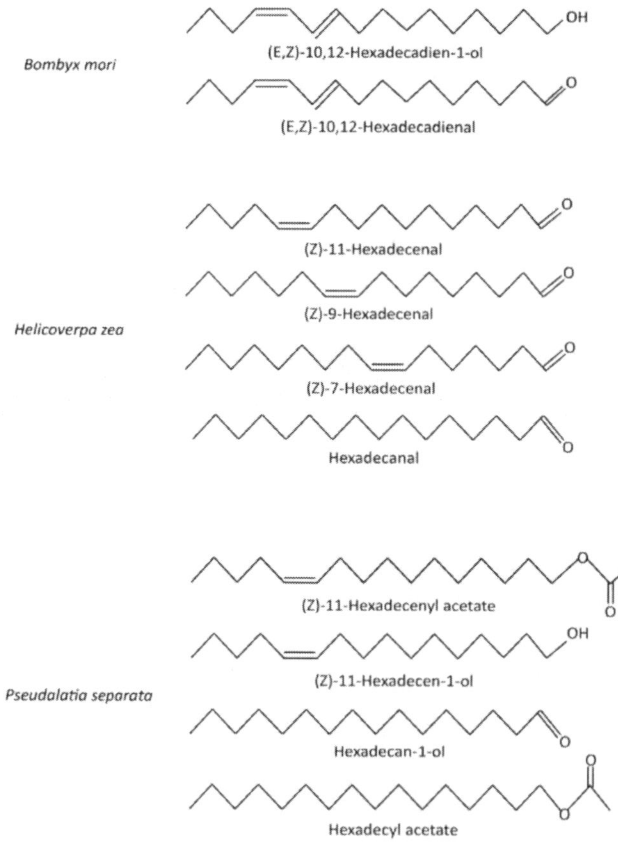

Figure 1. Examples of moth sex pheromones.

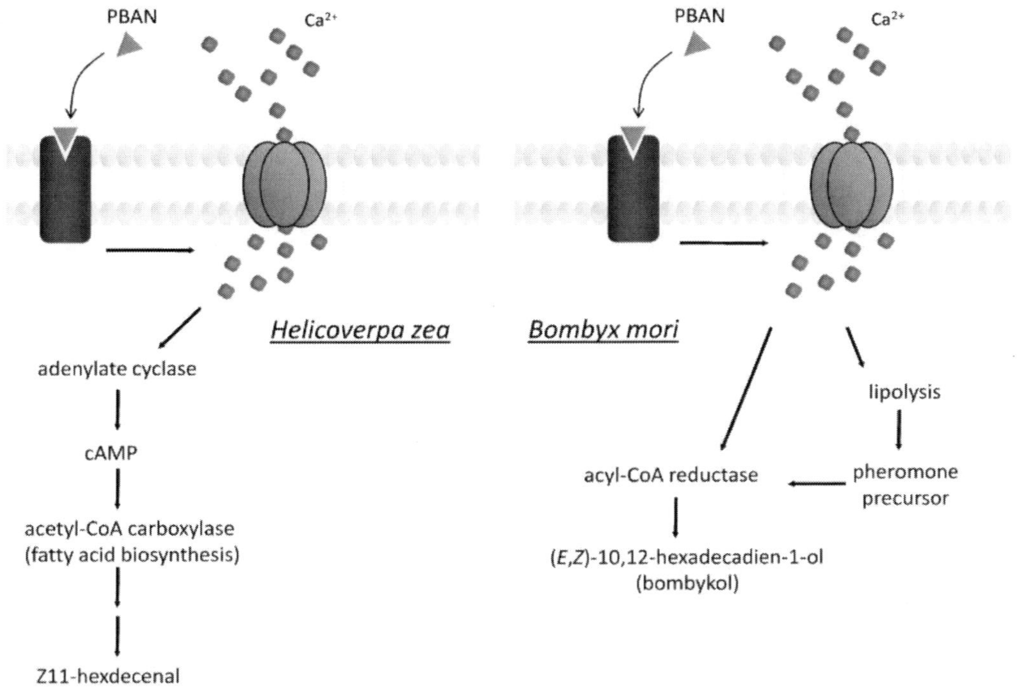

Figure 2. Intracellular pathways activated downstream of PBAN-mediated extracellular Ca^{2+} influx in *Helicoverpa zea* (corn earworm) and *Bombyx mori* (silkmoth). The PBAN signal cascade in *H. zea* proceeds through a cAMP-dependent pathway that activates a step (acetyl-CoA carboxylase) early in fatty acid biosynthesis that generates pheromone precursors. In contrast, the *B. mori* PBAN cascade proceeds independent of cAMP to activate the terminal step in sex pheromone production (reduction of the fatty acyl group) as well as lipolysis of stored sex pheromone precursors.

Diel patterns of female pheromone production are regulated by circadian cues that trigger release of pheromone biosynthesis activating neuropeptide (PBAN), a 33-34 amino acid neuropeptide produced in the subesophageal ganglion. Initially identified in *Helicoverpa zea* (corn earworm) and *Bombyx mori* (silkmoth) [3,4], PBAN has subsequently been identified in numerous other species in distantly-related insect orders [5,6]. The minimal sequence required for PBAN activity is defined by a conserved carboxyl terminal pentapeptide motif, FxPRLamide (where x is S, T, G, or V), that characterizes this family of insect neuropeptides. Interestingly, post-translational proteolytic processing of the PBAN gene product generates four additional FxPRLamide peptides. In addition to pheromonotropic effects, the FxPRLamide family of peptides have been implicated in the induction of melanization in larvae, induction of embryonic diapause and seasonal polyphenism in moths, termination of pupal diapause in heliothine moths, ecdysteroidogenesis in *B. mori* prothoracic glands, visceral muscle contraction in cockroaches, and the acceleration of puparium formation in flies (for review see [5]).

PBAN acts directly on the PG with the pheromonotropic effects mediated by the PBAN receptor (PBANR). PBAN binding triggers an influx of extracellular Ca^{2+} that drives species-specific biosynthetic processes (Figure 2) that affect either a step (most likely acetyl coenzyme A carboxylase) early in fatty acid synthesis via a cAMP-dependent pathway (ex. *H. zea*) or the terminal reductive step in the sex pheromone biosynthetic pathway via a cAMP independent pathway (ex. *B. mori*) [5,7]. PBAN has also been shown to stimulate lipolysis of the stored pheromone precursor in *B. mori* [8,9]. Given the economic and ecological importance of many lepidopteran species, a more precise understanding of the molecular processes underlying sex pheromone production is of major interest.

The aim of this chapter is to provide a short review on the molecular identification and characterization of PBANRs identified to date. Emphasis will be placed on recent significant developments regarding alternative splicing of the PBANR gene and how that relates to our understanding of PBANR functionality. The final major topic will examine the potential of PBANR as a target, both chemically and genetically, for novel pest control strategies.

1. PBANR Identification

Early *in vitro* studies demonstrated that PBAN exerts its pheromonotropic effects directly on the PG [5] indicating the presence of a cell surface receptor. Pharmacological profiling suggested the putative receptor was a G-protein coupled receptor (GPCR) since NaF, a potent G-protein activator, was able to mimic the pheromonotropic effects of PBAN [10]. The use of a biotinylated photoaffinity analog of PBAN (*N*-[*N*-(4-azido-tetrafluorobenzoyl)-biocytinyloxyl-succinimide (Atf-Bct-NHS-PBAN) labeled a ~ 50 kDa cell membrane protein in *Helicoverpa armigera* pheromone producing cells [11,12]. Despite these tantalizing results, molecular identification of the PBANR would not come about until after publication of the *Drosophila melanogaster* genome.

In 2001, based on sequence similarities between certain mammalian receptors and putative GPCRs in the *Drosophila* genome, Hewes and Taghert [13] suggested that peptide ligands with a carboxyl terminal PRxamide motif co-evolved with their respective receptors. They postulated that PBAN-like peptides displaying FxPRLamide motifs would interact with receptors exhibiting sequence similarity to mammalian neuromedin U receptors that are activated by peptides with the carboxyl terminal LxxPRxamide sequence. Park *et al.* [14] built on this hypothesis and demonstrated that *H. zea* PBAN (HzeaPBAN) activated GPCR CG8795, one of the four *Drosophila* GPCRs (CG8784, CG8795, CG9918, and CG14575) with similarity to neuromedin U receptors. Reciprocal experiments performed measuring the *in vivo* effects of neuromedin U in *H. zea* revealed that the mammalian peptide stimulated pheromone production [15]. Taken together these findings strongly implicated neuromedin U-like receptor homologs as PBANR candidates. Using degenerate primers designed to conserved regions of the *Drosophila* neuromedin U receptors, Choi *et al.* [15] identified a 346 amino acid GPCR (Figure 3) in the PGs of *H. zea* that exhibited significant sequence identity with *Drosophila* CG8795.

Functional analysis of this receptor in cultured Sf9 insect cells (an ovarian-derived cell line from *Spodoptera frugiperda*) revealed dose-dependent Ca^{2+} mobilization in response to HzeaPBAN stimulation with an EC_{50} value of 25 nM [15], supporting its identification as the HzeaPBANR. Working independently, we used similarly designed degenerate oligonucleotide primers to clone a 413 amino acid receptor (Figure 3) from PGs of the silkmoth *B. mori* [16].

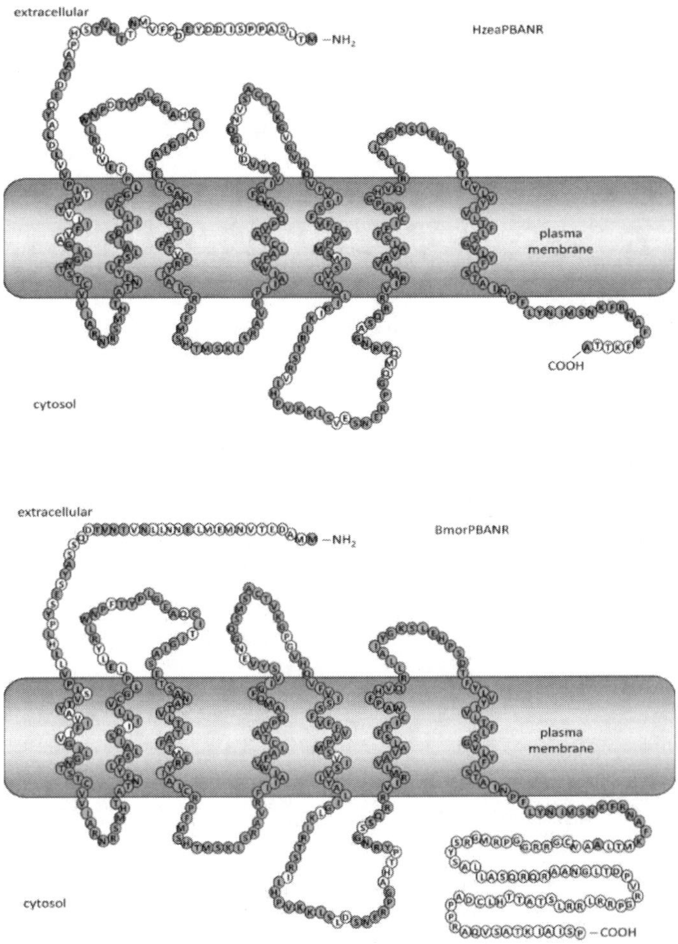

Figure 3. Illustration of the *H. zea* and *B. mori* PBAN receptors. Both PBANRs are seven-transmembrane domain receptors with extracellular amino terminal ends and intracellular carboxyl terminal ends. The *H. zea* PBANR (HzeaPBANR) contains 346 amino acids. The *B. mori* PBANR (BmorPBANR) contains 413 amino acids. Amino acids conserved between the two PBANRs are depicted in dark grey, while unique residues are in white.

Similar to HzeaPBANR, the *B. mori* receptor mobilized extracellular Ca^{2+} in response to PBAN stimulation, had significant sequence similarity with CG8795, and exhibited features consistent with GPCRs. Based on Northern blots, expression of the *B. mori* receptor was specific to PGs and underwent significant upregulation one day prior to adult pupal emergence [16], a time period that coincides with *B. mori* pheromonogenesis [17,18]. This receptor was consequently designated BmorPBANR. Dose-dependent reduction in *B. mori* sex pheromone production following RNA interference (RNAi)-mediated knockdown of PG-expressed BmorPBANR later confirmed its role in mediating the phermonotropic effects of PBAN [9].

Intriguingly, the ~50 kDa protein detected with the photoaffinity-labeled Atf-Bct-NHS-PBAN in the epidermis of the intersegmental membranes of *H. armigera* [11] was closer in size to BmorPBANR (~ 46 kDa) than the smaller HzeaPBANR (38.6 kDa). Furthermore, despite presumably mediating similar biological responses and significant sequence identity, BmoriPBANR and HzeaPBANR are structurally different (Figure 3). BmoriPBANR has a 67 amino acid carboxyl terminal extension that is absent in HzeaPBANR. We found that removal of this extension abolished ligand-induced internalization [16,19], a defined endocytotic mechanism that allows for the fine-tuning of ligand signal magnitude and duration associated with GPCR feedback regulation and desensitization [20,21]. Similar variation in carboxyl terminal length and function has been reported for the type I gonadotropin-releasing hormone receptors [22]. The non-mammalian type I receptors have a carboxyl terminal tail and undergo rapid ligand-induced internalization, whereas the mammalian type I receptors lack an extended carboxyl terminus and have significantly different internalization kinetics [22,23]. We thus concluded that the mechanisms underlying receptor-based regulation of the PBAN signal were different in *B. mori* and *H. zea*. The potential biological significance of this disparity led to early speculation that the respective receptor forms reflected the different signal transduction pathways utilized in the two species (cAMP dependent in *H. zea* vs. cAMP independent in *B. mori*).

Following the identification of PBANR homologs from *H. armigera* [24], *Spodoptera littoralis* (African cotton leafworm) [25], *Heliothis virescens* [26], *Spodoptera exigua* (beet armyworm)[27], and *Plutella xylostella* (diamondback moth) [28], however, it became apparent that PBANR expression and functionality likely mirrored the pleiotropic complexity of its ligand. The identification of PBANR homologs in *S. littoralis* and *H. virescens,* which were cloned from larval tissues and not adult female PGs as described in *H. zea* and *B. mori*, suggested a possible role in mediating PBAN induction of larval melanization and/or pupal diapause. The putative pleiotropic nature of the receptor was further underscored with reports that PBANR transcripts could, in addition to PGs, be amplified from brain, subesophageal ganglion, ventral nerve cord, thoracic ganglion, ovary, and, quite unexpectedly, tissues comprising the male abdominal tip [24,27,29]. Despite being cloned from diverse tissues, and presumably mediating different functions, the PBANRs identified in *H. armigera, S. littoralis, S. exigua* and *P. xylostella* all exhibited a truncated carboxyl terminus similar to HzeaPBANR, suggesting that feedback regulation of these receptors was also atypical. Moreover, the prevalence of the truncated carboxyl terminus amongst the various PBANRs raised questions concerning the evolutionary significance of the longer carboxyl terminus in *B. mori*. Kim and co-workers [26] began to address this issue with their identification of three structurally similar PBANR isoforms (referred to as HvirPBANR A-C) that diverged only at their carboxyl ends and which were concomitantly expressed in *H. virescens*. Similar to BmorPBANR, the HvirPBANR-C carboxyl terminus is extended and contains a defined internalization motif (see below), while the carboxyl terminal end of HvirPBANR-A is more like the shorter HzeaPBANR. Furthermore, HvirPBANR-C was preferentially amplified from PGs and, when expressed in a mammalian cell line, generated robust Ca^{2+} mobilization responses following PBAN stimulation. In contrast, the other two HvirPBANR isoforms were amplified from larval central nervous tissues and failed to respond to PBAN [26].

Sequence information (FJ240221-FJ200224) has also been deposited for multiple *Manduca sexta* (tobacco hornworm) PBANR isoforms that likewise exhibit variation only within the carboxyl terminus. Kim *et al.* [26] postulated that this variation arose from alternative splicing of the 3' exons in the PBANR gene, which would provide an appealing explanation for the observed diversity of PBANR expression, and presumably function, across multiple tissues/developmental stages.

Alternative splicing is one of the principal means by which organisms generate functional protein diversity in a temporal and/or tissue dependent manner, and involves a larger set of genes than initially surmised [30-32]. In *Drosophila*, approximately 61% of the expressed multi-exon genes have at least one alternative splice product [33], this number is thought to be even higher in humans [34,35]. Alternative splicing of GPCRs has been extensively documented [36,37] with splicing events frequently affecting/altering ligand binding, G-protein coupling, protein-protein interactions, intracellular signaling pathways, internalization, and receptor desensitization. Building on the work of Kim *et al.* [26], which looked at PBANR variants across multiple tissues, we sought to explore the potential role of alternative splicing in generating the functionally diverse PBANR variants initially cloned from PGs (i.e., HzeaPBANR and BmorPBANR). We found that multiple splice variants that differ only in the sequence and length of their respective carboxyl terminal ends are concomitantly expressed in the PGs of *B. mori*, *H. zea*, *H. armigera*, and *Pseudaletia (=Mythimna) separata* (Oriental armyworm) [38]. In the *B. mori* PG, we identified PBANR variants of 475 amino acids (BmorPBANR-B), 413 amino acids (BmorPBANR-C, the original BmoriPBANR identified), 345 amino acids (BmorPBANR-A), and 306 amino acids (BmorPBANR-As).

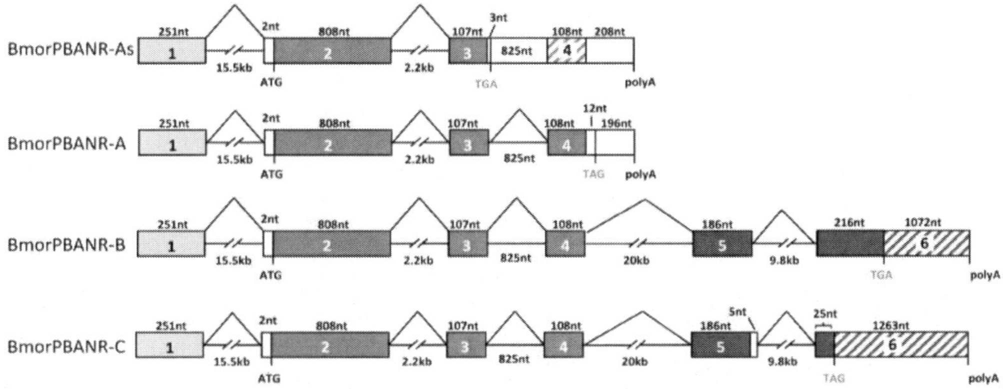

Figure 4. Schematic diagram of the *B. mori* PBANR genomic structure and alternative splice sites. The four BmorPBANR variants (As, A, B, and C) are depicted. Filled boxes represent the 6 exons, while white boxes represent non-spliced introns. Exon 1 (light grey) corresponds to the 5' untranslated region, exons 2-4 (medium grey) are common to the four variants, and exons 5-6 (black) encode the carboxyl terminal region specific to the B and C variants. Hatching indicates portions of exons that are not translated. Initiation (ATG) and stop sites (TGA or TAG) are indicated by their respective codons. Figure adapted from [38].

Sequence variation was limited to the carboxyl terminus with BmorPBANR-As transcripts encoding a GPCR truncated within transmembrane domain 7 (TM7).

The six exons and five introns comprising the *B. mori* PBANR gene span >50 kb (Figure 4) with the 5'untranslated region corresponding to exon 1 and a significant portion of the BmorPBANR coding sequence (start codon through TM7) encoded by exons 2-4. The carboxyl terminal portion of BmorPBANR is encoded by exons 5 and 6. BmorPBANR-As and BmorPBANR-A are generated by the retention of introns 3 and 4 respectively, both of which introduce premature stop codons. BmorPBANR-B is generated from exons 2-6, while

BmorPBANR-C arises from a 5 nucleotide frame shift insertion at the 3' end of exon 5 that changes the identity of the remaining ten amino acids (residues 404–413) and introduces a stop codon that generates a carboxyl terminus truncated by 62 amino acids compared to BmorPBANR-B. Variants of similar sizes were identified in *H. zea*, *H. armigera*, and *P. separata* (Figure 5). PCR-based expression analyses (Figure 6) revealed that the PBANR-C variant was preferentially amplified from the PGs of all four species [38]. Apparent preferential amplification of the shorter variant (designated as the PBANR-A variant) from PGs in *H. zea* [15], *H. armigera* [24], *S. exigua* [27], and *P. xylostella* [28] likely resulted from poor amplification of the B and C variants due the high GC content (55-80%) their extended carboxyl terminal ends. Ligand-induced internalization assays that examined the degree of internalization of a red fluorescent PBAN analog (RR-C10PBANR2K), revealed endocytosis occurred only in cells expressing the longer PBANR variants. Consistent with previous reports [16,19], no internalization was seen with the A variant (Figure 7). Furthermore, when stably transformed into insect cells, the BmorPBANR-A variant failed to mobilize extracellular Ca^{2+} in response to a range of PBAN concentrations (Figure 8) [39]. Similar inability to mobilize Ca^{2+} in response to PBAN stimulation was reported for the HvirPBANR-A variant [26]. Taken together, we suggest that the PBANR-C variant is the principal receptor molecule involved in mediating the pheromonotropic effects of PBAN. This conclusion is supported by specific binding of the photoaffinity-based biotin-labeled PBAN to a membrane-bound protein of ~50 kDa in *H. armigera* PGs [40], which is more consistent with the molecular weight of HarmPBANR-C (51.1kDa) than HarmPBANR-A (38.7 kDa). However, given their conservation across multiple insect orders, it is likely that the other variants are also biologically relevant.

We suggest that the coincident expression of multiple PBANR variants within individual PG cells represents a regulatory mechanism for fine-tuning cellular responsiveness to the PBAN signal. Nominally non-responsive PBANR-A receptors expressed at the cell surface could potentially function as ligand sinks that compete with PBANR-C for ligand binding. The net result would be less bioactive peptide available to trigger the cellular response thus decreasing overall sensitivity. Alternatively, heterodimerization of PBANR-As or PBANR-A with the longer variants could impede trafficking to the cell surface, thereby decreasing the pool of available receptors for ligand binding, which would likewise decrease overall cellular sensitivity.

When co-expressed in cultured cells with their predominant full-length receptor forms, truncated variants of the calcitonin receptor, corticotropin releasing factor receptor type 1, and growth hormone secretagogue receptor type 1a exerted dominant negative effects on signaling [41-43]. On a global scale, the presence of multiple PBANR variants expressed in diverse tissues across varying developmental stages likely reflects variations in receptor functionality. Many receptor variants have been reported to exhibit distinct spatial and temporal expression profiles, altered ligand binding, atypical feedback regulation, and differential activation of downstream effector pathways [36].

Similar to that observed for the PBANR-A variants, a naturally occurring variant of the neurokinin 1 receptor severely truncated at the TM7-cytosol junction is differentiated from the full-length receptor by an inability to mobilize extracellular Ca^{2+} and impaired ligand-induced internalization [44,45].

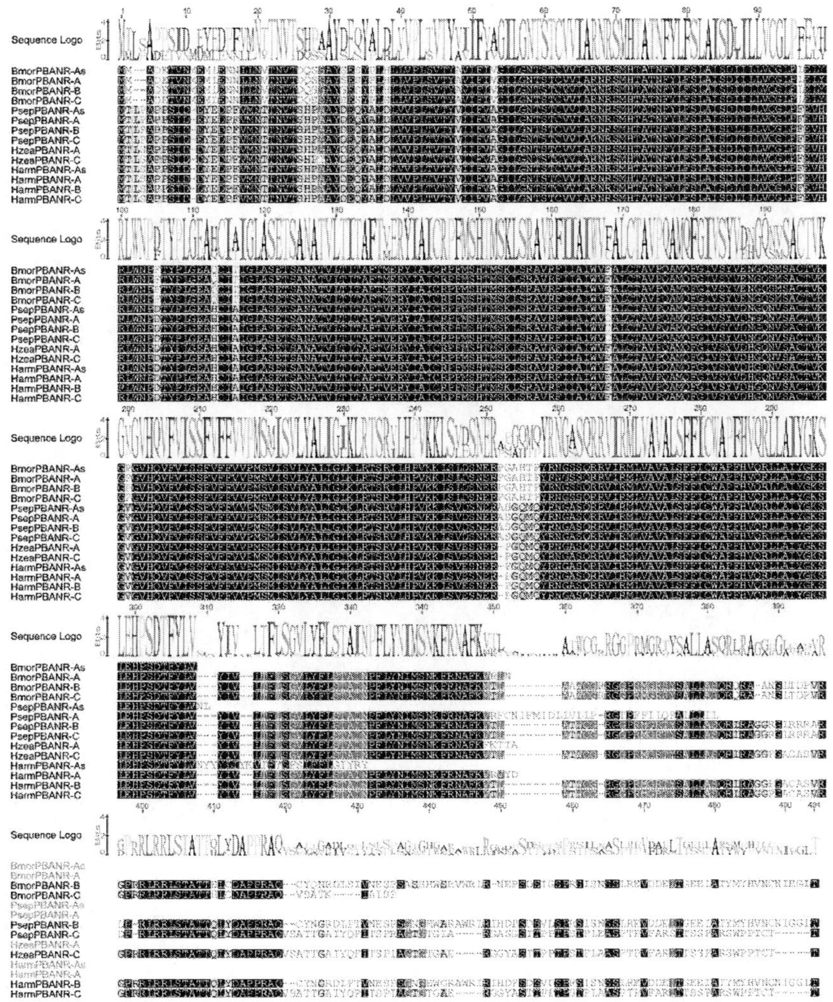

Figure 5. Multiple sequence alignment of PBANR variants from various species of moths. Sequences were aligned using the L-INS-I strategy in MAFFT [113]. Sequence abbreviations are: Bmor (*Bombyx mori;* accession #s for As, A, B, and C variants - AEX31546, AEX15646, AEX15643, BAD44726), Psep (*Pseudalatia separata*; accession #s for As, A, B, and C variants – AEX31548, AEX15648, AEX15645, AEX15642), Hzea (*Helicoverpa zea*; accession #s for As and C variants – AAP93921 and AEO17028), Harm (*Helicoverpa armigera*; accession #s for As, A, B, and C variants – AEX31547, AEX15647, AEX15644, AEX15641). In the sequence logo stacks [114], the height of each stack corresponds to the degree of sequence conservation at that position.

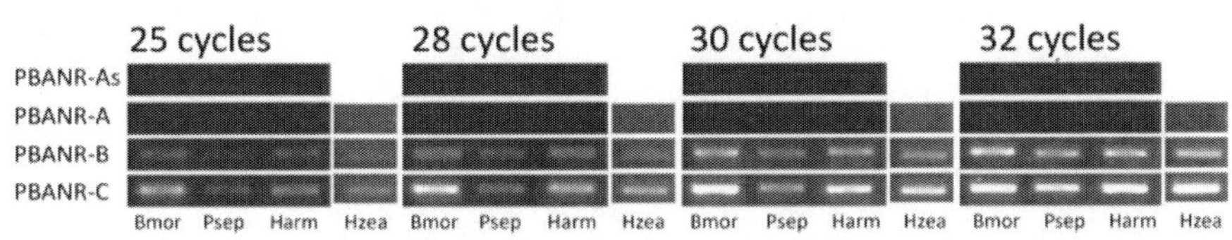

Figure 6. RT-PCR transcript expression profile of PBANR variants in the PGs of various moth species. RT-PCR was performed using cDNAs generated from *B. mori* (Bmor), *P. separata* (Psep), *H. armigera* (Harm), or *H. zea* (Hzea) PG total RNAs. PCR was performed using a GC-optimized DNA polymerase for 25–32 PCR cycles as indicated. Figure adapted from [38].

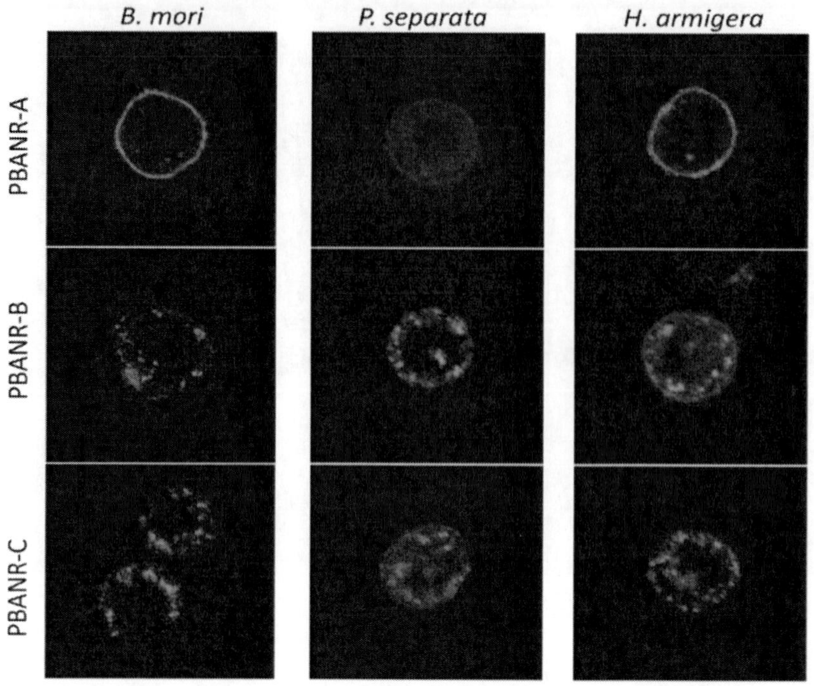

Figure 7. Ligand-induced internalization of PBANR variants. Cultured insect Sf9 cells transiently expressing EGFP-chimeras of the *B. mori*, *P. separata*, or *H. armigera* PBANR variants were challenged with 50 nM RR-C10PBANR2K (red fluorescent PBAN analog). Images obtained using a Leica TCS NT laser confocal microscope and processed in Adobe Photoshop 6.0. Note – no internalization observed in the cells expressing the PBANR-A variants. Figure adapted from [38].

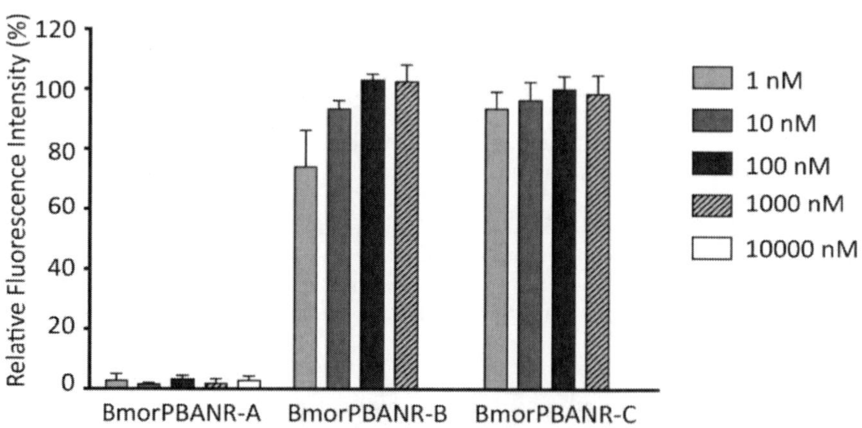

Figure 8. Dose–response profile of Sf9 cell lines stably expressing EGFP chimeras of the BmorPBANR A-C variants. Fluorescence intensity was measured on a scanning laser confocal microscope using FuraRed in response to increasing concentrations (1 nM –10000 nM) of C10PBAN$_{R2K}$ (truncated PBAN analog). Fluorescence intensity was normalized to BmorPBANR-C–EGFP activation with 100 nM C10PBAN$_{R2K}$. Error bars represent mean values + SEM of five replicates. Figure adapted from [39].

2. Structure-Function Analyses

2.1. Extracellular Domains

Because of the central role of PBANR in mediating sex pheromone biosynthesis, a process crucial to moth mating, there is significant interest in developing methods that disrupt PBANR functionality (see below). However, our understanding of the molecular mechanisms underlying the activation and regulation of insect peptide hormone GPCRs lags behind that of vertebrate GPCRs. To address this issue, a number of studies were initiated using various methodologies (e.g., chimeric receptors, site-directed mutagenesis, and molecular modeling) to begin to elucidate the structural determinants driving PBAN binding, PBANR activation, and propagation/termination of the PBAN signal.

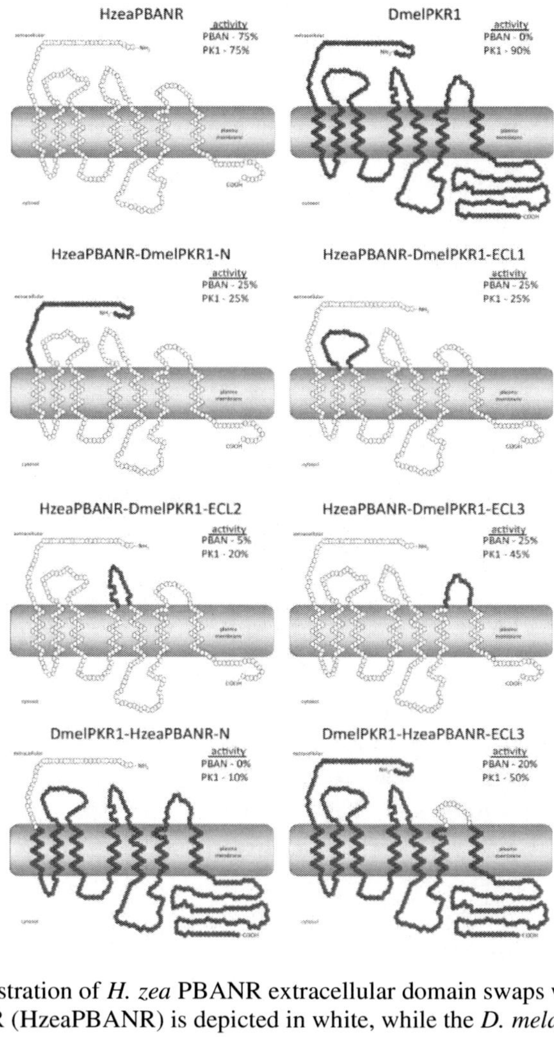

Figure 9. Schematic illustration of *H. zea* PBANR extracellular domain swaps with a distantly related receptor. *H. zea* PBANR (HzeaPBANR) is depicted in white, while the *D. melanogaster* pyrokinin 1 receptor (DmelPKR1) is depicted in black. Individual extracellular domain swaps are shown along with functional effects of swaps on mobilization of Ca^{2+} in response to PBAN and PK. Figure adapted from [48].

The extracellular loops (ECLs) and amino terminus of many GPCRs have been implicated in peptide ligand binding/discrimination [46]. Consequently, the generation of chimeric receptors that incorporate these domains from distant but related receptors can provide insights into the molecular determinants that govern ligand binding and signal transduction [47]. The ligand for the *Drosophila* pyrokinin receptor 1 (DmelPKR1; also referred to as CG9918) has a similar carboxyl terminal sequence (WGFPRLamide) as PBAN (FSPRLamide). However, while the DmelPKR1 is strongly activated by low nanomolar concentrations of pyrokinin 1 (PK1), it is largely unresponsive to PBAN [48]. The reciprocal is not true though, as PBANR can be activated by both PBAN and PK1.

To begin to elucidate the structural determinants that govern this ligand discrimination, Choi and co-workers generated a series of chimeric GPCRs in which the extracellular domains of HzeaPBANR and DmelPKR1 were swapped for one another (Figure 9). They then assessed the effectiveness of the respective chimeras at mobilizing extracellular Ca^{2+} in response to PBAN and PK1. Ligand discrimination was found to largely reside in ECL3, and to a lesser extent in the amino terminus [48]; regions that have been implicated in peptide ligand binding [46,49,50].

The impaired activity of chimeras containing the ECL2 swaps was attributed to altered positioning of the cysteine pairs that generate the disulfide bridge connecting ECL2 and TM3; a linkage that appears critical for GPCR folding and ligand binding [51]. To further explore the role of ECL3 in ligand discrimination, three separate point mutations in the HzeaPBANR ECL3 were made to residues (E297, S300, and F303) with side group functionalities (i.e., charged, polar, aromatic) that could potentially interact with a peptide ligand (Figure 10) [52]. Alanine substitution of S300 and F303 both reduced the efficiency of Ca^{2+} mobilization 50% compared to non-mutated controls in response to PBAN stimulation, suggesting that these residues may comprise potential contact points or that they contribute to the overall stabilization of the ligand binding domain.

Choi *et al.* [48] also examined the role of *N*-glycosylation, which is frequently associated with efficient cell surface trafficking [53], on HzeaPBANR function. Glutamine substitution of the two consensus *N*-glycosylation sites, N19 and N22, in the HzeaPBANR amino terminus (Figure 10) reduced the efficiency of Ca^{2+} mobilization 50% compared to non-mutated controls in response to PBAN stimulation. This reduction in receptor activity was attributed to disruption of forces stabilizing the overall HzeaPBANR structure [48]. Intriguingly, the BmorPBANR amino terminus also has two consensus *N*- glycosylation sites (N18 and N21). Deletion of the first 27 residues, including the two *N*- glycosylation sites, from the BmorPBANR amino terminus, however, had no effect on receptor trafficking, ligand binding, or ligand-induced internalization of a red fluorescent PBAN analog [54]. This apparent discrepancy in function could simply be an artifact of the differing assays, or could reflect differences in the respective receptors as the affects of *N*-glycosylation on trafficking and activity are known to be largely receptor-dependent [53]. Alternatively, glycosylation has been shown to play a role in stabilizing the position of the rhodopsin amino terminus [55]. Because considerable evidence suggests that the amino terminus of many peptide GPCRs is involved in ligand binding [49], it could be that glycosylation likewise stabilizes the positioning of the HzeaPBANR amino terminus in relation to the PBAN binding domain.

In silico methods have also been adapted to explore the structural determinants that govern ligand interactions with PBANR. While overall low sequence identity with GPCRs (HzeaPBANR is 15% identical) is a limitation to using the rhodopsin structure as a template for GPCR homology building [56-58], this approach has proven useful for identifying potential ligand contact points [56].

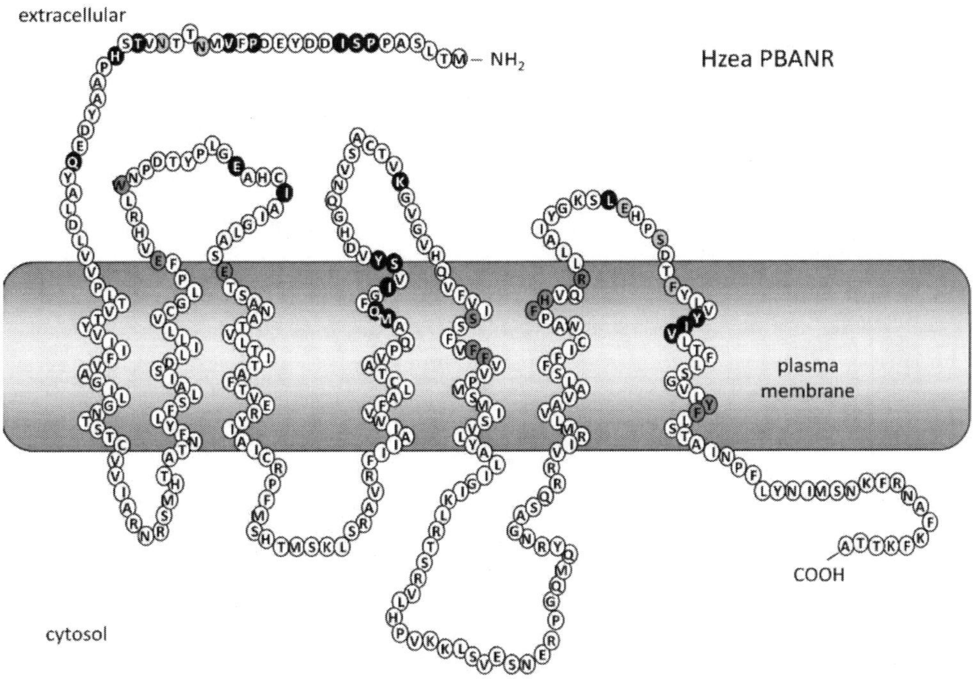

Figure 10. Schematic illustration of amino acids in the *H. zea* PBANR targeted for mutational analyses or predicted to comprise a portion of the ligand pocket. Residues depicted in light grey are positions that have undergone site-directed mutagenesis [48, 52]; mutations of N19, N22, S300, and F303 decreased receptor activity. Positions depicted in medium grey correspond to residues predicted using an evolutionary trace model to encode a portion of the ligand pocket [6]. Positions depicted in black correspond to residues predicted from a three-dimensional HzeaPBANR model to have potential ligand contact points [59].

To overcome these limitations, Stern *et al.* [59] incorporated a multiple sequence alignment of seven PBANR-like receptors prior to modeling the structural features of HzeaPBANR on the bovine rhodopsin crystal X-ray structure [60]. Molecular docking of a PBAN hexapeptide analog (YFSPRLamide) to that model identified a number of residues that potentially comprise the inner side of the HzeaPBANR binding domain: P7, S8, I9, P15, V17, T24, H26, Q33, E110, I114, M177, Q178, I181, S183, Y184, K197, L296, Y307 I308, and V309 (Figure 10) [59]. Experimental verification of these sites, however, has yet to be demonstrated. Building on the same structure, Stern *et al.* [59] also modeled the conformational effects of the HzeaPBANR and DmelPKR1 extracellular domain swaps (see above). In agreement with the experimental data, they found that the number of potential ligand contact points decreased 37%, 84% and 58% in the ECL1, 2, and 3 swaps, respectively. The large reduction seen with the ECL2 swap resulted from displacement of the loop following misorientation of the cysteines composing the ECL2-TM3 disulfide bridge.

In a complementary approach, Jurenka and Nusawardani [6] combined multiple sequence alignments and phylogenetic analysis of related receptors to predict potential ligand binding and discrimination sites in PBANRs. Based on the observation that the spatial coordinates of ligand binding domains in GPCR structures tend to be evolutionary conserved, the authors suggested that potential PBAN interactions sites included: TM2 - E96 and W101, TM3 - E121; TM5 - S208, F212, and F213; TM6 – F283, H284, and R287; and TM7 – Y318 and F319 (Figure 10). Because ECLs have been linked to ligand specific conformational changes [46,49,50], it has been suggested that charged residues in GPCR ECLs could also be critical to ligand recognition. Alanine substitution of E297 in ECL3 of HzeaPBANR, however, was shown to have little affect on receptor activity [52].

Due to the distinctive focuses between the *in silico* approaches of Stern *et al.* [59] and Jurenka and Nusawardani [6], different aspects of the potential binding pocket were identified. The structural approach using extracellular domain swaps concentrated on the potential role of the ECLs, while the evolutionary trace approach focused on identifying the conserved GPCR binding pocket bounded by the TM helices. Regardless, both approaches have identified a number of potential ligand interaction points that await experimental verification.

2.2. Intracellular Domains

In contrast to the ligand binding function of the GPCR extracellular domains, the intracellular domains comprising the carboxyl terminus and intracellular loops (ICLs) are critical to propagation and termination of the ligand signal [61,62]. Following ligand binding, heterotrimeric guanine nucleotide-binding proteins (G-proteins) propagate the signal to other effector molecules that transmit the signal to downstream cellular processes, which then feedback to regulate and desensitize the receptor [61,62].

A pivotal mechanism underlying GPCR desensitization is removal of the receptor from the cell surface by endocytotic mechanisms, followed by post-endocytotic sorting of ligand-free receptor back to the plasma membrane or to lysosomes for degradation and eventual down-regulation of receptor expression [61,62].

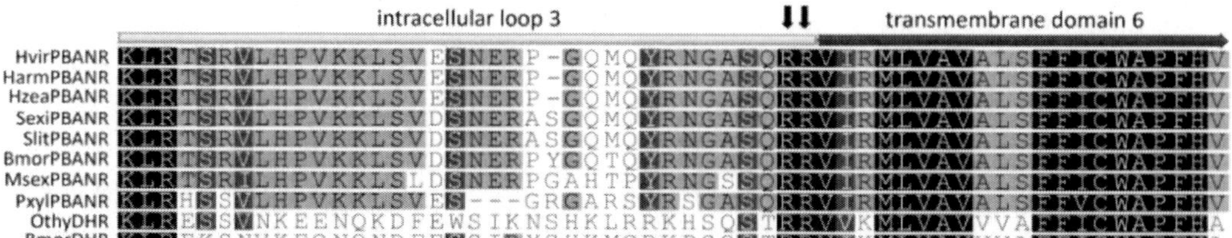

Figure 11. Sequence alignment of amino acids comprising the ICL3-TM6 junction in PBANRs and closely related receptors. Sequence abbreviations are: HvirPBANR (*Heliothis virescens*, ABV58013); HarmPBANR (*Helicoverpa armigera*, AAW47417); HzeaPBANR (*Helicoverpa zea*, AAP93921); SexiPBANR (*Spodoptera exigua*, ABY62317.2); SlitPBANR (*Spodoptera littoralis*, ABD52277); BmorPBANR (*Bombyx mori*, BAD44726); MsexPBANR (*Manduca sexta*, ACQ90219); PxylPBANR (*Plutella xylostella*, AAY34744); OthyDHR (*Orgyia thyellina* diapause hormone receptor, BAF36977); BmorDHR (*Bombyx mori*, BAE93495); AgamPKR1 (*Anopheles gambiae* pyrokinin receptor 1, AAZ43087). Arrows indicate the conserved Arg residues. Figure from [54].

Recruitment of proteins involved in mediating the desensitization/endocytotic process is dependent on GPCR phosphorylation. While components in the signal propagation and receptor desensitization pathways have been identified, less is known concerning the specific structural motifs within the GPCRs that mediate these processes, especially in insect receptors. It is clear, however, that the residues that constitute these motifs reside in the ICLs and carboxyl terminus. To begin to address these structure-function questions as they relate to PBANR, we used both knowledge-based and empirically driven site-directed mutagenesis to identify functionally important residues in PBAN signal propagation and PBANR regulation.

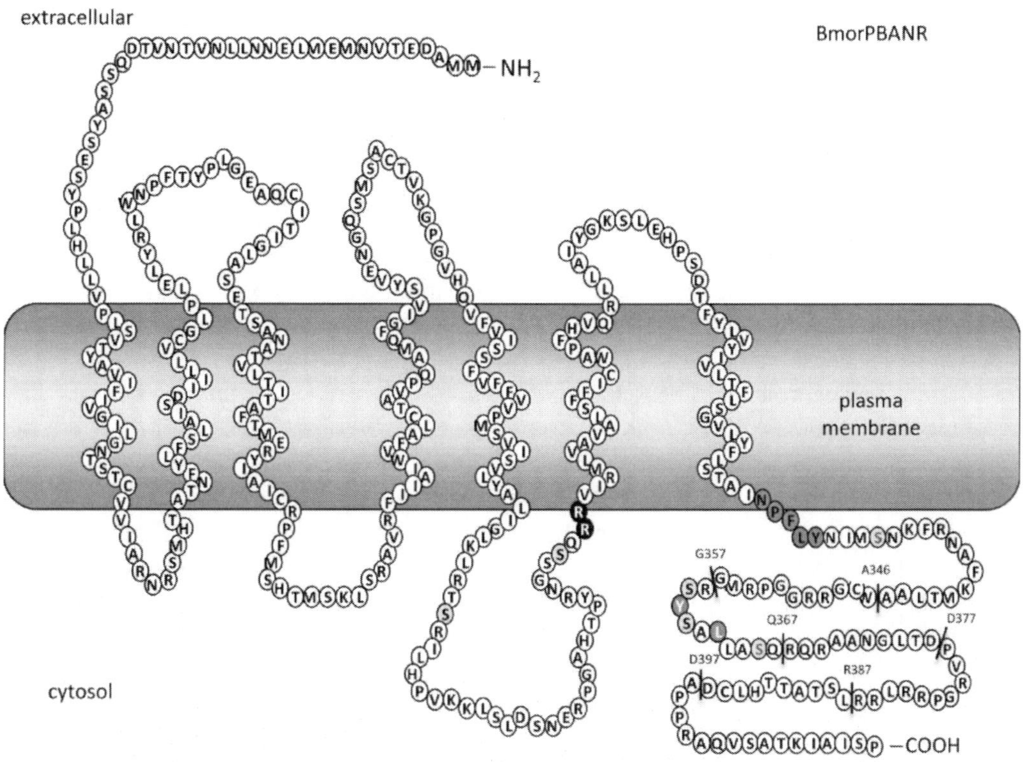

Figure 12. Schematic illustration of amino acids in the *B. mori* PBANR targeted for mutational analyses. Residues depicted in black (R263 and R264) are potential G-protein interaction sites [54]. Positions depicted in light grey correspond to potential phosphorylation sites that underwent Ala substitution [54]. Mutation of S333 and S366 (both of which are consensus PKC sites) impaired ligand-induced internalization. Potential internalization motifs (NPxxY and YxxΘ□□are shown in dark grey. The essential residues in the YxxΘ internalization motif (Y360 and L363) are shown in white font [19]. The locations of the carboxyl terminal truncations (A346, G357, Q367, D377, R387, and D397) are shown. Ligand-induced internalization was effectively abolished following truncation at position G357 [19].

Sex pheromone production in many moths is dependent on extracellular Ca^{2+} [5], suggesting PBANR activation triggers the opening of cell surface ion channels. We demonstrated that this event in *B. mori* is mediated by a Gaq-linked pathway [63]. Cationic residues at the cytosolic junction between ICL3 and TM6 of Gaq-linked GPCRs form ionic interactions with oppositely charged residues in the carboxyl terminus of Gaq [64-67]. Two conserved Arg residues (R263 and R264 in BmPBANR) are at this junction in all PBANRs identified to date, as well as closely related receptors (Figure 11). As indicated above, a critical element in GPCR biology is internalization of the ligand- receptor complex [61,62]. Using fluorescent confocal microscopic imaging in conjunction with a rhodamine red-tagged PBAN analog (RR-PBAN) and fluorescently tagged receptors in cultured insect cells, we were able to quantify ligand internalization relative to cell surface fluorescence, which we used as a indicator of PBANR activation. Site-directed mutagenesis (Figure 12) of R263 and R264 in BmorPBANR with either Ala (neutral amino acid) or Gln (anionic amino acid) reduced ligand-induced internalization 34% and 58% respectively compared to wildtype BmorPBANR [54]. The substitutions had no effect on targeting of the mutated receptors to the cell surface. We suggested that the neutral charge associated with the Ala mutations likely weakened coupling but did not completely disrupt it, thus causing the reduction in internalization. The charge-charge repulsions that would arise with the anionic Gln mutations, in contrast, would be expected to destabilize Gaq coupling and thus prevent propagation of the PBAN signal necessary for activating the regulatory feedback loop that would remove the receptor from the cell surface.

Taken together, these data provide a strong indication that these Arg residues are critical for PBAN signal transduction.

A number of conserved motifs, many of which reside in the carboxyl terminus, have been shown to play critical roles in GPCR internalization [61,62]. BmorPBANR has two tyrosine-based endocytotic signals (NPxxY and YxxΘ) in the carboxyl terminus/TM7 (Figure 12) that have been described as internalization signals in some GPCRs. The NPxxY motif is utilized in internalization of many vertebrate GPCRs [68-73], but is not universal as mutations of the critical residues had little effect on internalization of the gastrin-releasing peptide receptor or the type I angiotensin II receptor [74,75].

The YxxΘ motif (Y = tyrosine, x = any amino acid, and Θ = amino acid with a bulky hydrophobic sidechain) was implicated in ligand-induced internalization of protease activated receptor-1 [76] among others [77], and is present in the carboxyl terminus of numerous peptide GPCRs. To define the site(s) important for BmorPBANR internalization, we constructed a series of carboxyl terminal truncations with a fluorescent reporter protein fused in-frame with the last PBANR residue (Figure 13A) [19]. Ligand-induced internalization in cells expressing BmorPBANR truncated at G357 was significantly reduced compared to controls (Figure 13B). We thus concluded that the BmorPBANR internalization motif resided in the amino acid region spanning G357 and Q367. This region contains a putative YxxΘ endocytotic motif, YSAL (residues 360-363), which when mutated to ASAA impaired BmorPBANR internalization, albeit not to the same degree as the G357 truncation. We concluded that the YxxΘ motif is necessary but not essential for BmorPBANR internalization. This suggests that, like other receptors [78-80], multiple endocytotic signals are utilized. BmorPBANR also has a NPxxY motif (NPFLY, residues 325–329) near the TM7-carboxyl terminus cytosolic junction (Figure 12). The severely impaired internalization in the G357 truncation, which contains this motif, though suggests that it is not utilized. Examination of other PBANRs and closely related insect receptor sequences revealed that the carboxyl terminal YxxΘ□sequence is conserved consisting of YSAL, YSSL, YLVL, YLIL, YSAV, YNGV, and a closely related sequence, FSAL (Figure 14). Intriguingly, aside from a few exceptions, the presence of the motif sorts with the PBANR/PK2R clade and is not found in the PK1R/DHR group, suggesting that regulation of those receptors either utilizes a different internalization signal or proceeds via a non-endocytotic pathway. Whether or not this sequence is sufficient in and of itself to promote internalization of these PBANRs has yet to be experimentally determined.

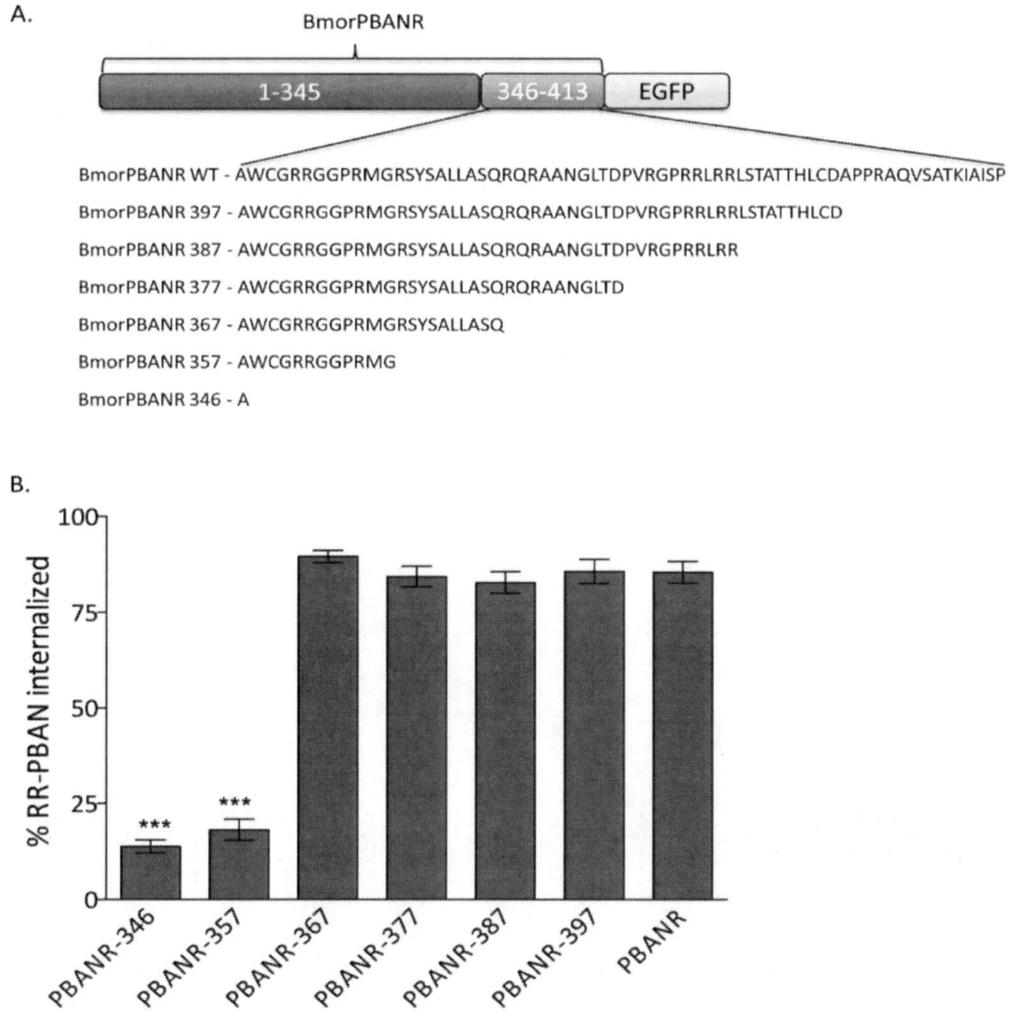

Figure 13. Effect of *B. mori* PBANR carboxyl terminus on ligand-induced internalization. A. Schematic representation of *B. mori* PBANR-EGFP carboxyl terminal truncations. The location and sequence of the truncations are indicated. B. Effect of the carboxyl terminal truncations on rhodamine red-PBAN (RR-PBAN)-induced internalization. Cells transiently expressing the respective truncations were incubated with 50 nM RR-PBAN for 1 h at 4 °C and then transferred to 27 °C for 30 min prior to fixation. The percentage of RR-PBAN internalized *vs.* cell surface localized was determined using multiple confocal microscopy images and represent means ± SEM from three separate transfections. Figure adapted from [19].

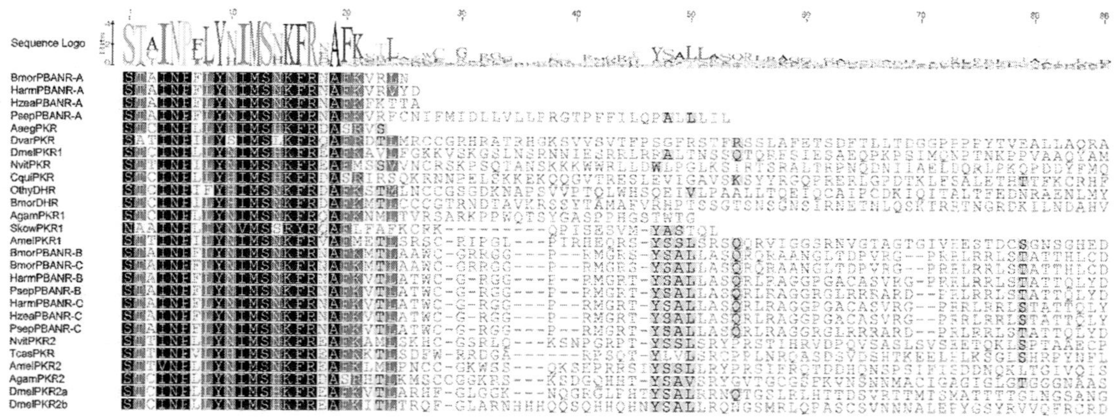

Figure 14. Conservation of the YxxΘ internalization motif among PBANRs. A portion of the carboxyl terminal end sequences from PBANRs and related receptors were aligned using the L-INS-I strategy in MAFFT [113]. The YxxΘ internalization motif is conserved in long variant PBANRs as well as PK2 receptors from a number of species. The DHR and PK1 receptors, in contrast, lack this motif, suggesting that feedback regulation of those receptors utilizes different mechanisms. Sequence abbreviations are: BmorPBANR-A (*B. mori*, AEX15646); HarmPBANR-A (*H. armigera*, AEX15647); HzeaPBANR-A (*H. zea*, AAP93921); PsepPBANR-A (*P. separata*, AEX15648); AaegPKR (*Aedes aegypti*, XP_001657210); DvarPKR (*Dermacentor variabilis*, ACC99623); DmelPKR1 (*D. melanogaster*, NP_001014620, CG9918); NvitPKR (*Nasonia vitripennis*, XP_001606201); CquiPKR (*Culex quinquefasciatus*, XP_001861460); OthyDHR (*O. thyellina*, BAF36977); BmorDHR (*B. mori*, BAE93495); AgamPKR1 (*A. gambiae*, AAZ43087); SkowPKR1 (*Saccoglossus kowalevskii*, ZP_002738336); AmelPKR1 (*Apis mellifera*, NP_001157480); BmorPBANR-B (*B. mori*, AEX15643); BmorPBANR-C (*B. mori*, BAD44726); HarmPBANR-B (*H. armigera*, AEX15644); PsepPBANR-B (*P. separata*, AEX15645); HarmPBANR-C (*H. armigera*, AEX15641); HzeaPBANR-C (*H. zea*, AEO17028); PsepPBANR-C (*P. separata*, AEX15642); NvitPKR2 (*N. vitripennis*, XP_001600587); TcasPKR (*Tribolium castaneum*, XP_968501); AmelPKR2 (*A. mellifera*, NP_001091688); AgamPKR2 (*A. gambiae*, AAX84798); DmelPKR2a (*D. melanogaster*, AAF54929, CG8795); DmelPKR2b (*D. melanogaster*, AAF54930, CG8784). In the sequence logo stacks [114], the height of each stack corresponds to the degree of sequence conservation at that position.

Ligand-induced internalization of GPCRs is triggered in response to phosphorylation of sites in the ICLs and carboxyl terminus. Consistent with this paradigm, we demonstrated that ligand-induced internalization of BmorPBANR could be blocked following general kinase inhibition [19]. We thus sought to determine which sites are utilized in BmorPBANR internalization. Of the 30 potential intracellular phosphorylation sites, we targeted six Ser residues (S233, S261, S333, S359, S361 and S366) for Ala substitution (Figure 12) and assessed their affect on RR-PBAN internalization. Despite close proximity to regions implicated in G-protein coupling, Ala substitution of S233 and S266 had no effect. In contrast, double mutation of S333 and S336 significantly impaired internalization, while double mutation of S359 and S361, which bound the critical tyrosine (Y360) in the YxxΘ motif, had a moderate effect [54]. Consequently, the modest impairment in internalization following Ala substitution of these Ser residues may be a consequence of perturbing the conformational space surrounding the motif rather than an indication of phosphorylation. S333 and S366A are consensus protein kinase C (PKC) sites (i.e., S/T-X-R/K). The strong negative effect on internalization following Ala substitution of these residues suggests that they are the principal phosphorylation sites used to initiate desensitization and endocytosis of BmorPBANR. While typically associated with the actions of GPCR kinases, feedback regulation can also be triggered by second messenger-dependent kinases such as PKC [61,62]. Reduced BmorPBNAR internalization following pharmacological-based kinase inhibition [19] and targeted PKC RNAi-mediated knockdown [54] are consistent with the role of S333 and S366 in BmorPBANR regulation. These Ser sites are also evolutionarily conserved in other PBANRs, further demonstrating their functional importance.

3. PBANR As a Potential Target

Given the critical importance of sex pheromones for moth mating, it is clear that any disruption in pheromone perception and/or production would be useful in reducing the incidence of mating. Indeed, integrated pest management strategies have already benefitted from synthetic pheromone-mediated mating disruption [81,82]. Targeted disruption of moth sex pheromone biosynthetic pathways, or the myriad other physiological functions (cuticular melanization, smooth muscle contraction, induction of embryonic diapause, etc.) regulated by the FxPRLamide family of peptides likewise offer attractive avenues for pest control. Peptide ligands, however, are unsuitable insect control candidates for a number of reasons (environmentally unstable, poor solubility making transition across the insect cuticle difficult, rapid proteolytic degradation). Insect GPCRs, in contrast, have been proposed as promising targets for the next generation of insecticides [83-86]. As such, significant effort has gone towards developing peptidomimetic approaches and identifying the chemical modifications needed to generate PBAN agonists and antagonists. Because this topic has been extensively reviewed elsewhere [83,87-90], we provide only a brief overview of some of the more intriguing developments.

While not suitable for receptor disruption, PBANR agonists (small molecules that activate the receptor in the absence of the endogenous ligand) provide valuable insights into the structural requirements and chemistries necessary for ligand binding and cuticular penetration. This information is necessary for designing antagonistic PBANR mimetics.

Nachman *et al.* [91] demonstrated the importance of Phe in the FTPRLamide PBAN sequence for interacting with PBANR. Replacing the associated phenyl ring with a carboranyl moiety increased the overall surface area and hydrophobic character of the Phe sidechain and was 10-fold more potent than native PBAN in stimulating pheromone biosynthesis in *H. virescens*. Furthermore, unlike native PBAN, topical application of the analog in aqueous solution also triggered pheromone production, suggesting that similar peptidomimetics may be able to penetrate the hydrophobic insect cuticle. Replacement of the Phe residue with either 1-pyrenebutyric acid or 9-fluoreneacetic acid induced continuous pheromone production for 18 hours in *H. virescens* following topical application [92], demonstrating that addition of hydrophobic moieties to the PBAN active core could also yield peptide mimetics that effectively penetrate the insect cuticle and have prolonged physiological effects. Attachment of 2-amino-7-bromofluorene to the terminal Phe in the PBAN FxPRLamide sequence likewise induced pheromone production when injected and topically applied. Intriguingly, while stimulating prolonged pheromone production, 500 pmol of the analog also induced mortality within 24 hours of injection. Mortality was not an effect of the bromofluorene itself, but was rather linked to bromine and the PBAN active core, suggesting direct involvement of the PBAN receptor [93]. The *trans*-orientation of the Pro in the core PBAN FxPRLamide sequence has also been identified as crucial for PBANR interactions [94].

It has been reported that replacement of naturally occurring L-amino acids with D-isomeric aromatic amino acids such as D-Phe or D-Trp can convert PBAN agonists to antagonists [95,96]. Zeltser and co-workers used this approach to develop a linear PBAN antagonist in which the 4th position in the peptide RYFTPRLamide was replaced with D-Phe to generate RYF[dF]PRLamide. Injection of this PBAN-mimetic significantly inhibited sex pheromone production in *Heliothis peltigera* (bordered straw moth) [97]. Building on this compound, Altstein *et al.* [98] increased peptide stability and enhanced bioactivity by developing a series of cyclic backbone-constrained PBAN analogs (see [99] for review of this method). Injections revealed that the inhibitory effects for some of the analogs persisted for several hours. Recently, topical application in an aqueous solution revealed that two of the more potent analogs (BBC-25 and BBC-28) inhibited endogenous sex pheromone production for up to 9 hours post-application [100]. In a parallel study, the linear RYF[dF]PRLamide analog was modified by incorporating an aliphatic amine to enhance cuticular penetration while retaining specific inhibition of sex pheromone production [101]. These cuticular permeant antagonists represent major advances in the development of novel, ecologically-responsible pest control tactics and will likely serve as structural scaffolds for future studies.

As indicated previously, in addition to regulating moth sex pheromone production, the PBAN/pyrokinin family of peptides also mediate a number of other physiological/biological responses (see [5]). A number of studies examining the effects of agonists and/or antagonists have reported differing inhibitory and stimulatory patterns as well as assay-specific differences in the efficacies of some compounds [89,102,103]. While these discrepancies may reflect differences inherent to the individual assays, we suggest that, given the concomitant expression of multiple PBANR variants in multiple tissues, the observed variation is linked to PBANR heterogeneity. As discussed above, clear differences have been reported for the PBANR-A and PBANR-C variants in terms of Ca^{2+} mobilization in response to PBAN [26,39]. Furthermore, heterodimerization amongst the PBANR variants would be expected to influence receptor behavior including responses to agonists/antagonists, G-protein signaling, cell surface trafficking, and internalization [104].

RNAi – The New Frontier?

As an ecologically responsible system that can be specifically tailored to individual pest species, RNAi-mediated knockdown of essential genes holds great promise for the future of insect pest management [105,106]. While still in its infancy, the efficacy of the methodology in transgenic plants has been effectively demonstrated in targeting the vacuolar ATPase of *Diabrotica virgifera virgifera* (western corn rootworm) [107], a cytochrome P450 involved in detoxification of a plant metabolite in *H. armigera* [108,109], and a gene critical to aphid-plant interactions in *Myzus persicae* (green peach aphid) [110]. While similar targeted disruption of PBANR activity in terms of sex pheromone production currently lacks a reliable delivery mode, it does present an intriguing possibility. To date, PBANR transcripts have been knocked down in three moth species – *B. mori*, *P. xylostella*, and *H. armigera*. We showed that injection of dsRNAs corresponding to a 417 nucleotide fragment of the BmorPBANR open reading frame into 1-day-old pupae would trigger receptor knockdown in adult females [9]. The observed transcript knockdown was dose dependent and severely impaired sex pheromone production (Figure 15). While effective for defining the *in vivo* role of BmorPBANR in sex pheromone biosynthesis, the high amount of dsRNAs (10 μg) required for knockdown suggests this specific fragment is a poor candidate target for development as a pest control tactic. Lee and co-workers [28], however, reported that injecting 100 ng dsRNA corresponding to a portion of the PxylPBANR sequence into pupae prior to adult emergence was sufficient to induce PBANR knockdown for three days. Using this method, they were able to achieve ~50% reduction in sex pheromone production with a 20-40% reduction in the female mating rate. They also reported that PxylPBANR knockdown decreased the expression of two desaturases thought to be involved in the *P. xylostella* sex pheromone pathway [111]. That study, however, failed to include a nonsense dsRNA control to account for non-specific RNAi effects. Unlike PBANR knockdown in the other two species, knockdown in *H. armigera* was performed using adult male moths. A previous study [24] revealed HarmPBANR expression in the male abdominal tip, a body region in lepidopteran species associated with male-derived sex pheromone-like compounds.

Bober and Rafaeli [112] explored the role HarmPBANR might have in mediating the biosynthesis of the male-derived compounds by injecting 10 μg dsRNA corresponding to a 880 base pair fragment of the HarmPBANR open reading frame into the hemocoel between the sixth and seventh abdominal segments of 1-day-old adult male *H. armigera*. They observed reduction in a number of the male-derived compounds with 59-74% inhibition. While the behavioral relevance of these compounds remains to be demonstrated, similar compounds have been linked to stimulation of female receptivity and inhibition of male competition. Regardless, these results demonstrate that PBANR functionality is not restricted to females and further underscores the pleiotropic nature of the receptor and its multifunctional ligand.

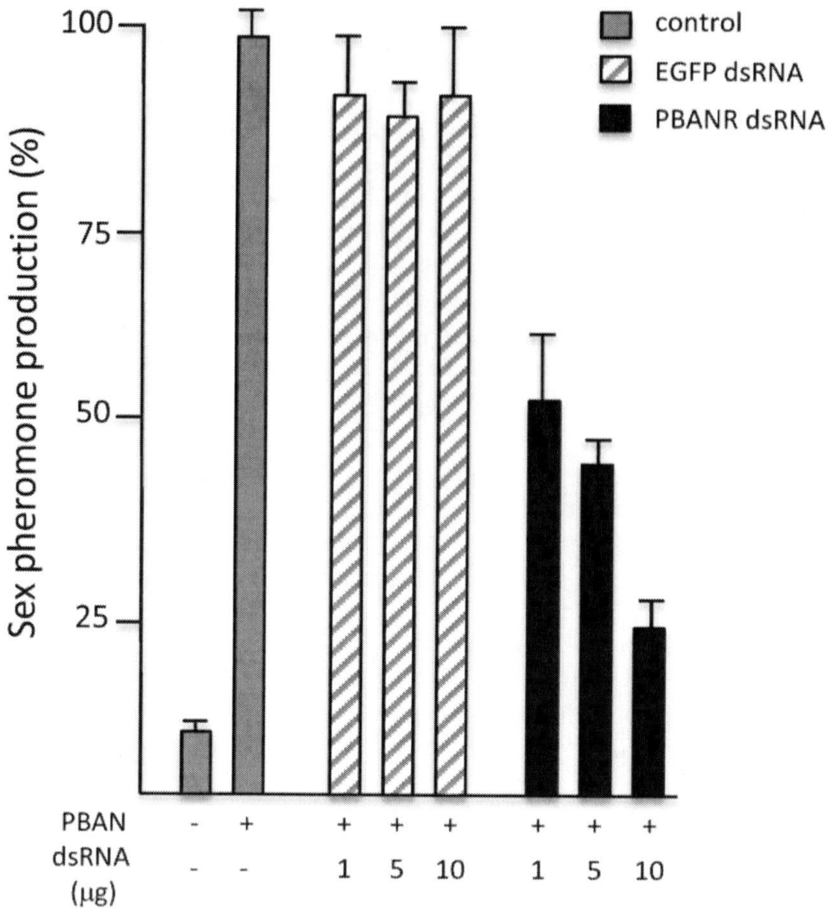

Figure 15. Dose-dependent effects of RNAi-mediated *B. mori* PBANR knockdown on *in vivo* sex pheromone production. One-day-old pupae were injected with dsRNAs corresponding to a portion of BmorPBANR or EGFP (nonsensical RNAi control). Results are expressed as percent reduction in sex pheromone levels compared with non-RNAi-treated females after intra-abdominal injections of 5 pmol of PBAN (n > 9). Figure adapted from [9].

CONCLUSION

Further research into the mechanisms underlying PBANR function in moths, as well as related receptors in other species, will help answer questions regarding the biological significance of the FxPRLamide family and how alternative splicing plays a role in mediating that biology. This knowledge is expected to provide insights into the complexities of GPCRs, and potentially can be applied towards the development of novel biorationally designed insect control agents.

REFERENCES

[1] Ando T, Inomata S-I, Yamamoto M (2004) Lepidopteran sex pheromones. *Topics Current Chem.* 239: 51-96.

[2] Tillman JA, Seybold SJ, Jurenka RA, Blomquist GJ (1999) Insect pheromones--an overview of biosynthesis and endocrine regulation. *Insect. Biochem. Mol. Biol.* 29: 481–514.

[3] Raina A, Jaffe H, Kempe T, Keim P, Blacher R (1989) Identification of a neuropeptide hormone that regulates sex pheromone production in female moths. *Science*. 244: 796-798.

[4] Kitamura A, Nagasawa H, Kataoka H, Inoue T, Matsumoto S, *et al.* (1989) Amino acid sequence of pheromone-biosynthesis-activating neuropeptide (PBAN) of the silkworm, *Bombyx mori. Biochem. Biophys. Res. Commun.* 163: 520–526.

[5] Rafaeli A (2009) Pheromone biosynthesis activating neuropeptide (PBAN): regulatory role and mode of action. *Gen. Comp. Endocrinol.* 162: 69–78.

[6] Jurenka R, Nusawardani T (2011) The pyrokinin/ pheromone biosynthesis-activating neuropeptide (PBAN) family of peptides and their receptors in Insecta: evolutionary trace indicates potential receptor ligand-binding domains. *Insect Mol. Biol.* 20: 323–334.

[7] Matsumoto S (2010) Molecular mechanisms underlying sex pheromone production in moths. *Biosci. Biotech. Biochem.* 74: 223-231.

[8] Matsumoto S, Fónagy A, Yamamoto M (2002) Chemical characterization of cytoplasmic lipid droplets in the pheromone-producing cells of the silkmoth, *Bombyx mori. Insect Biochem. Mol. Biol.* 32: 1447-1455.

[9] Ohnishi A, Hull JJ, Matsumoto S (2006) Targeted disruption of genes in the *Bombyx mori* sex pheromone biosynthetic pathway. *Proc. Natl. Acad. Sci. USA* 103: 4398–4403.

[10] Rafaeli A, Gileadi C (1996) Down regulation of pheromone biosynthesis: Cellular mechanisms of pheromonostatic responses. *Insect Biochem. Mol. Biol.* 26: 797–807.

[11] Rafaeli A, Gileadi C (1997) Neuroendocrine control of pheromone production in moths. *Invert. Neurosci.* 3: 223–229.

[12] Rafaeli A, Gileadi C (1999) Synthesis and biological activity of a photoaffinity-biotinylated pheromone-biosynthesis activating neuropeptide (PBAN) analog. *Peptides* 20: 787–794.

[13] Hewes RS, Taghert PH (2001) Neuropeptides and neuropeptide receptors in the *Drosophila melanogaster* genome. *Genome. Res.* 11: 1126–1142.

[14] Park Y, Kim Y-J, Adams ME (2002) Identification of G protein-coupled receptors for *Drosophila* PRXamide peptides, CCAP, corazonin, and AKH supports a theory of ligand-receptor coevolution. *Proc. Natl. Acad. Sci. USA* 99: 11423–11428.

[15] Choi M-Y, Fuerst E-J, Rafaeli A, Jurenka RA (2003) Identification of a G protein-coupled receptor for pheromone biosynthesis activating neuropeptide from pheromone glands of the moth *Helicoverpa zea. Proc. Natl. Acad. Sci. USA* 100: 9721–9726.

[16] Hull JJ, Ohnishi A, Moto K, Kawasaki Y, Kurata R, *et al.* (2004) Cloning and characterization of the pheromone biosynthesis activating neuropeptide receptor from the silkmoth, *Bombyx mori*. Significance of the carboxyl terminus in receptor internalization. *J. Biol. Chem.* 279: 51500–51507.

[17] Matsumoto S, Hull JJ, Ohnishi A, Moto K, Fónagy A (2007) Molecular mechanisms underlying sex pheromone production in the silkmoth, *Bombyx mori*: Characterization of the molecular components involved in bombykol biosynthesis. *J. Insect Physiol.* 53: 752–759.

[18] Matsumoto S, Ohnishi A, Lee J, Hull JJ (2010) Unraveling the pheromone biosynthesis activating neuropeptide (PBAN) signal transduction cascade that regulates sex pheromone production in moths. *Vitam. Horm.* 83: 425–445.

[19] Hull JJ, Ohnishi A, Matsumoto S (2005) Regulatory mechanisms underlying pheromone biosynthesis activating neuropeptide (PBAN)-induced internalization of the *Bombyx mori* PBAN receptor. *Biochem. Biophys. Res. Commun.* 334: 69–78.

[20] Moore CAC, Milano SK, Benovic JL (2007) Regulation of receptor trafficking by GRKs and arrestins. *Annu. Rev. Physiol.* 69: 451–482.

[21] Marchese A, Paing MM, Temple BRS, Trejo J (2008) G protein–coupled receptor sorting to endosomes and lysosomes. *Annu. Rev. Pharmacol. Toxicol.* 48: 601–629.

[22] McArdle CA, Franklin J, Green L, Hislop JN (2002) Signalling, cycling and desensitisation of gonadotrophin-releasing hormone receptors. *J. Endocrinol.* 173: 1-11.

[23] Pawson AJ, Katz A, Sun YM, Lopes J, Illing N, *et al.* (1998) Contrasting internalization kinetics of human and chicken gonadotropin-releasing hormone receptors mediated by C-terminal tail. *J. Endocrinol.* 156: R9–R12.

[24] Rafaeli A, Bober R, Becker L, Choi M-Y, Fuerst E-J, *et al.* (2007) Spatial distribution and differential expression of the PBAN receptor in tissues of adult *Helicoverpa* spp. (Lepidoptera: Noctuidae). *Insect Mol. Biol.* 16: 287–293.

[25] Zheng L, Lytle C, Njauw C-N, Altstein M, Martins-Green M (2007) Cloning and characterization of the pheromone biosynthesis activating neuropeptide receptor gene in *Spodoptera littoralis* larvae. *Gene.* 393: 20–30.

[26] Kim Y-J, Nachman R, Aimanova K, Gill S, Adams, ME (2008) The pheromone biosynthesis activating neuropeptide (PBAN) receptor of *Heliothis virescens*: Identification, functional expression, and structure-activity relationships of ligand analogs. *Peptides* 29: 268–275.

[27] Cheng Y, Luo L, Jiang X, Zhang L, Niu C (2010) Expression of pheromone biosynthesis activating neuropeptide and its receptor (PBANR) mRNA in adult female *Spodoptera exigua* (Lepidoptera: Noctuidae). *Arch. Insect Biochem. Physiol.* 75: 13–27.

[28] Lee D-W, Shrestha S, Kim AY, Park SJ, Yang CY, *et al.* (2011) RNA interference of pheromone biosynthesis-activating neuropeptide receptor suppresses mating behavior by inhibiting sex pheromone production in *Plutella xylostella* (L.). *Insect Biochem. Mol. Biol.* 41: 236–243.

[29] Watanabe K, Hull JJ, Niimi T, Imai K, Matsumoto S, *et al.* (2007) FXPRL-amide peptides induce ecdysteroidogenesis through a G-protein coupled receptor expressed in the prothoracic gland of *Bombyx mori*. *Mol. Cell Endocrinol.* 273: 51–58.

[30] Nilsen TW, Graveley BR (2010) Expansion of the eukaryotic proteome by alternative splicing. *Nature* 463: 457–463.

[31] Venables J, Tazi J (2012) Regulated functional alternative splicing in *Drosophila*. *Nucleic Acids Res.* 40: 1-10.

[32] Hartmann B, Valcárcel J (2009) Decrypting the genome's alternative messages. *Curr. Opin. Cell Biol.* 21: 377–386.

[33] Graveley BR, Brooks AN, Carlson JW, Duff MO, Landolin JM, *et al.* (2011) The developmental transcriptome of *Drosophila melanogaster*. *Nature* 471: 473–479.

[34] Pan Q, Shai O, Lee LJ, Frey BJ, Blencowe BJ (2008) Deep surveying of alternative splicing complexity in the human transcriptome by high-throughput sequencing. *Nat. Genet.* 40: 1413–1415.

[35] Wang ET, Sandberg R, Luo S, Khrebtukova I, Zhang L, *et al.* (2008) Alternative isoform regulation in human tissue transcriptomes. *Nature* 456: 470–476.

[36] Markovic D, Challiss RAJ (2009) Alternative splicing of G protein-coupled receptors: physiology and pathophysiology. *Cell Mol. Life Sci.* 66: 3337–3352.

[37] Minneman KP (2001) Splice variants of G protein-coupled receptors. *Mol. Interv.* 1: 108–116.

[38] Lee JM, Hull JJ, Kawai T, Goto C, Kurihara M, *et al.* (2012) Re-Evaluation of the PBAN receptor molecule: Characterization of PBANR variants expressed in the pheromone glands of moths. *Front. Endocrinol.* 3: 6.

[39] Lee JM, Hull JJ, Kawai T, Tsuneizumi K, Kurihara M, *et al.* (2012) Establishment of Sf9 transformants constitutively expressing PBAN receptor variants: Application to functional evaluation. *Front. Endocrinol.* 3: 56.

[40] Rafaeli A, Zakharova T, Lapsker Z, Jurenka RA (2003) The identification of an age- and female-specific putative PBAN membrane-receptor protein in pheromone glands of *Helicoverpa armigera*: possible up-regulation by Juvenile Hormone. *Insect Biochem. Mol. Biol.* 33: 371–380.

[41] Seck T, Pellegrini M, Florea AM, Grignoux V, Baron R, *et al.* (2005) The delta e13 isoform of the calcitonin receptor forms a six-transmembrane domain receptor with dominant-negative effects on receptor surface expression and signaling. *Mol. Endocrinol.* 19: 2132–2144.

[42] Zmijewski MA, Slominski AT (2009) CRF1 receptor splicing in epidermal keratinocytes: Potential biological role and environmental regulations. *J. Cell Physiol.* 218: 593–602.

[43] Chow KBS, Sun J, Chu KM, Tai Cheung W, Cheng CHK, *et al.* (2012) The truncated ghrelin receptor polypeptide (GHS-R1b) is localized in the endoplasmic reticulum where it forms heterodimers with ghrelin receptors (GHS-R1a) to attenuate their cell surface expression. *Mol. Cell Endocrinol.* 348: 247–254.

[44] DeFea KA, Vaughn ZD, O'Bryan EM, Nishijima D, Déry O, *et al.* (2000) The proliferative and antiapoptotic effects of substance P are facilitated by formation of a beta -arrestin-dependent scaffolding complex. *Proc. Natl. Acad. Sci. USA* 97: 11086–11091.

[45] Lai J-P, Lai S, Tuluc F, Tansky MF, Kilpatrick LE, *et al.* (2008) Differences in the length of the carboxyl terminus mediate functional properties of neurokinin-1 receptor. *Proc. Natl. Acad. Sci. USA* 105: 12605–12610.

[46] Peeters MC, van Westen G, Li Q, Ijzerman AP (2011) Importance of the extracellular loops in G protein-coupled receptors for ligand recognition and receptor activation. *Trends Pharmacol. Sci.* 32: 35–42.

[47] Yin D, Gavi S, Wang H-Y, Malbon CC (2004) Probing receptor structure/function with chimeric G-protein-coupled receptors. *Mol. Pharmacol.* 65: 1323–1332.

[48] Choi M-Y, Fuerst E-J, Rafaeli A, Jurenka RA (2007) Role of extracellular domains in PBAN/pyrokinin GPCRs from insects using chimera receptors. *Insect Biochem. Mol. Biol.* 37: 296–306.

[49] Gether U (2000) Uncovering molecular mechanisms involved in activation of G protein-coupled receptors. *Endocr. Rev.* 21: 90–113.

[50] Gether U, Asmar F, Meinild AK, Rasmussen SGF (2002) Structural basis for activation of G-protein-coupled receptors. *Pharmacol. Toxicol.* 91: 304–312.

[51] de Graaf C, Foata N, Engkvist O, Rognan D (2008) Molecular modeling of the second extracellular loop of G-protein coupled receptors and its implication on structure-based virtual screening. *Proteins* 71: 599–620.

[52] Choi M-Y, Jurenka RA (2010) Site-directed mutagenesis and PBAN activation of the *Helicoverpa zea* PBAN-receptor. *FEBS Lett.* 584: 1212–1216.

[53] Duvernay MT, Filipeanu CM, Wu G (2005). The regulatory mechanisms of export trafficking of G protein-coupled receptors. *Cell Signal* 17: 1457–1465.

[54] Hull JJ, Lee JM, Matsumoto S (2011) Identification of specific sites in the third intracellular loop and carboxyl terminus of the *Bombyx mori* pheromone biosynthesis activating neuropeptide receptor crucial for ligand-induced internalization. *Insect Mol. Biol.* 20: 801–811.

[55] Tam BM, Moritz OL (2009) The role of rhodopsin glycosylation in protein folding, trafficking, and light-sensitive retinal degeneration. *J. Neurosci.* 29: 15145–15154.

[56] Congreve M, Langmead CJ, Mason JS, Marshall FH (2011) Progress in structure based drug design for G protein-coupled receptors. *J. Med. Chem.* 54: 4283–4311.

[57] Sabio M, Jones K, Topiol S (2008) Use of the X-ray structure of the beta2-adrenergic receptor for drug discovery. Part 2: Identification of active compounds. *Bioorg. Med. Chem. Lett.* 18: 5391–5395.

[58] Mobarec JC, Sanchez R, Filizola M (2009) Modern homology modeling of G-protein coupled receptors: which structural template to use? *J. Med. Chem.* 52: 5207–5216.

[59] Stern PS, Yu L, Choi M-Y, Jurenka RA, Becker L, *et al.* (2007) Molecular modeling of the binding of pheromone biosynthesis activating neuropeptide to its receptor. *J. Insect Physiol.* 53: 803–818.

[60] Okada T, Sugihara M, Bondar A-N, Elstner M, Entel P, *et al.* (2004) The retinal conformation and its environment in rhodopsin in light of a new 2.2 A crystal structure. *J. Mol. Biol.* 342: 571–583.

[61] Ferguson S (2001) Evolving concepts in G protein-coupled receptor endocytosis: the role in receptor desensitization and signaling. *Pharmacol. Rev.* 53: 1-24.

[62] Kristiansen K (2004) Molecular mechanisms of ligand binding, signaling, and regulation within the superfamily of G-protein-coupled receptors: molecular modeling and mutagenesis approaches to receptor structure and function. *Pharmacol. Ther.* 103: 21–80.

[63] Hull JJ, Lee JM, Matsumoto S (2010) Gqalpha-linked phospholipase Cbeta1 and phospholipase Cgamma are essential components of the pheromone biosynthesis activating neuropeptide (PBAN) signal transduction cascade. *Insect Mol. Biol.* 19: 553–566.

[64] Wonerow P, Schöneberg T, Schultz G, Gudermann T, Paschke R (1998) Deletions in the third intracellular loop of the thyrotropin receptor. A new mechanism for constitutive activation. *J. Biol. Chem.* 273: 7900–7905.

[65] Yang M, Wang W, Zhong M, Philippi A, Lichtarge O, *et al.* (2002) Lysine 270 in the third intracellular domain of the oxytocin receptor is an important determinant for G alpha(q) coupling specificity. *Mol. Endocrinol.* 16: 814–823.

[66] Claus M, Neumann S, Kleinau G, Krause G, Paschke R (2006) Structural determinants for G-protein activation and specificity in the third intracellular loop of the thyroid-stimulating hormone receptor. *J. Mol. Med.* 84: 943–954.

[67] Kleinau G, Jaeschke H, Worth CL, Mueller S, Gonzalez J, *et al.* (2010) Principles and determinants of G-protein coupling by the rhodopsin-like thyrotropin receptor. *PLoS ONE* 5: e9745.

[68] Barak LS, Tiberi M, Freedman NJ, Kwatra MM (1994) A highly conserved tyrosine residue in G protein-coupled receptors is required for agonist-mediated beta 2-adrenergic receptor sequestration. *J. Biol. Chem.* 269: 2790-2794.

[69] Barak LS, Ménard L, Ferguson SS, Colapietro AM, Caron MG (1995) The conserved seven-transmembrane sequence NP(X)2,3Y of the G-protein-coupled receptor superfamily regulates multiple properties of the beta 2-adrenergic receptor. *Biochemistry* 34: 15407–15414.

[70] Le Gouill C, Parent JL, Rola-Pleszczynski M (1997) Structural and functional requirements for agonist-induced internalization of the human platelet-activating factor receptor. *J. Biol. Chem.* 272: 21289-21295.

[71] Gripentrog JM, Jesaitis AJ, Miettinen HM (2000) A single amino acid substitution (N297A) in the conserved NPXXY sequence of the human N-formyl peptide receptor results in inhibition of desensitization and endocytosis, and a dose-dependent shift in p42/44 mitogen-activated protein kinase activation and chemotaxis *Biochem. J.* 352: 399-407.

[72] He R, Browning DD, Ye RD (2001) Differential roles of the NPXXY motif in formyl peptide receptor signaling. *J. Immunol.* 166: 4099–4105.

[73] Bouley R, Sun T-X, Chenard M, McLaughlin M, McKee M, *et al.* (2003) Functional role of the NPxxY motif in internalization of the type 2 vasopressin receptor in LLC-PK1 cells. *Am. J. Physiol. Cell Physiol.* 285: C750–C762.

[74] Slice LW, Wong HC, Sternini C, Grady EF, Bunnett NW, *et al.* (1994) The conserved NPXnY motif present in the gastrin-releasing peptide receptor is not a general sequestration sequence. *J. Biol. Chem.* 269: 21755–21761.

[75] Hunyady L, Bor M, Baukal AJ, Balla T, Catt KJ (1995) A conserved NPLFY sequence contributes to agonist binding and signal transduction but is not an internalization signal for the type 1 angiotensin II receptor. *J. Biol. Chem.* 270: 16602–16609.

[76] Paing MM, Temple BRS, Trejo J (2004) A tyrosine-based sorting signal regulates intracellular trafficking of protease-activated receptor-1: multiple regulatory mechanisms for agonist-induced G protein-coupled receptor internalization. *J. Biol. Chem.* 279: 21938–21947.

[77] Pandey KN (2009) Functional roles of short sequence motifs in the endocytosis of membrane receptors. *Front. Biosci.* 14: 5339–5360.

[78] Nussenzveig DR, Heinflink M, Gershengorn MC (1993) Agonist-stimulated internalization of the thyrotropin-releasing hormone receptor is dependent on two domains in the receptor carboxyl terminus. *J. Biol. Chem.* 268: 2389–2392.

[79] Thomas WG, Baker KM, Motel TJ, Thekkumkara TJ (1995) Angiotensin II receptor endocytosis involves two distinct regions of the cytoplasmic tail. A role for residues on

the hydrophobic face of a putative amphipathic helix. *J. Biol. Chem.* 270: 22153–22159.

[80] Johnson KF, Chan W, Kornfeld S (1990) Cation-dependent mannose 6-phosphate receptor contains two internalization signals in its cytoplasmic domain. *Proc. Natl. Acad. Sci. USA* 87: 10010–10014.

[81] Witzgall P, Kirsch P, Cork A (2010) Sex pheromones and their impact on pest management. *J. Chem. Ecol.* 36: 80–100.

[82] El-Sayed AM, Suckling DM, Byers JA, Jang EB, Wearing CH (2009) Potential of "lure and kill" in long-term pest management and eradication of invasive species. *J. Econ. Entomol.* 102: 815–835.

[83] Scherkenbeck J, Zdobinsky T (2009) Insect neuropeptides: structures, chemical modifications and potential for insect control. *Bioorganic. Med. Chem.* 17: 4071–4084.

[84] Hiel MB, Loy T, Poels J, Vandersmissen HP, Verlinden H, *et al.* (2010) Neuropeptide receptors as possible targets for development of insect pest control agents. In: Neuropeptide Systems as Targets for Parasite and Pest Control. Geary TG and Maule AG, editors. Landes BioScience and Springer Science + Business Media Boston, MA. pp. 211-226.

[85] Bai H, Palli SR (2013) G protein-coupled receptors as target sites for insecticide discovery. In: Advanced Technologies for Managing Insect Pests. Ishaaya I, Palli SR, Horowitz AR, editors. Springer Netherlands. pp. 57–82.

[86] Grimmelikhuijzen CJP, Hauser F (2013) Arthropod genomics and pest management targeting GPCRs. In: Advanced Technologies for Managing Insect Pests. Ishaaya I, Palli SR, Horowitz AR, editors. Springer Netherlands. pp. 165–177.

[87] Altstein M (2001) Insect neuropeptide antagonists. *Biopolymers* 60: 460–473.

[88] Altstein M (2004) Novel insect control agents based on neuropeptide antagonists: The PK/PBAN family as a case study. *J. Mol. Neurosci.* 22: 147–157.

[89] Nachman RJ, Ben-Aziz O, Davidovitch M, Zubrzak P, Isaac RE, *et al.* (2009) Biostable beta-amino acid PK/PBAN analogs: agonist and antagonist properties. *Peptides 30*: 2174–2181.

[90] Nachman RJ (2009) Agonists/antagonists of the insect kinin and pyrokinin/PBAN neuropeptide classes as tools for rational pest control. In: Biorational Control of Arthropod Pests. Ishaaya I, Horowitz AR, editors. Dordrecht: Springer Netherlands. pp. 21-48.

[91] Nachman RJ, Teal PE, Radel PA, Holman GM, Abernathy RL (1996) Potent pheromonotropic/myotropic activity of a carboranyl pseudotetrapeptide analogue of the insect pyrokinin/PBAN neuropeptide family administered via injection or topical application. *Peptides* 17: 747–752.

[92] Teal PE, Nachman RJ (1997) Prolonged pheromonotropic activity of pseudopeptide mimics of insect pyrokinin neuropeptides after topical application or injection into a moth. *Regul. Pept.* 72: 161–167.

[93] Teal PEA, Nachman RJ (2002) A brominated-fluorene insect neuropeptide analog exhibits pyrokinin/PBAN-specific toxicity for adult females of the tobacco budworm moth. *Peptides* 23: 801–806.

[94] Nachman RJ, Wang XJ, Etzkorn FA, Aziz OB, Davidovitch M, *et al.* (2009) Evaluation of a PK/PBAN analog with an (E)-alkene, trans-Pro isostere identifies the Pro orientation for activity in four diverse PK/PBAN bioassays. *Peptides* 30: 1254–1259.

[95] Hruby VJ, al-Obeidi F, Kazmierski W (1990) Emerging approaches in the molecular design of receptor-selective peptide ligands: conformational, topographical and dynamic considerations. *Biochem. J.* 268: 249–262.

[96] Hruby VJ (2002) Designing peptide receptor agonists and antagonists. *Nat. Rev. Drug Discov.* 1: 847–858.

[97] Zeltser I, Gilon C, Ben-Aziz O, Schefler I, Altstein M (2000) Discovery of a linear lead antagonist to the insect pheromone biosynthesis activating neuropeptide (PBAN). *Peptides* 21: 1457–1465.

[98] Altstein M, Ben-Aziz O, Daniel S, Schefler I, Zeltser I, *et al.* (1999) Backbone cyclic peptide antagonists, derived from the insect pheromone biosynthesis activating neuropeptide, inhibit sex pheromone biosynthesis in moths. *J. Biol. Chem.* 274: 17573–17579.

[99] Gilon C, Halle D, Chorev M, Selinger Z, Byk G (1991) Backbone cyclization: A new method for conferring conformational constraint on peptides. *Biopolymers* 31: 745–750.

[100] Hariton A, Ben-Aziz O, Davidovitch M, Altstein M (2010) Bioavailability of backbone cyclic PK/PBAN neuropeptide antagonists--inhibition of sex pheromone biosynthesis elicited by the natural mechanism in *Heliothis peltigera* females. *FEBS J.* 277: 1035–1044.

[101] Nachman RJ, Teal PEA, Aziz OB, Davidovitch M, Zubrzak P, *et al.* (2009) An amphiphilic, PK/PBAN analog is a selective pheromonotropic antagonist that penetrates the cuticle of a heliothine insect. *Peptides* 30: 616–621.

[102] Altstein M, Ben-Aziz O, Zeltser I, Bhargava K, Davidovitch M, *et al.* (2007) Inhibition of PK/PBAN-mediated functions in insects: discovery of selective and non-selective inhibitors. *Peptides* 28: 574–584.

[103] Nachman R, Ben A, Davidovitch M, Kaczmarek K, Zabrocki J, *et al.* (2010) A novel dihydroimidazoline, trans-Pro mimetic analog is a selective PK/PBAN agonist. *Front. Biosci.* 2: 195.

[104] Haack KKV, McCarty NA (2011) Functional consequences of GPCR heterodimerization: GPCRs as allosteric modulators. *Pharmaceuticals* 4: 509–523.

[105] Price DRG, Gatehouse JA (2008) RNAi-mediated crop protection against insects. *Trends Biotech.* 26: 393–400.

[106] Burand JP, Hunter WB (2012) RNAi: Future in insect management. *J. Invertebr. Pathol.* http://dx.doi.org/10.1016/ j.jip.2012.07.012

[107] Baum JA, Bogaert T, Clinton W, Heck GR, Feldmann P, *et al.* (2007) Control of coleopteran insect pests through RNA interference. *Nat. Biotechnol.* 25: 1322–1326.

[108] Mao Y-B, Cai W-J, Wang J-W, Hong G-J, Tao X-Y, *et al.* (2007) Silencing a cotton bollworm P450 monooxygenase gene by plant-mediated RNAi impairs larval tolerance of gossypol. *Nat. Biotechnol.* 25: 1307–1313.

[109] Mao Y-B, Tao X-Y, Xue X-Y, Wang L-J, Chen X-Y (2010) Cotton plants expressing CYP6AE14 double-stranded RNA show enhanced resistance to bollworms. *Transgenic Res.* 20: 665-673.

[110] Pitino M, Coleman AD, Maffei ME, Ridout CJ, Hogenhout SA (2011) Silencing of aphid genes by dsRNA feeding from plants. *PLoS ONE* 6: e25709.

[111] Lee D, Kim Y (2011) RNA interference of PBAN receptor suppresses expression of two fatty acid desaturases in female *Plutella xylostella*. *J. Asia-Pacific Entomol.* 14: 405-410.

[112] Bober R, Rafaeli A (2010) Gene-silencing reveals the functional significance of pheromone biosynthesis activating neuropeptide receptor (PBAN-R) in a male moth. *Proc. Natl. Acad. Sci. USA* 107: 16858–16862.

[113] Katoh K, Kuma K, Toh H, Miyata T (2005) MAFFT version 5: improvement in accuracy of multiple sequence alignment. *Nucleic Acids Res.* 33: 511-518.

[114] Crooks GE, Hon G, Chandonia J-M, Brenner SE (2004) WebLogo: A sequence logo generator. *Genome. Res.* 14: 1188-1190.

In: New Developments on Signal Transduction Research
Editor: Masayoshi Yamaguchi
ISBN: 978-1-62618-230-1

Chapter 4

CDK5 ACTS AS A SURVEILLANCE SYSTEM IN THE NERVOUS SYSTEM

Jared S. Rosenblum*
College of Medicine and the Department of Chemistry,
Drexel University, Philadelphia, PA, US

ABSTRACT

Cyclin dependent kinase-5 (cdk5) belongs to the family of cyclin dependent kinases (cdks) based on its sequence homology, varying between 60-70 percent, with the other cdks. However, its activity is primarily restricted to neuronal cells. It is activated by neuron-specific molecules p35 and p39 with molecular weights of 35kD and 39kD, respectively, which have similar cyclin structures. Cdk5 has been shown to be a multifunctional protein kinase that regulates a large number of neuronal processes including synaptic activity, neural system development, and memory and learning. Cdk5 is not only necessary for nervous system development but is also essential for survival. The knockout of cdk5 is embryonically lethal. It has been shown that cdk5 not only phosphorylates its substrates such as NF, tau, and other cytoskeletal molecules, but also regulates the activity of kinases and phosphatases by kinase-phosphatase cross-talk. In this review, the cross-talk of cdk5 with different kinases and phosphatases and their role in the nervous system function and development is described.

Keywords: Cdk5, neuronal cell survival, signal transduction, neurofilament regulation

INTRODUCTION

Cyclin-dependent kinase 5 (Cdk5) is unique amongst the cyclin-dependent kinases for several reasons. First, the majority of these protein kinases are mitotically active, whereas Cdk5 is active post-mitotically during the proliferative cell cycle [1]. Second, Cdk5, unlike the other members of this family, does not require a cyclin compound to be activated, though

* Corresponding Author: Email: jared.s.rosenblum@drexel.edu

it can bind to cyclin D/E. It is instead activated by an interaction with either p35 or p39. These noncyclin molecules have molecular weights of 35 kDa and 39 kDa, respectively [1]. Finally, Cdk5 is a neuron-specific kinase [1]. Thus, Cdk5 has important role in neuronal survival, neurite outgrowth, and synaptic transmission.

The relationship between the activator proteins, p35 and p39, and Cdk5 has been demonstrated, and it has been found that Cdk5 is critical for neuronal physiology while the activator proteins are only supplementary. Mice lacking the activator proteins, p35 or p39, are viable; however, lamination defects present in the cerebral cortex [1]. Mild disruption of the hippocampus and cerebellum are also found in the absence of the activator proteins [1]. The deleterious effects of p39 absence were not nearly as significant as those of p35, suggesting that p39 holds a lesser role in activating Cdk5 and that it may do so when p35 is absent in early brain development [1].

The functional necessity of Cdk5 in physiology is demonstrated by the observations of mice lacking it. Cdk5 deficiency is a fatal mutation. Mice lacking Cdk5 die perinatally with defects in neuronal migration. The double knockout of p35/p39 showed the same deleterious effects as Cdk5 knock-outs, showing that p39 is compensatory and Cdk5 is critical for neuronal function [1].

The actions of Cdk5 are related to its substrate. As a proline-directed serine/threonine kinase, Cdk5 acts on neurofilaments and tau in the nervous system, components of the neuronal cytoskeleton [1]. The restriction of Cdk5 action to neurons is not due to the local production of Cdk5, which is expressed in a number of tissues. Rather, Cdk5 activators p35 and p39 are neuronally specific. The phosphorylation of neurofilaments by Cdk5 is topographically regulated. A persistent question remains unanswered regarding this topographic regulation: the cell bodies of physiologic neurons show little or no phosphorylation of neurofilaments while the axons of these neurons contain extensively phosphorylated neurofilaments. The pathology of neurons is thus related to the abberant phosphorylation of neurofilaments in the cell bodies [1].

Evidence of the relationship of Cdk5 to the pathology of neurons was demonstrated in Cdk5 knockout mice. These mice not only showed the neuronal migration patterns discussed above but also showed phosphorylation of cytoskeletal proteins such as tau and neurofilaments in the brain stem and spinal cord neurons [1]. It was thus concluded that Cdk5 activity regulates the activity of other kinase pathways and that a cross-talk system might be present and responsible for neuronal regulation.

INTEGRIN A_1B_1 REGULATES CDK5 ACTIVITY

The activity of cdk5, *i.e.* axonal cytoskeleton protein phosphorylation, is regulated by exogenous signals including integrin $\alpha_1\beta_1$ [1]. Not only are integrins a family of proteins responsible for cellular adhesion to the extracellular matrix, but these α_1/β_1 heterodimeric receptors also transduce signals from the cell surface. The result is intracellular activation of signal pathways. In neuronal cells, the transduction of extracellular signals in this fashion mediates processes such as outgrowth through the activation of cdk5 [1]. The activation of Cdk5 causes phosphorylation of particular domains of neurofilaments, particularly hNF-H, and affects the neuronal cytoskeleton [1].

PHOSPHORYLATION OF NEUROFILAMENTS

Neurofilaments are neuron-specific intermediate filaments [1]. The current model for neurofilament association is dependent on the structure of the neurofilaments. The heavy, middle , and light neurofilaments (NF-H 190 kDa, NF-M 115 kDa, and NF-L 68 kDa) are all comprised of a central alpha helical coil-coiled domain [1]. This central domain is flanked by a globular amino-terminal head domain [1]. NF-L differs from the other neurofilaments in that it does not contain a carboxyl-terminal tail domain. This tail domain in NF-H and NF-M is highly or hyper-variable. Neurofilament assembly is accomplished primarily by the golubular head domain while the coil-coiled domain is involved in the formation of the filamentous structure [1]. The carboxyl-terminal domain forms side-arms. These side arms extend from the neurofilament and form cross links. These side chains are notable in their containing lysrine/serine/proline (KSP) repeat motifis [1]. These KSP repeats are phosphorylated specifically by proline-directed serine-threonine kinases [1]. Thus the axonal cytoskeleton is simultaneously stable and dynamic and composed of neurofilaments, microtubules, actin, and mictrotubule-associated proteins (MAPs).

The proline-directed kinases, mitogen-activated kinases, stress activated protein kinase, and GSK3 are known to be responsible for the phosphorylation of KSP repeats found in the carboxyl-terminal domains on NF-H and NF-M [1]. The NF tail domain consensus sequence for phosphorylation based on site-specific substitution studies is X(S/T)PXK [1]. NF-H is the preferred target for Cdk5 phosphorylation, as demonstrated by COS cell co-transfection studies with p35/Cdk5 or GSK3 [1]. The two kinases phosphorylate different sites on the neurofilament. Equivalent phosphorylation of neurofilaments by extracellular signal-regulated kinase 1/2 (ERK1/2) demonstrated that Cdk5 phosphorylation of neurofilaments is specified by KSP repeats in the tail domain [1]. It was determined that human NF-H (hNF-H) is the preferred substrate for Cdk5 due to the specific KSPXK motif in NF-H where X is a basic or neutral residue [1]. This was further demonstrated by the phosphorylation of a synthetic peptide containing 14 KSPXK multiple repeats derived from hNF-H by Cdk5 and by recovery of phosphorylation of the synthetic peptide following its dephosphorylation [1].

CDK5 PHOSPHORYLATES NEUROFILAMENTS IN RESPONSE TO OXIDATIVE STRESS

NF dynamics are regulated in several ways by phosphorylation. This includes stabilizing subunits and increasing their resistance to proteolysis, increasing the spacing of NFs and thus axonal caliber, dissociating NFs from the motor protein kinesin, slowing transport rate, and promoting NF-NF interactions leading to bundles within axons [2, 3]. Neurofilament axonal transport is inhibited by overexpression of Cdk5 [2]. This then induces perikaryal accumulations of disordered cables of phosphorylated neurofilaments (phospho-NF cables) [2, 4]. The accumulation of these cables perikaryally, or in proximal axons, is a characteristic feature of experimental or clinical motor neuron disease found with oxidative stress and increased Cdk5 activity [2, 4]. Thus, the possible relationship between Cdk5, oxidative stress, and phospho-NF cables was tested by treating cultured neurons with hydrogen peroxide and observing the transport of green fluorescent protein (GFP) [2].

Previous experiments demonstrated that phosphorylation of NF-H by Cdk5 leads to the inhibition of NF axonal transport and induction of perikaryal accumulations of thick and disordered NF cables [4]. These accumulations demonstrate the phosphoepitopes seen in those observed in motor neurons in Alzheimer's disease and amyotrophic lateral sclerosis (ALS) [4]. It was also suggested that mutations of side-arm domains on NF-H affect known phosphorylation sites and act as risk factors for motor neuron disease [5, 6, 7]. Finally, mutations in Cu/Zn superoxide dismutase (SOD1) are also concomitant with ALS [8, 9]. This mutation also promotes oxidative stress, and such transgenic mice show an increased p25:p35 ratio [10].

The effective increase in p25 leads to the abnormal localization and hyperactivity of Cdk5 [10]. Evidence therefore existed for the physiologic Cdk5-dependent formation of perikaryal accumulations by phosphorylation and for the pathological SOD1-dependent increase in abberant Cdk5 activity by increase of the p25:p35 ratio. Moreover, axonal transport was perturbed in mutant SOD1 transgenic mice, and said mice displayed abberant perikaryal NF accumulations [2].

While Cdk5 is localized at the cell membrane by p35, NFs are relatively inaccessible [2]. Under oxidative stress conditions, as induced by SOD1 mutations, p35 is cleaved to p25 and Cdk5 is localized to the cytosol [10]. In these conditions, NFs are readily accessible targets of abberantly active p25/Cdk5 [10]. In spinal tissue from ALS patients, NF-H phosphorylation patterns were identical to that of normal individuals [11]. This suggests the mislocation of NF accumulations rather than the misphosphorylation. In other words, the NFs are normally phosphorylated but are improperly localized [2]. This makes the involvement of the kinase (Cdk5) normally responsible for phosphorylation of NFs in the abberant accumulation plausible. Perikaryal accumulation is thus likely to occur by either impaired translocation of NFs into axons or from inhibition of proteolysis, both secondary to aberrant phosphorylation [3].

The experiments involving GFP mentioned above concluded that perikaryal accumulation after oxidative stress was in part due to impaired translocation of NFs into the axon [2]. This followed observation that the levels of newly synthesized subunits within the distal axonal neurites were reduced upon treatment with hydroxen peroxide [2]. However, an additional explanation was necessitated by the observation that, in response to oxidative stress, the ratio of phospho-dependent immunoreactivity exceeded the total in perikarya. This indicated that Cdk5 both impairs transport and fosters a specific increase in C-terminal phosphorylation of NF subunits retained within perikarya [2].

The direct mechanism for inhibition of NF transport by Cdk5 was not therein determined and may include phosphorylation of NFs directly, NF transport proteins, NF linker proteins, or cross talk [2]. p42/44 MAPK activity promotes anterograde transport. Cdk5 down-regulates the MAPK pathway and might act in this way on NF transport. It was substantiated, however, that oxidative stress, aberrant Cdk5, and disturbed NF dynamics are linked [2].

NF inclusions contain kynesin and dynein, suggesting an inability of the NFs to release these motor proteins [2]. This alone would impair axonal transport by the fact that the motors for transport are trapped. Accumulation of NF spheroids by destabilization of the axonal cytoskeleton and compromising transport may occur and represent the last occurrence leading to neuronal degeneration in neurons compromised early on by oxidate stress [2].

High Molecular Weight Neurofilament (NF-H) Phosphorylation Modulated by Pin1 via Cdk5

Under conditions of oxidative stress, Pin1, a peptidoprolyl cis/*trans*-isomerase has been shown to induce NF-H phosphorylation [12]. It is important to keep in mind that many cellular processes are controlled by reversible phosphorylation on Ser/Thr-Pro motifs by proline-directed kinases. The phosphorylation state of the protein indicates the balance between kinases and phosphatases [13]. Regulation of the latter may occur by conformational changes imposed by enzymes such as isomerases. Moreover, NF-H SP moiety repeats are physiologically phosphorylated *in vivo* and the further phosphorylation seems to require reconfiguration by peptidylprolyl isomerization [13]. The structure of NF-H is remarkably suited for Pin1 action [13]. Pin1 (protein interacting with NIMA—never in mitosis A) is likely involved based on its ability to isomerize Ser and Thr bonds to Pro [13].

Pin1 was found in neurofibrially tangles [14]. It had been previously hypothesized that Pin1 may be involved in the regulation of hypoerphosphorylation of Tau by facilitating dephosphorylation via protein phosphatase 2A (PP2A) [14]. It was found that Pin1 increased phosphorylation of NF-H, as mentioned, at KSP repeats by proline directed kinases [12]. These include ERK1/2, p35/Cdk5, and JNK3, the major stress activated protein kinase of the nervous system, specifically in cell bodies, and is activated under glutamate stress [13]. The induction of phosphorylation is seen in a concentration dependent manner. It was also shown by small interfering RNA (siRNA) inhibition or by knockout of Pin1 that epidermal growth factor-induced NF-H phosphorylation is enhanced by Pin1. It should be noted that a number of phosphorylated proteins such as RNA polymerase II and CDC25 are regulated through Pin1 induced prolyl-isomerizations [13].

Pin1 was reported to affect target proteins in two conflicting roles [13]. First, it was observed to facilitate Tau dephosphorylation by PP2A [14]. Second, it was shown that Pin1 facilitates hyperphosphorylation and inhibits dephosphorylation of RNA polymerase II [15]. Moreover a number of processes are up-regulated while others are down-regulated. The effect of oxidative stress as it relates to Pin1 was explored in cortical neurons and it was found in both peroxide-induced stress and heat shock that NF-H phosphorylation was induced [13]. This was rescued by knockout of siRNA inhibition of Pin1 [13]. It was also noted that the stress conditions caused perikaryal motor neuron inclusions typical of ALS of phospho-NF and that the associated neuronal apoptotis was similarly resuced by Pin1 inhibition [13]. Finally, it was demonstrated that inhibition of Pin1 inhibits JNK3 pathway, which is activated under oxidative stress [13]. Pin1 does, however, appear to have different roles in different neurodegenerative disorders, such as inclusions in Lewy bodies. It appears based on these findings that Pin1 is a regulator of the stress response and promotes NF-H phosphorylation and perikaryal accumulation [13].

Regulation of Glucocorticoid Receptor Transcriptional Activity by Cdk5

Glucocorticoids are major effectors of the stress response involved in homeostasis of the central nervous system [16. 17]. They impact many neuronal cell functions and are critical to homeostasis in the central nervous system. Electrical activity, synthesis or secretion of neurotransmitters and/or neuropeptides, and cell death regulation are all effects of glucocorticoids at the neuronal cellular level [16, 17]. These hormones are important in the regulation of memory, mood, and sleep [17]. It has also been found that the anatomic structure of the brain is influenced by glucocorticoids; hippocampal volume is reduced while ventricles are enlarged and reversible cortical atrophy occurs [17]. Most of these effects are mediated by action of glucocorticoids are the associated receptors, which are of the nuclear receptor superfamily [18].

Glucocorticoid receptors (GR) act as transcription factors in a substrate-dependent fashion. The responsive genes are acted on by these transcription factors following binding of the ligand that results in changes and interactions in the cytoplasm [18]. Translocation of the GR transcription factor complex occurs in an energy-dependent manner and follows liberation of the receptor from the heterooligomer that includes heat shock proteins [18]. When bound to the promotor region of glucocorticoid-responsive genes, the GR transcription factor attracts coactivator complexes as well as chromatin-remodeling factors [18]. RNA polymerase II and the associated factors are also recruited [18, 19].

These recruitments occur by the two transactivation domains of the GR transcription factor: transactivation function (AF)-1 and -2 in the N-terminal and ligand-binding domains, respectively [18, 19].

It was found that Cdk5 ineracts with the ligand binding-domain of the GR [20]. Regulation of GR transcriptional activity at many levels makes glucocorticoids critical for survival. A common mechanism of regulation of the GR transcriptional activity is by post-translational modification, *i.e.* phosphorylation of specific serine or threonine residues by cell-cycle and mitogen-activated kinases [21]. These are downstream of cell surface receptors for growth factors and cytokines [22-25].

The interaction of Cdk5 with the GR was explored via cyto-trap yeast two-hybrid screening assays using human GRα ligand binding domain as bait. It was found to be through the Cdk5 activator, p35, or the aberrant form, p25, at the C-terminal domain from residues 190-262. The binding is ligand-dependent and p25 demonstrates a much stronger association. Cdk5 phosphorylated GR at several serine residues, including Ser^{203} and Ser^{211} at the N-terminal domain.

Phosphorylation suppressed the transcriptional activity of GR on glucocorticoid-responsive promoters by attenuating attraction of transcriptional cofactors. This occurs by preventing the interaction of transcriptional cofactors and DNA. Roscovitine, an inhibitor of Cdk5 activity, differentially modulated GR transcriptional activity on endogenous glucocorticoid-responsive genes [20].

It was suggested that Cdk5 is a substrate for GR and this may be a mechanism for the regulation of Cdk5 function in the brain. The effect of Cdk5 on glucocorticoid-induced transcriptional activity is specific to the promoter region and may be positive or negative. GR phosphorylation may determine target promoter specificity as well as cofactor interaction, strength and duration of signaling, and receptor stability. Subcellular localization is thus accomplished and regulated by various kinase phosphorylation of GR, including MAPK, c-JUN N-terminal kinase, and GSK-3 [20].

REGULATION OF MINERALOCORTICOID RECEPTOR TRANSCRIPTIONAL ACTIVITY BY CDK5

The homeostasis of the central nervous system is largely affected by glucocorticoids, as mentioned above and in regulating memory, cognition, mood, and anatomic structure [16, 17]. The regulation of glucocorticoid receptor transcriptional activity by Cdk5 was discussed above [20]. It was later found that Cdk5 also regulates mineralocorticoid receptor transcriptional activity [26]. Both the glucocorticoid and mineralocorticoid receptors belong to the nuclear hormone receptor superfamily [18]. As discussed, they are dependent on agonist hormone binding to dissociate from heat shock proteins with which they are complex, translocate to the nucleus in an energetically dependent manner, and actively modify transcriptional activity at the respective responsive genes [18].

Glucocorticoid and mineralcorticoid receptors share an evolutionary origin [27, 28] and major homologies [29, 30] at their DNA binding sites and ligand binding sites, 94% and 57%, respectively. In the absence of cortisol-inactiving 11ß-hydroxysteroid dehydrogenase type 2 (11ßHSD2), the mineralocorticoid receptor binds with a higher affinity to the endogenous glucocorticoid cortisol or corticosterone [29]. Regardless of 11ßHSD2, however, both glucocorticoid and mineralocorticoid receptors are stimulated by dexamethasone and aldosterone, respectively [29].

Transcription is initiated by association of the receptors with coactivator complexes and chromatin-remodeling factors at the N-terminal domain as well as RNA polymerase II and its ancillary factors at the ligand-binding domain through the receptor transactivation domains [18]. The N-terminal domains do not have as high a sequence homology (<15%) as the other domains discussed; this adds greatly to the transcriptional specificity [30]. The characteristics of these receptors described suggests that in areas of the brain without 11ßHSD2, the mineralocorticoid receptor acts as a second glucocorticoid receptor to regulate the expression of particular genes that may have different transcriptional potencies from the glucocorticoid receptor in response to cortisol or in association with aldosterone [26].

The regulation of the mineralocorticoid by Cdk5 occurs via phosphorylation of multiple serine (Ser^{128} and Ser^{250}) and one threonine (Thr^{159}) residues located in the activation domain of the N-terminal domain. This phosphorylation is mediated by physical interaction at the ligand-binding domain with the coactivator p35. This effect was examined with aldosterone and dexamethasone as well as the endogenous glucocorticoid corticosterone. It was suggested that Cdk5 differentially alters association of the mineralocorticoid receptor with cofactors to the activation domain to negatively or positively influence transcription [26].

It was also possible that the effect was tissue specific [31]. While aldosterone treatment yielded increased mRNA and protein expression of brain-derived neurotrophic factor (BDNF), dexamethasone treatment resulted in its suppression [26]. Moreover, corticosterone demonstrated a biphasic effect [26]. Physiologically, cortisol circulates at two log units higher than aldosterone in the blood of humans and rodents [29]. The mineralocorticoid receptor has a higher affinity for cortisol than does the glucocorticoid receptor, as mentioned. Thus, at basal levels the mineralocorticoid receptor is fully activated and the glucocorticoid receptor is not; it is activated under circadian rhythms and stress when cortisol levels are elevated [26].

While the glucocorticoid receptor is expressed throughout the entire central nervous system, the mineralocorticoid receptor is only found in the hippocampus and amygdala neurons, which also lack the glucocorticoid-inactivating enzyme 11ßHSD2 [32]. The effects of cortisol in the hippocampus/amygdala system and in the hypothalamic-pituitary-adrenal axis system are contradictory. This is based on the mediation of the cortisol response by these receptors. The mineralocorticoid receptor mediates enhancement of neuronal excitability, stabilization of synaptic transmission, and stimulation of long term potentiation in CA1 cells while simultaneous presence of both receptors attenuates this excitability and disrupts long term potentiation [33, 34]. Moreover, both excess of glucocorticoid via the glucocorticoid receptor and deficiency via the mineralocorticoid receptor in the hippocampus cause apoptosis [35].

BDNF stimulates synaptic transmission in hippocampal neurons and decreases the vulnerability to glucose deprivation as well as the glutamate toxicity seen in cultured hippocampal neurons [36]. It was found that Cdk5 enhanced the effects of dexamethasone on BDNF expression and that the effects of the aberrant p25/Cdk5 complex may affect neuronal viability, synaptic plasticity, consolidation of memory, and emotional changes through the corticosteroid receptor/BDNF relationship in hippocampal neurons [26].

Mediation of Kinesin-Driven Motility by Cdk5 Regulation of GSK3

The neuronal cytoskeleton is also responsible for the generation of compartments and the subsequent distribution of membrane proteins within the axonal and dendritic compartments. Membrane proteins are thus synthesized within membrane-bound organelles in the neuronal cell body and transported to the axon by microtubule-dependent motor proteins [37]. Such transportation and targeted delivery is observed in regions of axons such as nodes of Ranvier or at presynaptic terminals [38]. Proteins and membrane components necessary for axonal function must be transported from their point of origin or synthesis in the cell body. The mature axon does not have synthetic machinery to a significant degree. The motor protein, kinesin, is an essential component of fast anterograde transport of membrane bound organelles in neurons [39]. Fast anterograde transport is a form of specialized delivery within neurons to biochemically and structurally distinct domains [40].

Kinesin, a phosphoprotein *in vivo*, is a heterotetramer composed of two heavy and two light chains, KHC and KLC respectively. While the KHC is responsible for generating force by virtue of its ATPase and microtubule-binding domain [41], KLC are essential for the binding of and specificity for cargo [42]. The release of membrane-bound organelles from kinesin is mediated by phosphorylation by GSK3, a ubiquitous kinase, on KLCs [43]. GSK3 can inhibit fast anterograde transport by interacting with kinesin. However, GSK3 was not shown to act on dynein to affect retrograde transport in extruded axoplasm from the giant axon of the squid [43]. These findings suggested, as was found, that GSK3 is inactive in the axon and mechanisms for its upregulation exist and are localized [38]. Inhibition of Cdk5 activity in axons similarly demonstrated reduced anterograde transport rates without affecting retrograde transport [38].

The lack of a consensus site for Cdk5 on kinesin suggested a common pathway between Cdk5 and GSK3 leading to regulation of kinesin and thus anterograde transport. A novel regulatory pathway for kinsein-driven motility in axons was determined wherein inhibition of

Cdk5 activity leads to protein phosphatase 1 (PP1) activation of glycogen synthase kinase 3 (GSK3). Finally, the phosphorylation of kinesin light chains is accomplished by GSK3, and kinesin is detached from its successfully transported cargo. The definition of distinct domains within axons may be a result of this pathway [38].

Inactivation of GSK3 may occur by phosphorylation on serine resiudes: Ser^9 on GSK3ß and Ser^{21} on GSK3α. This may be a result of autophosphorylation [44] or the activity of other kinases such PKB/Akt [45]. Cdk5 does not directly act on GSK3. Inhibition of Cdk5, however, resulted in Ser^9 dephosphorylation and subsequent activiation of GSK3 in cortical neurons [38]. Phosphatases were implicated in the activation of GSK3 by the observation that inhibition of serine/threonine phosphatases by pharmacological agents increased phosphorylation at the same site on GSK3 [38]. It was determined *in vitro* that PP1 was capable of phosphorylation of GSK3 at the same site. Specific peptide inhibition of GSK3 prevented inhibition of kinesin-driven motility in axoplasm [38].

GSK3 may be compartmentally or substrate-localized by proteins without intrinsic activity, *i.e.* scaffolds such as AKAP220 and axin in the Wnt signaling pathway [46]. Temporal and spatial localization of GSK3 is particularly important in neurons, wherein the constitutive activity results in neurite retraction [47], apoptosis, behavioral abnormalities [48], and embryonic lethality [49].

GSK3, PP1, and Cdk5 can interact *in vivo* and have overlapping distributions in neuritic growth cones, for example. The three components of the pathway are found in a high molecular weight complex in the brain. It is thought that Cdk5 keeps a certain population of PP1 inactive; however, Cdk5 does not directly phosphorylate PP1 [38]. Thus, it is thought that Cdk5 might phosphorylate one of associated regulatory proteins of PP1. The inactivation of GSK3 by Cdk5 is consistent with the observation that sustained Cdk5 activity is required for normal fast anterograde transport in neurons and that tau and GSK3 activity are increased in p35 knockout mice [38].

The pathway described relies on a balance of kinase and phosphatase activities to regulate fast anterograde transport within specific cellular domains.

Local changes affecting this balance include phosphorylation of neurofilaments at Nodes of Ranvier, wherein Na channels must be delivered to the cell membrane. Most important, however, is the local inhibition of Cdk5 and local activation of GSK3. Cdk5 activity can be regulated in a number of ways, including tumor suppressor protein APC, p35 binding proteins, or modification by other kinases [38].

Inhibition of Anterograde Axonal Neurofilament Transport by MAPK Activity Inhibition

Axonal cytoskeletal dyanmics are regulated primarily by phosphorylation of the constituent proteins. This includes NFs and the axon-specific microtubule associated protein tau [50, 51, 52]. Phosphorylation of the C-terminal regions of NF-H and NF-M begins in the cell bodies and continues during axonal transport [51, 53]. It has been previously shown that other kinases phosphorylate NFs at distinct sites possibly through distinct or combined mechanisms aimed at different regulation. These include GSK-3α, GSK-3ß, casein kinase I and II, and stress-activated protein (SAP) kinases [54. 55]. Cdk5 regulates NF phosphorylation and axonal transport [56]; however, the mechanisms by which both are accomplished was questioned.

Cdk5 activity down-regulates p42/44 mitogen-activated protein kinase activity by deactivation of an upstream activator (MEK-1) via phosphorylation [57]. Anterograde axonal transport of NFs is promoted by p42/44 MAPK and inhibited by Cdk5 activity [58]. The inhibition of NF transport by Cdk5 may be through inhibition of MAPK activity, an established cross-talk of Cdk5. It was shown that inhibition of Cdk5 increased anterograde axonal transport, as did loss of MAPK activity [59]. Constitutive expression of MAPK enhanced axonal transport. None of these treatments affected tau transport. Inhibition of transport by transfection with p25/Cdk5 was prevented by co-transfection with constitutively active MAPK. However, tau transport was not recovered; p25/Cdk5 still inhibited tau transport. Thus, MAPK activity is required for anterograde NF transport but not tau transport. While Cdk5 inhibits translocation of tau along axonal neurites, it is not by MAPK activity [59].

It was also found that the inhibition of axonal transport by inhibition of MAPK activity was not additive to that seen with p35/Cdk5. Moreoever, constitutively activated MAPK did not show an additive increase to that of inhibition of Cdk5 by roscovitine. This suggests that these two affectors of NF transport are in the same pathway, not separate ones. Thus, while NF transport is regulated by two distinct kinase cascades, the two overlap at the regulation of MAPK activity by Cdk5 [59].

It was concluded that manipulation of MAPK activity can oppose Cdk5 activity on anterograde axonal transport of NFs. Further, it was suggested that this influence is downstream from Cdk5. The observation that NF subunits can be selectively assembled in the distal-most region of axonal neurites, *i.e.* the growth cone, was substantiated when it was shown that cells treated with roscovitine or transfected with constitutively active MAPK were brighter in GFP studies than the rest of the axon [59]. It was suggested that an interaction between tau and NFs might exist and be regulated by MAPK and/or Cdk5. Moreover, it was said that tau interferes with axonal transport of NFs and induces oxidative stress that leads to neuronal degeneration [60]. During neurodegeneration, when oxidative stress occurs and Cdk5 is over-activated [61, 62], a feedback loop involving tau, NFs, Cdk5, and the redox state of the cell may be critical and deleterious [59].

Spatial and temporal localization is key to development of the nervous system in so far as neurite outgrowth, synaptic activity, and neuronal migration, actions mediated by Cdk5, as well as neurogenesis, neuronal differentiation, and neuronal remodeling, actions mediated by MAPK pathway. Cdk5 and MAPK contribute to the stabilization of developing axons via temporal regulation of NF accumulation and phospho-isoforms in growing axonal neurites [59]. Localization is further accomplished by the formation of cytoskeletal complexes containing NF in the brain.

Distribution of Neuronal Cytoskeletal Proteins: Notch and Cdk5

Both Notch signaling and p35/Cdk5 activity are critical to proper development of the nervous system. Notch signaling favors differentiation toward the glial cell line while inhibition of Notch results in neuronal differentiation [63]. The Notch response is mediated by its intracellular domain that is cleaved from the full-length receptor. This process occurs in two proteolytic steps and is mediated by a presenilin-secretase complex [64]. Lateral inhibitory interactions necessary for patterning of neural cells are mediated by Notch-Delta signaling [65]. In lateral inhibition, the canonical Notch signaling pathway is active and dependent on Delta/Serrate (Jagged)/Lag ligand-regulated binding of the extracellular domain of Notch [66]. Thus, a presenilin/γ-secretase complex gains access to Notch and cleaves it to release the cytosolic domain. This domain then translocates to the nucleus to form a transcriptional activation complex. The latter binds with CSL/RBP-jK and Mastermind to positively regulate transcription of Notch target genes such as *Hes* and negatively regulate *Ngn1* [67, 68].

As discussed, Cdk5 phosphorylates neuronal cytoskeletal proteins. The relationship between Notch and Cdk5 was invesitaged by observation of Cdk5 activity in the presence of *N*-[*N*(3,5-difluorophenacetyl)-1-alanyl]-*S*-phenylglycine *t*-butyl ester (DAPT), an inhibitor of Notch signaling by inhibition of γ-secretase [69]. This inhibitor also enhances neuronal differentiation of embryonic stem cell derived embryoid bodies as a consequence of its prevention of the Notch response. Cdk5 activity was attenuated with DAPT treatment despite an increase in Cdk5 protein level. This same phenomenon, herein described in cortical neurons, was found in transgenic mice. p35 levels, however, were not increased. Over-expression of p35 restored attenuation of Cdk5 activity by DAPT. This suggested that Notch interacts with Cdk5 to up-regulate its activity. In the presence of DAPT and decreased Cdk5 activity, phospho-NF and phospho-tau shifted from the axonal compartment to the cell body, which was reversed by over-expression of p35. DAPT was determined to not disrupt the interaction between p35 and Cdk5. Notch signaling was disrupted by DAPT, as demonstrated by up-regulation of *neurogenin* and down-regulation of *Hes1*. It was determined by RT-PCR that DAPT up-regulated Cdk5 expression at the transcriptional level.

The results agree with demonstrations of accumulated phospho-NF in the soma associated with reduced Cdk5 and ERK1/2 hyperactivation. This also agrees with redistributions of phosphorylated cytoskeletal proteins observed in p35 knockout mice [69]. It is important to note that though the consequences of inhibition of Cdk5 activity with DAPT and roscovitine are similar, the mechanisms are very different. Also, it is important to understand that the findings indicate the translocation of p-Tau and p-NF into the cell body acts as a primer for apoptosis. There remains no explanation for why elevated Cdk5 levels do not result in elevated activity, though it is hypothesized that inactive Cdk5 my bind to p35/Cdk5 and preclude the active site. This is supported by the rescue of activity with provision of ectopic p35. p-Tau and p-NF translocation was also reversed. Regulation of Cdk5 is affected by Notch signaling pathway and this affects localization of cytoskeletal components [69].

Organization and Interaction of Nestin with p35 by Cdk5

The localization of Cdk5 is critical to neuronal function and is accomplished by its interaction with its activator proteins, p35 and p39, as well as with scaffolds. Nestin is an intermediate filament protein expressed during the early stages of central nervous system development [70, 71]. The down-regulation of nestin in myotubes is followed by replacement by other tissue-specific IFs, specifically desmin, in differentiation [70, 71]. Nestin is of particular interest because of its involvement in pathology of the central nervous system. It is expressed in processes such as glial scarring after CNS injury [72]. Nestin is also expressed again during the regeneration of injured muscle tissue [72, 73]. Thus, it was concluded that nestin expression is found in morphologically dynamic cells, *i.e.* proliferating and migrating cells. It was further demonstrated in denervated myofibers that nestin is closely associated with neural transmission structures [73].

Cdk5 is critical for neuronal development by its involvement in neuronal migration [74] and for neuromuscular junctions by its regulation of differentiation, organization, and regeneration of muscle cells [75]. Both of these processes also, as mentioned, involve nestin and the Cdk5 activitator, p35. While nestin is present during development, it is the preferential substrate of Cdk5 over its co-expressed intermediate filament, vimentin, despite containing the proper consensus site [76]. This phosphorylation of nestin by Cdk5 at a particular threonine residue (Thr^{316}) induces reorganization of the nestin network in a spatiotemporally appropriate manner; the expression and activation of Cdk5 and nestin and subsequent association between Cdk5 and nestin are concomitant [76]. Phosphorylation by Cdk5 increases nestin solubility. This followed the known organization of Cdk5 in the nervous system by its association with cytoskeletal components in neurons. In differentiating myotubes, it was determined that Cdk5 is a modulator of nestin organization and dynamics and that nestin serves as a scaffold for Cdk5 and p35 and the related signaling [76].

Posttranslational regulation of nestin is necessary to reorganize the intermediate filaments in myotubes with indistinguishable networks of nestin, vimentin, and desmin during development. Mature myotubes have desmin and nestin localized to the Z bands until nestin expression decreases [71]. The association of nestin with p35 and Cdk5 allows for the organization and dynamics of nestin in the intermediate filament network [76]. This may have mild effects on other intermediate filaments indirectly. Not only does Cdk5 regulate nestin and its association with p35, but nestin also regulates the activity of Cdk5 [76]. Nestin accomplishes this by regulating the cleavage of p35 to p25, the protein that associates with Cdk5 and results in abberant activity. Nestin association with Cdk5 and p35 halts this process and allows continued differentiation of the myotube intermediate filament network. The activation of Cdk5 by association with p35 or p25 will allow its involvement in other pathways, such as the production of acetylcholine receptors at neuromuscular junctions [76].

NEUROMUSCULAR JUNCTION ACETYLCHOLINE RECEPTOR REGULATION BY CDK5

The activation of signaling cascades involving Ras, Raf, ERK1/2, JNK3, and phosphatidylinositol-2 (PIP_3) has been shown to increase transcription of genes producing acetylcholine receptors (AchR). Transcription is also regulated by the nerve-derived factor neuregulin (NRG) at the sub-synaptic nuclei.

The NRG family includes heregulin, acetylcholine receptor-inducing activity, neu differentiation factor, and glial growth factor. The actions of NRG are dictated by its being bound by members of the epidermal growth factor receptor family (ErbB). These tyrosine kinase receptors, particularly ErbB3 and ErbB4, are activated by this binding of NRG, and a homo- or heterodimerization occurs that results in tyrosine phosphorylation of intracellular substrates. Despite phosphorylation in response to NRG binding, the ErbB3 receptor does not have tyrosine kinase activity [77-79].

In muscle, ErbB receptors associated with Cdk5 and p35 [80]. Inhibition of Cdk5 prevents NRG binding to ErbB receptors and thus transcription of AchR [1, 81]. It can be concluded then that Cdk5 is involved in the transcription of AchR [81].

In cultured myotubes, removal of Cdk5 activity resulted in the loss of ErbB activitation [81]. NRG is involved in cell survival, differentiation, and cell migration. Cdk5 is both involved in regulating the expression of the AchR gene and in NRG signaling at the neuromuscular junction [1]. Disruption of the NRG gene results in abnormal heart and nervous system formation. NRG-signaling also mediates for the survival of Schwann cells through the PI-3K/Akt pathways.

This pathway is a critical survival pathway and is mediated by Cdk5 via phosphorylation of ErbB2 and ErbB3 and subsequent regulation of Akt activity [1]. The survivial pathway outlined is upregulated by Cdk5, as demonstrated in Cdk5 knockout mice wherein brain extracts and lysates demonstrated reduced PI-3K/Akt activities [81].

CDK5 MEDIATES THE NEUREGULIN-DEPENDENT PI-3K/AKT NEURONAL SURVIVAL PATHWAY

Neuregulins are involved in cell survival, mitosis, migration, and differentiation. The survival of Schwann cells is mediated by NRG signaling via the phosphoinositide 3-kinase (PI-3K)/Akt pathway [1]. This pathway, as mentioned, is a critical survival pathway in neurons and other cells. Cdk5 regulates this pathway by phosphorylation of NRG receptors ErbB2/ErbB3 [77-79]. This was demonstrated by the finding that brain extracts and cortical neurons from Cdk5 knockout mice contained reduced activites of PI-3K and Akt together with increased apoptosis. This pathway also presents a novel Cdk5 consensus sequence for the serine residue (Ser^{1120}) on the ErbB3 receptor: SPRXR [81].

Cell survival is favored by Akt kinase, which inhibits many apoptotic substrates, including Bad, Forkhead transcription factor, and caspase-9. Akt kinase is phosphorylated at Ser^{473} and Thr^{308} and thus activated by PDK1, which is activated by localization of phosphoinositides at the cytosolic side of the plasma membrane. This follows PI-3K activation caused by NRG binding to ErbB receptors. Cdk5 modulates this pathway in neurons via phosphorylation of ErbB2 and ErbB3 at consensus motifs containing Ser^{1176} and Thr^{871}, respectively. The novel consensus sequence was phosphorylated by Cdk5 on ErbB3 as a preferred site. Thus, Cdk5 is involved in the regulation of downstream signaling transduction pathways involving receptor protein-tyrosine kinases [81].

An important aspect of the regulation of this neuronal survival pathway is the idea that Cdk5 can regulate Akt activity independent of PI-3K activation. It has been suggested that, in Cdk5 -/- mice, several of the calcium dependent of cAMP dependent pathways involved in activation of Akt may be compromised to reduce Akt and PI-3K activities.

Most important, however, is the neuregulin-induced Schwann cell survival via the PI-3K/Akt pathway in which Akt and MAPK pathways are activated in parallel, resulting in the phosphorylation and inactivation of Bad and proving anti-apoptotic. There is, however, evidence that Cdk5 is both pro-apoptotic and anti-apoptotic. The cross-talk herein described demonstrates that Cdk5 activity is regulated both within Schwann cells and within axons by Scwhann cells [81].

Regulation of Cdk5 Activity and Neurofilament Phosphorylation by Myelin-Associated Glycoprotein

Myelination of axons is performed by oligodendrocytes in the central nervous system and by Schwann cells in the peripheral nervous system. It was suggested that axons might be affected by myelination by a number of possibilities because phosphorylation of neurofilament proteins was observed in axons within or near myelin sheaths. Axonal caliber was noted to increase in the presence of myelin and this was partly determined by neurofilament phosphorylation [82, 83].

Microtubule associated proteins (MAPs) phosphorylation was also caused by myelin [84]. Myelin-associated glycoprotein (MAG) is specific to myelin produced by Schwann cells and oligodendrocytes and is localized to the periaxonal membrane [85].

MAG thus is involved in glia/axon interactions and bi-directional signaling. Mice lacking MAG demonstrate decreased phosphorylation of neurofilaments related to decreased Cdk5 and ERK1/2 activities. While it has been shown MAG is not essential for myelin formation, the decreased activities of the enzymes affected by MAG results in subtle neuronal changes: disruption of the periaxonal junction, delayed myelination, and formation of dystrophic oligodendrocytes [86].

In vitro experiments demonstrated that MAG is a glial signaling molecule that acts at a receptor on the axonal membrane. This relationship subsequently affects cytoskeletal proteins and associated kinases. The presence of MAG in COS-7 cells co-cultured with primary dorsal root ganglion neurons upregulated the total amount of NF-M, MAP1B, MAP 2, and tau. Not only was the total amount of protein elevated, but the amount of phosphorylated NF-H, NF-M, and MAP1B was also elevated. Similar results were found in PC12 neurons.

The increased phosphorylation of neurofilaments and cytoskeletal proteins in the presence of MAG was found to be concomitant with increased activities of Cdk5 and ERK1/2. Thus it was concluded that the interaction of MAG with axonal receptors induces signal transduction cascade that affects cytoskeletal protein expression and phosphorylation. The phosphorylation is affected by the changed activities of the proline-directed serine/threonine kinases [87].

Cdk5 Phosphorylates MEK1 to Regulate the MAPK Neuronal Growth Pathway

Both ERK1/2 and Cdk5 are proline-directed serine/threonine kinases that phosphorylate neurofilaments following activation by an axonal receptor that, when bound to MAG, transduces the signal from the glial membrane. It has been demonstrated that these two kinase pathways, Cdk5 and mitogen activated protein kinase (MAPK pathway), are not only connected by similar activation but are also regulated and linked in their affect on neurofilaments by a cross-talk pathway [88].

Independent of the MAG activation, ERK1/2 is activated by an upstream regulator, modulator MAP kinase kinase-1 (MEK1) [89, 90]. This upstream regulator of ERK1/2 is activated by the interaction of extracellular signals such as nerve growth factor (NGF) with receptors at the membrane surface. The signal is transduced on the intracellular surface through an intermediate process. Interaction of NGF with its receptor causes activation of Ras, which then activates Raf. MEK1/2 is then phosphorylated by Raf. ERK1/2 is then activated by MEK1/2 by phosphorylation [89]. Regulation of this pathway is achieved by p35/Cdk5 by phosphorylation of MEK1/2 on a specific threonine residue (Thr^{286}). The phosphorylation of this upstream activator of ERK1/2 causes the down-regulation of the MAPK pathway [88].

The relationship between Cdk5 and the MAPK pathway was substantiated by p35/Cdk5 knockouts that demonstrated higher activity of ERK1/2 in the cerebral cortex with increased phosphorylation of neurofilaments [88]. Thus, a loss of Cdk5 activity would lead to the up-regulation of the MAPK pathway involving ERK1/2 and in increased neurofilament production and phosphorylation. This is accomplished by virtue of ERK1/2 having several roles. First, activated ERK1/2 translocates into the nucleus and acts as a transcription factor and causes the increased synthesis of cytoskeletal proteins in addition to p35 activator protein [89].

This will cause the increased phosphorylation of neurofilaments by Cdk5. Second, ERK1/2 actively phosphorylates NF and MAP once activated [90]. The action of Cdk5 cross-talk in the MAPK pathway serves thus to inhibit neurofilament production and phosphorylation by limiting the amount of substrate and amount of activator protein for its own pathway as well as by limiting the MAPK pathway directly [88].

p35/Cdk5 Phosphorylates Ras Guanine Nucleotide Releasing Factor 2 (RasGRF2) to Mediate ERK1/2 Activity

ERK1/2 activity is reduced by a decrease in Rac-guanidine exchange factor activity upon phosphorylation of Ras Guanine Nucleotide Release Factor 2 (RasGRF2), a member of the RasGFR family of GEFs. After determing by yeast two-hybrid system that p35 associates with RasGRF2, it was demonstrated that p35/Cdk5 phosphorylates RasGRF2 on a particular serine residue (Ser^{737}) [91]. By affecting the Rac signaling cascade and ERK1/2 activity, Cdk5 mediates the distribution of RasGRF2 and MAP1b within neurons [91]. Rac is a member of the Rho family of GTPases; it is implicated in cell migration and neutire outgrowth [92]. p35/Cdk5 co-localizes with Rac in neuronal growth cones, and p35 associates with Rac in a GTP-dependent manner [93].

In down-regulating cellular Rac activity, p35/Cdk5 does not affect Ras activity or RasGRF2 steady-state levels. Cdk5/p35 decreases RasGRF2-Rac dependent activation of ERK1/2. These findings further support the cross-talk relationship of Cdk5 in the Rac signal transduction pathway. Further, co-transfection of p35/Cdk5 and RasGRF2 in neurons led to the accumulation of MAP1b but not tau, tubulin, or MAP2. It was concluded that MAP1b distribution is affected by the involvement of Cdk5 in the Rac signaling pathway [91].

The Cdk5 cross-talk with the Rac signaling pathway places Cdk5 as a regulator of ERK1/2 activity at two points physiologically. First, Cdk5 phosphorylates MEK1 in the MAPK pathway to affect ERK1/2 activity [91]. Second, as described, Cdk5 regulates ERK1/2 through phosphorylation of RasGRF2 in the Rac signaling pathway. Regulation of RasGRF2-mediated Rac activity via Cdk5 suggests a physiological pathway originating at the membrane that produces downstream intracellular changes in kinase activities. These changes in kinase activities subsequently change the cytoskeletal distribution of proteins.

The Rho GTPases, including Rac and PAK kinases, are involved in actin polymerization [92, 94]. Rac effector PAK1 kinase is hyperphosphorylated by the active Cdk5 complex [91]. This not only modulates activity of this Rac effector protein but also changes its co-localization with Rac1 [91]. Thus, the modification of PAK1 activity and co-localization affects the dyamics of the actin cytoskeleton. RasGRF2 has 16 putative targets for proline directed kinases, *i.e.* proline-directed serine and threonine residues. The regulation of it by p35/Cdk5, however, is similar to the regulation of RasGRF1, which, phosphorylated by PKA, leads to Ras signaling activation following intracellular calcium concentration increase [95]. RasGRF1 is only expressed in the brain, while RasGRF2 is expressed in several tissues including brain and lung. RasGRF1, however, is involved, by tyrosine-phosphorylation by Cdc42-associated kinase 1 and Src, in the activation of ERK2 and induction of Rac-guanine nucleotide exchange activity [96, 97].

NEURONAL NUCLEAR ORGANIZATION CONTROL BY CDK5 ACTION ON RAS GUANINE NUCLEOTIDE RELEASING FACTOR-1 (RASGRF1)

RasGRF1 is another member of the Ras guanine nucleotide exchange factor (RasGEF) family. While RasGRF2 is ubiquitously present, RasGRF1 is mainly expressed in the nervous system [98]. As mentioned, these are responsible for direct activation of Ras and Rac GTPases by conversion from GDP-bound enzymes to GTP-bound enzymes. There is a high degree of homology between them [98]. RasGRF1 is a phosphoprotein that is phosphorylated by kinases including PKA and non-receptor tyrosine kinase Ack1 and Src [99-101].

RasGRF1 is localized to the plasma membrane via its pleckstrin homology (PH) domain [102]. It may bind directly to the membrane phospholipids or via associated proteins such as G-protein subunits. Which occurs is not yet known definitively. It is known, however, that RasGRF1 is localized to the membrane additionally through its myristoylation sequence found near its N-terminal [103]. Ras is localized to the inner leaflet of the plasma membrane as well. It is critical for the signaling pathway that Ras be in close proximity to RasGRF1 and to p35/Cdk5 [98]. It has also been found that the coupling of the calcium signal and signal transduction responses, dependent on duration, amplitude, and frequency of temporospatially local concentration, is key in activation of RasGRF1 and activity regulation of Cdk5.

Phosphorylation of RasGRF1 by p35/Cdk5 did not affect RasGRF1-mediated ERK1/2 activity despite reduced RasGRF1 levels in CHO cell system. In other cell systems, this was not the same. The overexpression of p35/Cdk5 caused a decrease in RasGRF1 levels, as mentioned, and a reduction in Ras activation that led to a reduction in active Akt [98]. The latter is critical to neuronal survival and its reduction was linked to neuronal nuclear disintegration. This agreed with ischemia-induced CREB activation reduction and neuronal damage enhancement in RasGRF1 knockout mice [104]. Overexpression of RasGRF1 in mitotic cells results in the disorganization of the nucleus. Transfection of a dominant negative Cdk5 construct with the expression of wild-type RasGRF1 results in condensed and fragmented nuclei. This can be resuced by co-transfection of RasGRF1 with Cdk5 [98].

It has been shown that RasGRF2 and RasGRF1 couple NMDA receptors to activation of the Ras/ERK pathway [105]. NMDARs are acted on by serine/threonine and tyrosine kinases, as described [106]. Another important interaction between RasGRF1 and Cdk5 has also been shown in the ß cells of the pancreas. In the pancreas, RasGRF1 is required for ß cell development and glucose homeostatic regulation [107]. RasGRF1 knockout mice demonstrate reduced body weight, glucose intolerance, and hypoinsulinemia, hallmarks of clinical diabetes type 2 [107].

In addition, active p35/Cdk5 has been found in the pancreatic ß cells [108]. It has been shown that RasGRF1 is also acted on by Cdk5, which phosphorylates it at Ser^{731}. This phosphorylation regulates the steady state levels of RasGRF1 in neurons by leading to its degradation via a calpain-mediated mechanism. Calcium influx is mediated by glutamate and causes activation of Cdk5 dependent on calmodulin. Cdk5 then autophosphorylates p35 which is subsequently ubiquinated and degraded by the proteosome. Past a certain threshold, it is possible calpain recognizes phospho-RasGRF1 and degrades it. This was substantiated by the observation that RasGRF1 levels were higher in cortical neurons from Cdk5 knockout mice and that this level was reduced with Cdk5 transfection into the same neurons. It was concluded that cdk5 tightly regulates RasGRF1 by phosphorylation to maintain proper cellular organization. Deletion of RasGRF1 is damaging to long-term potentiation. It was suggested that the absolute levels of the four proteins involved—RasGRF1, Ras, Cdk5, and p35—are not the determinant of function. Rather, a balance between them is most important for neuronal function and survival [98].

Regulation of JNK3 Neuronal Apoptosis Pathway by Cdk5

Whereas involvement of Cdk5 in the MAPK pathway regulates neuronal growth and survival, its involvement in the c-Jun N-terminal kinase (JNK3) pathway regulates apoptosis. Phosphorylation of JNK3 at a particular threonine residue (Thr^{131}) inhibits its kinase activity. Reduced activity of JNK3 by Cdk5 leads to a decrease in phosphorylation of its substrate, c-Jun. The JNK3 pathway is normally pro-apoptotic. In the presence of Cdk5/p35, the pathway is inhibited and survival is favored. Thus, Cdk5/p35 cross-talk in the JNK3 pathway is anti-apoptotic.

The effect of Cdk5 on the JNK3 pathway as pro-survival was demonstrated by the introduction of catalytically inactive Cdk5 to cells transfected with p35/Cdk5 and induced with JNK3 activity by UV damage. This resulted initially in survival of the cells and subsequently in apoptosis upon introduction of inactive Cdk5. Therefore, Cdk5 can again be said to be critical for neuronal survival [1, 109].

Cell Cycle Re-Entry Control via Cdk5 Phosphorylation of Retinoblastoma (Rb) Protein in Neurons

Cell cycle re-entry is particularly detrimental to post-mitotic neurons. Neuronal re-entry into the cell cycle results in cell death. Cdks play an essential role in regulating the cell cycle and do so by modulating the functions of target proteins at each step of the cycle. The ten known Cdks are divided into three groups; the first of which, G_1-S phase Cdks, consisting of Cdk4/CycD and Cdk2/CycE, phosphorylates retinoblastoma (Rb) protein [110,11]. Unphosphorylated or hypo-phosphorylated Rb protein results in suppression of G_1. This happens because Rb protein binds to E2F transcription factors in the absence of phosphorylation. Once hyper-phosphorylated, Rb protein dissociates from E2F transcription factor to allow it to become active. This drives gene expression critical to the G_1/S phase transition and initiation of DNA synthesis.

The second group of Cdks is concerned with the inhibition of DNA over-replication, while the third group phosphorylates lamins, condensing, separase, and other mitotic proteins [110-112]. Cdk5 has been discussed thus far as uninvolved in the cell cycle, *i.e.* post-mitotic. Moroever, Cdk5 has been localized thus far to the cytoplasm and cell membrane, not the nucleus, as opposed to the other Cdks. Finally, Cdk5, as mentioned, does not require a cyclin molecule for activation, but rather is activated by p35 or p39 proteins that take cyclin forms at the binding pocket. While most Cdk5 substrates discussed are cytoplasmic or membrane localized, there are substrates of Cdk5 in the nucleus [113-115]. It has been shown that Cdk5 may be involved in centromere mediation during the cell cycle [116, 117]. Aberrant activation of Cdk5 by p25 has been shown to result in cell cycle activity in neurons with DNA damage and neuronal cell death [118]. This same pattern as a result of cell cycle re-entry has been seen with a p25/Cdk5 activity and ß-amyloid protein [119]. It was demonstrated that p35/Cdk5 is not only located in nuclear fractions but that it also phosphorylates Rb protein. It does so at sites similar to those phosphorylated by Cdk4 and Cdk2. E2F transcription factor activity is observed with increased Cdk5 activity. This can trigger cell cycle re-entry [120].

Normal levels of p35/Cdk5 do not trigger this re-entry. Thus, it was suggested that Cdk5 regulates the cell cycle in post-mitotic neurons. Mature neurons are subject to death after re-entry into the cell cycle. It leads to pathological conditions. Cdk5 is specifically expressed in mature neurons. The physiological levels of p35/Cdk5, as mentioned, do not induce cell cycle re-entry, suggesting a threshold effect of phosphorylation of Rb on E2F activation that is determined by Cdk5 [120]. These observations suggest that the degree of Rb phosphorylation through this pathway could be elevated in pathological conditions.

Neuron-Like Apoptosis in Min6 ß-Pancreatic Cells Induced by p35 Overexpression

Cdk5 and p35 expression has been noted in pancreatic beta cells [121-123]. The first piece of evidence to Cdk5 involvement in pancreatic cell functioning was that glucose-stimulated insulin secretion was enhanced by decreased Cdk5 activity [123, 124]. To evaluate whether Cdk5 might regulate insulin secretion, a low level of p35 in pancreatic Min6 cells was tested. Next, a time course experiment was performed in high glucose (25 mM) [124].

In the latter experiment, it was shown that endogenous p35 increased gradually by a three-fold increase in Cdk5 activity in 16 hours. Doubling insulin secretion over 2 hours followed by progressive down regulation negatively correlated with Cdk5 activity. Within 2 hours, a three-fold increase in Cdk5 activity was seen, however, in response to p35 overexpression. This was coupled to a 50% reduction in insulin secretion in high and low (3mM) glucose conditions. The induction of p25, the truncated aberrant activator of Cdk5, was seen following 4 hours of high glucose level treatment of cells overexpressing p35. In these cells, insulin secretion was inhibited and apoptosis was observed. Roscovitine, a Cdk5 inhibitor, or co-infection of dominant negative Cdk5 with p35 resulted in increased insulin secretion and resultant inhibition of apoptosis. These observations suggest a paradigm similar to neurodegeneration for type 2 diabetes, in accord with the current understanding linking Alzheimer's and type 2 diabetes metabolically and pathologically [124].

Studies have yielded reported conflicting observations. Cdk5 has been seen to increase insulin secretion and overexpression of p39 in Min6 cells activated and augmented calcium induced insulin exocytosis [122, 125]. The latter was accomplished by Munc18 phosphorylation [125]. Conversely, olomoucine inhibition of Cdk5 activity enhanced insulin secretion in Min6 cells and islets under high but not low glucose concentrations [123]. Further, p35 siRNA treatment of ß cells or p35 removal by genetic manipulations results in enhanced insulin secretion [126].

It was also found that insulin gene expression is preserved by a mechanism that involves restoration of DNA binding and nuclear localization of the insulin transcription factor, PDX-1, in chronic elevation of glucose with inhibition of Cdk5 activity [126]. Pancreatic ß cell function is tightly regulated by glucose concentration. Insulin synthesis and secretion are two points of regulation and it appears Cdk5 is involved at both. By its involvement at two points of regulation, Cdk5 may inhibit or activate insulin secretion, as determined by glucose concentration conditions and the duration of exposure [124]. It has been suggested that p35/Cdk5 and p39/Cdk5 might function in ways similar to their neuronal actions and be tightly regulated multifunctional kinases in these pancreatic cells.

Both transcription and translation of the insulin gene appear to be sensitive to glucose conditions. Several signal transduction pathways in which Cdk5 can cross-talk in neurons are activated at high glucose concentrations: MAP and p38/SAP kinases, PI3K and PKC, kinases acting on Pdx-1, and transcription factors [127]. The transcription factors translocate to the nucleus and interact with coregulators at the insulin promoter site. Transcription and synthesis of p35 is upregulated in accordance with this after 1-2 hours to phosphorylate and translocate Pdx-1. This explains the apparent dual role of Cdk5 with short and long durations of glucose concentrations [124].

Glucotoxicity is a pathological feature of type 2 diabetes. Insulin secretion is inhibited; tissue response to insulin is negatively modulated. Thus, it was suggested the pathogenesis of type 2 diabetes might be via glucotoxic stimulation of Cdk5 activity in ß cells. Cdk5 activity affects insulin secretion directly through its effect on modulation of exocytosis [124]. This process in ß cells uses the same machinery as in neuronal cells. Glucose induced calcium influx into these cells is the primary trigger for insulin secretion. Exposure to the Cdk5 inhibitor, olomucine, potentiates insulin secretion, suggesting phosphorylation of a subunit of the L-VDCC channel [123]. Cdk5 was also found to affect secretion and cell survival, as mentioned, for chronic glucose toxicity [124]. Neurodegeneration and type two diabetes share impaired glucose metabolism, increased oxidative stress, amyloid deposition, and insulin resistance, lending further support to the paradigm described [128].

ATTENUATION OF COCAINE-MEDIATED DOPAMINE SIGNALING BY INCREASED P35/CDK5 ACTIVITY

Synaptic dopamine levels in the striatum are increased upon exposure to cocaine, a drug of abuse [129]. Dopamine reuptake at axon terminals is blocked and leads to this effect. Cocaine also alters gene expression in dopaminoceptive neurons. This occurs by the avtivation of intracellular pathways resulting in signal transduction from the D1 dopamine receptor to the nucleus [129].

Addiction is thought to occur by the up-regulation of several transcription factors underlying neuronal adaptation. It has been demonstrated that the expression of Cdk5 is up-regulated with its activator p35 through induction of the transcription factor ΔFosB in the striatum [130, 131].

Cdk5 has been implicated in postnatal brain dopaminergic transmission [132. 133]. Up-regulation of Cdk5 and p35 has been observed in chronic cocaine exposure. Physiologically, Cdk5 is thought to act as a presynaptic regulator of dopamine release because inhibition of Cdk5 results in increased dopamine release [132].

Moreover, the efficacy of postsynaptic dopamine signaling by phosphorylation of dopamine and cAMP-regulated phosphoprotein, molecular mass 32 kDa (DARPP-32) at Thr^{75} is modulated by Cdk5. This converts DARPP-32 into a cAMP-dependent kinase inhibitor [133].

It was thus hypothesized that chronic cocaine exposure and addiction involve the downstream regulation of Cdk5 and p35 in dopamine signaling. To assess the relationship between Cdk5, p35, and induction of striatal dopamine signaling, two transgenic mouse lines were generated. These mice had either p35 or Cdk5 overexpressed specifically in neurons. It was found that cocaine-mediated dopamine signaling was attenuated by the overexpression of p35, not Cdk5 [134].

It has been found that without p35, Cdk5 in its monomeric form increases and is not active. Thus, p35 presence is a rate-limiting factor for Cdk5 activity. In Tgp35 mice, CREB phosphorylation and c-fos in the striatum following acute cocaine injection were diminished. This suggests the cocaine dopamine signaling response is attenuated by Cdk5 activity. It is likely this involved the inhibition of multiple singaling cascades. Cdk5 attenuated PKA phosphorylation of DARPP-32 at Thr^{34}. This phosphorylation inhibits the activity of protein phosphatase 1 (PP1). PP1 is responsible for dephosphorylation of CREB at Ser^{133}. This aspect of PP1 activity is left unantagonized by Cdk5 involvement via DARPP-32. Additionally, extracellular signal regulated kinase kinase 1 phosphorylation by Cdk5 at Thr^{286} was increased, decreasing activation of extracellular signal-regulated kinase 1/2. It is also possible that Cdk5 phosphorylation of DARPP-32 at Thr^{75} could inhibit PKA and lead to loss of activation of MEK1/2. Finally, the MAPK pathway may be affected by phosphorylation of MEK1 at Thr^{286} resulting in a decreased catalytic function [134].

REGULATION OF NOCICEPTIVE SIGNALING BY CDK5 THROUGH MAPK PATHWAY ENZYMES

Pain has two aspects: sensory or discriminative and affective or emotional components. The sensory aspect is termed nociception; it is necessary for survival and maintenance of an organism [135]. Pain is relayed from primary sensory neurons in the periphery to secondary spinal cord neurons that synapse to tertiary afferent neurons in the brain for conscious perception. While acute sensitization to pain is advantageous, chronic pain can lead to secondary symptoms such as anxiety, depression, and a decreased quality of life [136]. Thus two aspects of understanding pain were undertaken. The sensory cell types and associated signal transduction molecules were determined and then the parallel pathways for distribution of nociceptive information to the limbic system or sensory forebrain were examined.

The number of genes and gene products associated with pain increased dramatically with the introduction of studies involving genetically modified mice, antisense knockdowns in cells, gene expression assays (DNA microarray-based expression profiling), and linkage mapping to the field.

The initial suggestion of Cdk5 involvement in nociception was made based on the discovery of several substrates of Cdk5 in primary afferent neurons responsible for the regulation of these transduced signals [137, 138]. For example, Cdk5 modulates the MAPK pathway, a critical pathway in the hypersensitivity through both transcription-dependent and -independent mechanisms. It was found in nociception that MEK1/2 was phosphorylated at Ser^{221} and Ser^{217}. Cdk5 also interacts with CaMKII, a signal transduction pathway enzyme critical to nociception. Moroever Cdk5 is a downstream target of ΔFosB, which is upregulated during nociception [139, 140].

Further exploration revealed that Cdk5 serves a role in nociception in addition to its roles in learning, addiction, and apoptosis. This involvement in pain was first suggested by the determination of expression of Cdk5 and p35 in nociceptive neurons. This expression appears to be modulated in the peripheral inflammatory response. Cdk5 phosphorylates the NMDA receptor and the P/Q type voltage dependent calcium channel. These receptors control calcium influx during sensation of pain. The inflammatory response regulation is accomplished by the increased activation of calpain, which is calcium dependent, in these nociceptive neurons after an inflammatory response [134].

The increased activation of calpain leads to the cleavage of p25 from p35 and subsequent aberrant activation of Cdk5 by formation of a more stable complex. Delayed responses to thermal pain stimuli were seen in p35 knockout mice when compared to wild type controls, most likely due to significantly decreased Cdk5 activity.

Further, roscovitine treatment administered intrathecally in rats attenuated formalin-induced nociceptive responses. It was postulated that the genes found in nociception were linked to Cdk5 by the following relationship: Cdk5/p35 levels increase due to phospho-ERK1/2 activation that regulates c-fos transcription factor and early growth response 1 (ERG1), which are responsible for Cdk5 and p35 transcription. Decreased p35 levels were not observed with increased p25 levels, possibly due to this mechanism of inflammatory response transcription of p35 [134].

The enhanced formation of p25 was found concomitant with increased calpain activity [134]. The calpain-dependent cleave of p35 has been implicated in neurodegenerative disorders. Many patients with these disorders not only experience memory loss and impaired neural functions but also experience chronic pain. Studies have shown this involvement of Cdk5 in nociception in dorsal root ganglion, trigeminal ganglion, and spinal cord neurons. Thus, regulation of Cdk5 by oxidative stress and the inflammatory response combined with its cross talk in many other pathways poise it as a potential drug target for pain reduction and for the treatment of neurodegenerative disorders simultaneously.

Phosphorylation of Transient Receptor Potential Vanilloid 1 (TRPV1) by Cdk5 in Nociceptive Signaling

Chronic pain may result from abnormalities in the molecular signal transduction mechanisms involved in the primary, secondary, or tertiary neurons in the afferent system. Transient receptor potential vanilloid 1 (TRPV1) is a polymodal ligand-gated cation channel [141, 142]. It is expressed most in small-diameter sensory neurons and is activated by heat, protons, capsaicin, leukotrienes, and anandamide [143]. TRPV1 knockout mice demonstrate reduced thermal hyperalgesia following inflammation [144].

Reduced responsiveness to nociception in bone cancer pain models is also observed [145]. The idea that TRPV1 is subject to many levels of control and that those levels of control are linked has gained favor considering the involvement of the receptor in the processes mentioned as well as bladder disease, gastro-esophageal reflux disease, bladder disease, pancreatitis, and asthma [146].

TRPV1 response to painful stimuli is dependent on phosphorylation, which regulates intracellular calcium levels. PKA, PKA, and CaMKII phosphorylate TRPV1 to modulate responses to various adverse stimuli [147, 148]. It has been noted that Cdk5 mediates nociceptive signaling. Computational analysis suggested three potential sites on TRPV1 for Cdk5-mediated phosphorylation. It was then found that Cdk5 phosphorylation of TRPV1 at Thr^{407}, which was the most likely based on computations, modulates agonist-induced calcium influx [149].

This residue is located on the large amino-terminal intracellular loop of TRPV1 near the first transmembrane domain. Conformational changes and activation of TRPV1 by overlapping PKA, PKC, and CaMKII phosphorylation have been observed. Cdk5 likely causes similar conformation change and allows ion passage upon ligand bonding or heat activation. Intrathecal roscovitine infusion yielded an antinociceptive effect and attenuated nociception induced by formalin [149].

The expression and specific activity of Cdk5/p35 was noted in DRG and trigeminal ganglia. This was demonstrated in dorsal root ganglion neurons where inhibition of Cdk5 activity yielded a significant reduction of calcium influx by TRPV1. Restoration of Cdk5 activity was sufficient to restore calcium influx.

Primary nociceptor-specific Cdk5 conditional knockout mice experiments validated the findings. These conditional knockout mice lacked Cdk5 in C-fiber neurons. These mice showed significant hypoalgesia in the basal state. The phenotype was similar to that of knockouts of p35 and TRPV1. This confirmed the direct involvement of Cdk5 in the nociceptive pathway [149].

TRANSCRIPTIONAL ACTIVATION OF P35 VIA TUMOR-NECROSIS FACTOR-A DURING NOCICEPTION REGULATES CDK5

Cdk5 is up-regulated with p35 in nociceptive signaling under peripheral inflammation. Further understanding the pro-inflammatory molecules regulating this process in response to the inflammatory response was critical. Responses to nociceptive heat stimuli are attenuated in p35 knockout mice [150]. In transgenic mice overexpressing p35, the response is potentiated [150]. The heat stimulus being discussed activated TRPV1 receptor [149]. As noted, phosphorylation of this receptor is an important regulator of intracellular calcium levels [149, 151]. In mice models, TNF- α mRNA was increased at the site of injection of carrageenan to cause hypersensitivity to heat stimuli [150].

In the absence of heat stimuli, chronic pain due to inflammation or sensory nerve damage is mediated by cytokines [152, 153, 155]. These cytokines are released by activated immune cells. Macrophages are recruited by chemotactic molcules to release these pro-inflammatory cytokines [153]. These include interleukin-6 (IL-6), interleukin-1ß (IL-1ß), and tumor-necrosis factor- α (TNF- α). It has been shown in animal models with neuropathic pain that levels of IL-6, IL-1ß, and TNF- α are elevated [1, 155, 150]. This was demonstrated in the spinal cord and at sites of injury in rats.

TNF- α is a pro-inflammatory cytokine [152, 153]. It is produced by astrocytes, microglia, and neurons [150]. Its action is transduced by two main receptors: TNFR-1 and TNFR-2 [2,7]. Both are expressed in dorsal root ganglia (DRG). The trigeminal ganglion was also tested [7]. It was found that TNF- α activates p35 promoter activity [150, 153]. This is in a dose and time dependent manner. TNF- α activates ERK1/2, p38 MAPK, JNK, and NF-κB pathways [150]. These were evaluated with regards to their role in activation of the p35 promoter [153].

Activation of the ERK1/2 pathway has been shown to contribute to the regulation of Cdk5 activity. p35 promoter activity was decreased in the presence of MEK inhibitor [150]. In the physiologic pathway, MEK inhibits ERK activity. P35 promoter activity was increased, however, by the inhibitors of p38 MAPK, JNK, and NF-κB. IFN-γ and IL-1ß did not show any significant effect in these cell types [150]. TNF- α treatment increased Egr-1 transcription factor mRNA and protein levels in an ERK signaling dependent manner [150, 153]. Although TNF- α modulates several signaling pathways listed above either positively or negatively, the end result is an overall increase in p35 transcription. The MEK-ERK1/2-Egr-1 pathway appears to dominate the direction of the resultant events [150]. Inhibitors of TNF- α significantly reduced mechanical allodynia and thermal hyperalgesia. Thus, it was concluded that TNF- α induces Cdk5 activity by increasing levels of p35 during peripheral inflammation and this sensitizes TRPV1-positive sensory neurons [150].

INHIBITION OF P25/CDK5 ASSOCIATION LEADING TO APOPTOSIS AND REDUCTION OF TAU HYPERPHOSPHORYLATION

The association of Cdk5 with p25, a truncated Cdk5 activator produced upon exposure of the cells to amyloid ß (Aß) peptides, causes the phosphorylation of tau at epitopes specific to Alzheimer's disease (AD) [156, 157]. Both the extracellular aggregation of Aß and the intracellular hyperphosphorylation of tau are characteristic hallmarks of neurodegeneration, including AD. It has been suggested that the abberant activity of Cdk5 induced by conversion of p35 to p25 is at the root of neurodegenerative pathology. This includes diseases such as AD, ALS, and Neimann's Pick Disease (NPD) [156-158].

Under these conditions, with increases in intracellular calcium, proteolysis of p35 is calpain-directed and releases an N-terminal myristoylated membrane tether [159]. The dissociated p25/Cdk5 is thus induced to activate numerous cytosolic targets [159]. Abberant tau phosphorylation in neurons has been suppressed by the infection with Cdk5 inhibitory peptide (CIP), which is derived from p35. Moreover, apoptosis caused by introduction of Aß peptides to these neurons was reduced by treatment with CIP [1]. This inhibitory peptide not only attenuated apoptosis and the hyperphosphorylation of tau, but it also failed to inhibit endogenous or transfected p35/Cdk5 activity and other cdk activities.

In previous experiments, transgenic mice were made to overexpress p25. These mice demonstrated AD cytoskeletal disruptions and hyperphosphorylation of tau in areas of the brain such as the hippocampus. Inhibitors of Cdk5 and calpain stabilized p35 and reduced p25 neurotoxicity induced by $Aß_{1\text{-}42}$. This suggested that p25 is a downstream effector of Aß peptides [160, 161]. The abnormal phosphorylation of tau related to the Aß-induced p25/Cdk5 aberrant activity destabilizes microtubules [162]. Thus, neurite degeneration and the formation of paired helical filaments (PHFs) containing neurofibrillary tangles (NFTs) are observed.

CIP, a 125-residue peptide composed of $p35^{154\text{-}279}$, has a much higher affinity for Cdk5 than p25. Therefore, *in vitro*, it was found that Cdk5 activity is specifically and effectively inhibited by CIP. It was demonstrated in HEK293 cells that CIP reduces tau phosphorylation by the inhibition of Cdk5 activity in this manner while sparing Cdc2 kinase activity. Physiologic p35/Cdk5 was also unaffected. This was further evidence to the causative role of the pathway described above in the pathogenesis of AD [1].

CIP additionally promotes neuronal survival by inhibiting apoptotic pathways. Overexpression of cleaved caspase-3, a marker of apoptosis increased in cortical neurons coinfected with p25/Cdk5, is prevented by coinfection with CIP. $Aß_{1\text{-}42}$ similarly caused cleaved capsase-3 level increases. This was prevented by CIP and roscovitine, suggesting involvement of Cdk5 as an intermediate in amyloid-induced cleaved caspase-3 activation. Reduction of this apoptotic pathway was further observed by TUNEL assays. CIP represents a promising therapeutic avenue considering the extent to which Cdk5 is involved in other pathways such as maintaining the cytoskeleton or vesicle transport. It is thus key to be able to spare its physiologic function and that of other Cdks [163].

ASSOCIATION OF P35/CDK5 WITH CDK5/P35-REGULATED KINASE (CPRK)

Protein packaging and secretion via the Golgi apparatus is regulated by Cdk5/p35-regulated kinase (Cprk). Cdk5 regulates Cprk. This was demonstrated via isolation of Cprk by its association with p35 in a yeast two-hybrid screen. Cprk co-localizes with p35 in neurons in the Golgi apparatus. Thus Cdk5 is involved in cross-talk at each level of synaptic function ranging from transcription of the receptors in neuromuscular junctions to protein packaging and vesicle formation and finally cytoskeletal arrangement [1].

EXOCTYSOSIS REGULATION BY CDK5 VIA MUNC18-1 AND NUCLEAR LOCALIZATION

In regulated exocytosis, neurotransmitter release is accomplished by vesicle priming and fusion with the membrane. In the final stages of this process, target-soluble N-ethylmaleimide sensitive fusion protein (SNARE) complexes are essential. Munc18-1, also known as p67, has a high affinity for a particular SNARE, syntaxin, and is essential to the regulation of the final stages of exocytosis. Association of vesicle SNARE proteins with syntaxin 1a, a process required for assembly of the core protein complex that mediates vesicle fusion, is inhibited by the high affinity binding of Munc18-1 to syntaxin 1a [164-166].

Munc18-1 was determined to co-localize with Cdk5 and phosphor-NF-H in fiber tracts and the cerebellum by Western blot and immunohistochemistry analysis. Moreover, Munc18-1 is phosphorylated by and co-purifies with Cdk5 [167]. Thus, the interaction of Munc18-1 with syntaxin 1a is regulated by the activity of Cdk5. It is by this regulation of Munc18-1 interaction that Cdk5 modulates the level of vesicle SNARE association with syntaxin 1a and subsequent secretion of vesicles [167]. Cdk5 is therefore a regulator of Munc18-1 mediated exocytosis.

Munc18-1 was also shown to co-localize with components of the cytoskeleton, suggesting involvement in the dynamics of the cytoskeleton in addition to its role in vesicle docking and fusion described above. It was found in both PC12 neurons and rat brain neurons that Munc18-1 is nuclearly localized and binds to double stranded (ds) DNA [167]. The lack of DNA binding domains on Munc18-1 and its ability to bind dsDNA suggested binding via protein-protein interactions with other DNA binding proteins. Nuclear localization is substantiated by the nuclear import and export signals in the Munc-18-1 primary sequence. Thus, it was suggested Munc-18-1 might be a nuclear shuttle [167].

These potential additional functions of Munc18-1 were suggested by observation of the structure: two clusters of basic residues (20-31 and 457-467) characteristic of bipartite nuclear localization signal (NLS) sequence [167]. This implied the transport of Munc18-1 into the nucleus, as mentioned. A cluster of hydrophobic residues (405-414) represented the nuclear export signal. Amino acid sequences are important for sub-cellular distribution, a determinant of function.

MODULATION OF EXOCYTOSIS VIA CDK5-MEDIATED PHOSPHORYLATION OF SETPIN SEPT5 (HCDCREL-1)

Cdk5 regulates the cellular function of presynaptic proteins such as synapsin, Munc18, and dynamin 1 to mediate its role in synaptic transmission [167]. Cdk5 and its activators are more present at the nerve terminals and are localized to vesicular fractions [168]. Cdk5 forms a complex with syntaxin, p35, and nsec-1 (Munc18) [169]. These proteins are of central importance to synaptic fusion and release in exocytosis. Cdk5 in this complex phosphorylates dynamin 1 and regulates endocytosis [170].

To determine possible interactions with other proteins at the presynaptic terminal, a yeast-two hybrid screen of a human cDNA library with p35 as bait and isolated human septin 5 (SEPT5 or hCDCrel-1) as an interacting clone was used. It was through this system that Cdk5 targets such as p35/Cdk5-regulated membrane-associated kinase, RasGRF2, α-actinin-1, and CaMKII were found. SEPT5 can directly bind to syntaxin-1 but is readily displaced by α-SNAP. Adult-type septin 5 isoform (SEPT5_v1) at the synapse was phosphorylated by p35/Cdk5 at Ser^{17} in the N-terminal domain. This phosphorylation appears to inhibit its association with syntaxin, a key part of the SNARE complex. It has been suggested that SEPT5 forms tethers, networks, or meshes and that these prevent vesicle association with the membrane [171].

It was found through glutathione-S-transferase pull down and coimmunoprecipitation assays that Cdk5 associates with SEPT5 and is colocalized at synaptic vesicles and then in subsequent experiments that p35/Cdk5 phosphorylates SEPT5 *in vitro* and *in vivo* [171]. It was determined that the major phosphorylation site was Ser^{327} adjacent to the C-terminal coiled-coil domain and that the minor phosphorylation site was Ser^{161} in the GTP-binding domain [171].

These findings were comfirmed by observation of metabolic labeling studies in which mutant SEPT5 did not significantly take up orthophosphate where wild type did. Finally, a phospho-polyclonal antibody and mass spectrometric phosphoproteomic sequencing were used to confirm the major phosphorylation site [171].

SEPT5 association with Cdk5 was concomitant with a reduction in syntaxin-1 binding to SEPT5. Mutation studies in which this serine residue was converted to alanine resulted in stronger bonding of SEPT5 to syntaxin. Interestingly, the mutant potentiates secretion of human growth hormone excessively. Further study in synaptic vesicle fractions confirmed that the phosphorylation of SEPT5 by p35/Cdk5 decreased binding to syntaxin-1. Constitutively active nonphosphorylated SEPT5 conducted regulated exocytosis more than the wild type.

Thus, Cdk5 serves to phosphorylate SEPT5 to modulate exocytotic secretion by suppressing neurotransmitter release and enhancing endocytosis. Phosphorylation of SEPT5 may result in negative modulation of exocytosis by preventing SNARE association or by promoting tethering or networking or even by inducing translocation to the cytosol, as is known to occur with PKG phosphorylation of SEPT3. These processes may contribute to the rate of endocytotic recycling of vesicles [171].

Modulation of the SNARE secretion cycle occurs at various places via different components and proteins. Phosphorylation is the key regulatory mechanism and provides the balance between endocytosis and exocytosis. PKC and PKA phosphorylate SNAP-25 to potentiate vesicle recruitment and release [172].

As described, Cdk5 phosphorylates Munc18 to reduce syntaxin binding. Overall, the phosphorylation of the same protein machinery at different sites results in the ability to modulate the same cycle differently and effectively and septin isoforms may be an added level of regulation. This would be highly effective considering Cdk5 is a major contributor and its actions are dependent on the sub-cellular localization of its activators and substrates under particular conditions.

Endocytosis Regulation by Cdk5 via Association with Amphiphysin and Dynamin

Cdk5 regulates neurotransmission at exocytosis and its opposing process: endocytosis. Neurite outgrowth and endocytosis involves high levels of the phosphoprotein amphiphysin in neurons [1, 173, 174]. Amphiphysin is also a substrate of Cdk5 and a common constituent of a high molecular weight complex in the brain including p35/Cdk5. This association and co-localization of amphiphysin with p35/Cdk5 has been shown to occur in neuronal growth cones and at actin-rich lamellipodia of transfected fibroblasts [175, 176].

Endocytosis is a highly regulated process involving p35/Cdk5 and amphiphysin at the leading edges of neuronal processes. This suggests involvement of this complex or these co-localized enzymes in neuronal migration and growth cone navigation. Endocytosis is critical in sustaining synaptic activity following its opposing process of exocytosis. Calcineurin-mediated dephosphorylation activates endocytosis [1]. This occurs following an influx of calcium preceded by depolarization. This dephosphorylation acts on the eight proteins called the dephosphins that are physiologically phosphorylated in their resting state. Dynamin, synaptojanin, amphiphysin I and II, epsin, eps15, AP180, and PIP kinase Iγ are known as the dephosphins [177-179]. Thus rephosphorylation of these enzymes is required to perpetuate the cycle. Cdk5 is a dephosphin kinase and, as such, is essential for the process of endocytosis and the synaptic transmission cycle.

Synaptic Activity, Cdk5, and Long-Term Potentiation

The involvement of Cdk5 in synaptic transmission is further demonstrated by the observation of synaptic proteins as its substrates. N-methyl-D-aspartate (NMDAR) glutamate receptors contain intracellular and extracellular phosphorylation sites. These receptors are critical to learning, development, memory, and synaptic transmission. They are multimeric complexes formed from NR1 and NR2 subunits, produced by a single gene and four genes with alternatively spliced variants, respectively. The phosphorylation sites mentioned have been identified on the NR1, NR2A, and NR2B subunits [180-182]. Cdk5 was found associated with the NR2A subunit of these receptors. It was found that Cdk5 phosphorylates this subunit at Ser^{1232} and affects NMDAR activity [1]. This was shown by the NMDA-induced regulation of long-term potentiation (LTP) in hippocampal neurons.

NMDAR activity is upregulated by phosphorylation of the NR2A subunit as mentioned by Cdk5. This effect is reversed by roscovitine, a specific Cdk5 inhibitor. The result is down-regulation of NMDAR activity [183]. p25/Cdk5 was also found to phosphorylate the intracellular loop of the P/Q type voltage-dependent calcium channel (VGCC). Cdk5, p35, and the P/Q VGCC are found in abundance in synaptosomes. Therein, the inhibition of Cdk5 activity by roscovitine induced neurotransmitter release and an enhanced influx of calcium through the channel [184]. Phosphorylation within the intracellular loop of the channel also causes interaction of the channel with the SNARE proteins, SNAP25 and synaptotagmin I. These two SNARE proteins are required for neutotransmitter release. Thus, p35/Cdk5 activity down-regulates P/Q VGCC channel activity to inhibit neurotransmitter release. Therefore, Cdk5 affects both long-term potentiation and long-term depression in neurotransmission [1].

∂-Catenin Phosphorylation Regulates Its Localization and GluR2-Mediated Synaptic Activity

Exocytosis, as discussed, is regulated by direct Cdk5 phosphorylation of synapsin I, Munc-18, and voltage dependent calcium channel and the indirect phosphorylation of Pctaire-1. Endocytosis if also regulated via Cdk5 phosphorylation and modulation of the phosphoinositide pathway [185]. Post-synaptically, the neuregulin receptor ErbB3 is phosphorylated by Cdk5, as is Post-synaptic density protein-95 (PSD-95) [186, 187]. Multimerization of the post-synaptic scaffold is regulated by PSD-95 phosphorylation and results in a decrease in NMDA receptor and potassium channel clustering [187].

The Cdk5 activator, p35, interacts with ∂-catenin, a synaptic adherens junction protein comprising part of the anchorage complex of the AMPA receptor on the post-synaptic membrane [188]. This was determined by a yeast two-hybrid screening system with p35 as the "bait". ∂-catenin was first identified by its association with presenilin-1, a protein most frequently found mutated in familial Alzheimer's disease [189]. Moreover, ∂-catenin deletions and allele hemizygosity correlate with some of the mental retardation observed in Cri-du-Chat syndrome [190]. Mice lacking it demonstrate cognitive impairments, particularly involving hippocampal-mediated long- and short- term plasticity and spatial learning. ∂-catenin belongs to a family of proteins, the p120-catenin family, which coordinates cadherin-cytoskeletal interactions at the cell membrane through its Armadillo (ARM) repeats that bind it to cadherin [191].

The PDZ binding domain of ∂-catenin is important in its interaction with GRIP, ABP, and PSD-95 [192]. The phosphorylation site for Cdk5 is at the opposite end compared to the PDZ domain. The proteins that bind here form the scaffold for AMPA receptors GluR2 and GluR3 [188].

Phosphorylation of ∂-catenin by Cdk5 regulates the subcellular localization of ∂-catenin that is found with changes in dendritic morphogenesis and synaptic activity. Cdk5 phosphorylates ∂-catenin at two sites *in vivo* in the brain: Ser^{300} and Ser^{357}. The phosphorylation results in increased affinity between ∂-catenin and the PDZ-binding proteins mentioned. It was suggested that the interaction at the ARM domain was not affected by Cdk5 phosphorylation. Phosphorylation by Cdk5 regulates subcellular localization of ∂-catenin to a cytoplasmic or intraneurite pool as well as a membrane-localized pool. It has been reported that the phosphorylation of ∂-catenin by p35/Cdk5 increases its association with Pin1 [188]. Absence of Cdk5 phosphorylation results in localization of ∂-catenin to the membrane. Mutation studies to mimic the nonphosphorylated protein increased dendritic protrisions and AMPA receptor subunit GluR2 tethering at the membrane [188]. GluR1 levels at the membrane were also increased the presence of phospho-deficient ∂-catenin. While this is not a PDZ associated subunit, GluR1 has been shown to be involved in heterodimerization to upregulate GluR2 expression at the membrane [193]. It was concluded, as expected, that the AMPA/NMDA ratio was also increased [188]. This leads to the regulation of synaptic function by virtue of the increased AMPA-mediated response involved in synaptic activity [188]. Neurons demonstrated increased dendritic spines in the hippocampal neurons with knocked-down ∂-catenin. It has been suggested that kinases such as Akt and Abl also resulted in morphogenesis and dendritic spine changes through ∂-catenin. Loss of Cdk5 activity and resultant stabilization of ∂-catenin at the membrane results in sustained AMPA-mediated potentiation and enhanced sodium flux that may be detrimental to the neuron. These findings of AMPA-mediated neuronal changes related to post-translational modifications of ∂-catenin represent a developmental paradigm for this relationship to Cdk5.

NEURAL STEM CELL DIFFERENTIATION IS MODULATED BY CDK5 PHOSPHORYLATION OF P27^{KIP1}

Studies of Cdk5 in embryonic corticogenesis have demonstrated that the neurons of the initial cortical preplate form but subsequent migration of post-mitotic neurons into respective cortical layers fails [194]. Cdk5 is involved in neural migration rather than early neural commitment, as evidenced by experiments in *Xenopus* in which it was noted that p35/Cdk5 activity was induced following expression of neural differentiation transcription factors, neurogenin and neuroD [195]. It has been suggested that the importance of Cdk5 extends further into preventing neuronal cell re-entry into the cell cycle to avoid cell death. The point at which cell cycle exit occurs appears to be dictated by regulation of Cdk5 subcellular localization, not by Cdk5 activity [196].

Cdk5 is also crucial to adult neurogenesis. Conditional Cdk5 knockout in the dentate gyrus of the adult hippocampus results in granule cell neurogenesis inhibition and shows fewer immature neurons without affecting levels of proliferation. A major function of Cdk5 appears to be maintaining the survival of neurons in the subgranular zone in the hippocampus [197].

The role of Cdk5 in neuronal differentiation was further investigated in relation to p27^{Kip1}, a protein crucial to neuronal differenation [198]. Under this model, protection from cell cycle re-entry occurs when p35/Cdk5 is localized in nuclei bound to p27^{Kip1} without kinase activity. Mammals express three members of the Cip/Kip family that regulate cell cyle exit. These are p21^{Cip1}, p27^{Kip1}, and p57^{Kip1}. In particular, p27^{Kip1} has been implicated in neurogenesis processes [199]. The coupling of these processes may be due to the overlapping yet different roles of the N- and C-terminal domains of p27^{Kip1} [200]. Overlapping domains in the N-terminal region of p27^{Kip1} independently terminate progenitor cell cycles and determine neuronal cell fate. This is accomplished by stabilization of neurogenin2, an upstream bHLH transcription factor [201]. Levels of p27^{Kip1} are regulated at translation and protein turnover. It is phosphorylated by Cdk2-cyclin E at Thr^{187} in preparation for ubiquination during the S-G1 phase of the cell cycle preceding replication [202.

The question remained as to whether Cdk5 regulates neural differentiation by modulating the cell cycle or by regulating proneural gene expression. Therefore, homogeneous populations of uncommitted stem cells from E13 telencephalon of Cdk5 +/+ and -/- mice were obtained. Under stress conditions, Cdk5 is translocated into the cytoplasm and neurons undergo a delayed apoptosis. It was determined that phosphorylation of p27^{Kip1} by Cdk5 is crucial to differentiation of neuronal stem cells by the following observations. Frist, neurogenesis is specifically suppressed by transfection of p27^{Kip1} siRNA into Cdk5 +/+ neuronal stem cells. Second, neuronal differentiation in Cdk5 -/- stem cells is reduced compared to +/+ stem cells. Third, Cdk5 +/+ neuronal stem cells with inhibited differentiation due to a nonphosphorylatable mutant, p27/Thr^{187}A, were rescued by cotransfection of a phosphorylation mimicking mutant, p27/Thr^{187}D. Finally, differentiation was inhibited by the introduction of mutant p27^{Kip1} (p27/Thr^{187}A) into Cdk5 +/+ stem cells [198].

It was concluded that Cdk5 regulates neural differentiation of neuronal stem cells by p27^{Kip1} phosphorylation at Thr^{187} [198]. This correlates with cell cycle exit and the progressive decline of Cdc2 activity. Moreover, it was found that phosphorylation at Ser^{10} promotes neurite outgrowth as neurons differentiate with Thr^{187} phosphorylation [198]. Both sites need to be phosphorylated. p27^{Kip1} may regulate neuronal differentiation from cell cycle arrest through control of differentiation, neurite outgrowth, and migration. p35/Cdk5 is protected from inhibition by p27^{Kip1} in neurons, in contrast to cell cycle Cdks [203]. This allows the function of Cdk5 in normal corticogenesis. It is possible that p27^{Kip1} can be phosphorylated at a different site by another kinase to compensate in the absence of Cdk5. This allows the initiation of neurogenesis but not the later process of corticogenesis. Ultimately, it was suggested that a combination of niche factors from radial glial, ependymal, and endothelial cells, including extracellular matrix, are critical to signaling compensatory phosphorylation to allow an alternative path for neurogenesis [198]. Despite the presence of an alternative pathway, Cdk5 activity is necessary for corticogenesis.

Glutamatergic Regulation of p35 Stability and Cdk5 Activity

Cdk5 may be involved in the synaptic activity of mature neurons. Presynaptically, Cdk5 has been shown to phosphorylate synapsin I [204] and the α-subunit of the P/Q type voltage-dependent calcium channel [205]. This phosphorylation of a protein related to exocytotic release may regulate neurotransmitter release. Further evidence for this relationship of Cdk5 in presynaptic events is found in its regulation of endocytotic events via phosphorylation of dynamic I and amphiphysin I, as discussed. Post-synaptically, Cdk5 has been shown to phosphorylate the NR2A subunit of the NMDA receptor as well as PSD95 and protein phosphatase inhibitor-1 [206] and its homologue in the striatum, DARPP-32 [207]. Cdk5 activity may be changed post-synaptically by changing the association of p35. It has been found that NMDA stimulation increases the association of p35 with Ca^{2+}-calmodulin-dependent protein kinase IIα (CaMKII), which contributes greatly to post-synaptic density [208].

Cdk5 is only active when bound to its regulatory activation subunit, p35, and, therefore, understanding the synthesis or degradation of p35 is a strong indicator of Cdk5 activity [209]. Neurotrophic factors can stimulate the synthesis of p35 [210]. Extracellular matrix components also stimulate p35 synthesis [211]. Glutamate, the major excitatory neurotransmitter of the central nervous system, induces p35 degradation. Glutamate receptors are either ionotropic or metabotropic. Both classifications of families have subgroups. The function of each glutamate receptor is related to neuronal processes during stage of development and localization in the brain [212].

Cdk5 activity can be regulated by autophosphorylation as a result of downstream events of glutamatergic signaling via the NMDA receptor. It was found that the binding of glutamate to NMDA or kainate ionotropic receptors, but not to metabtropic receptors, leads to activation of Cdk5 [212]. This transient activation is dependent on calcium and calmodulin. The same signaling cascade results in the proteasome-depdent degradation of p35, the Cdk5 activator. These events result in decreased Cdk5 activity. The regulation of Cdk5 activity this way is complex and involves p35 binding in a calcium dependent manner to CaMKII and α –actinin, both of which are enriched in the post-synaptic region. The impact of these findings was examined in theta burst stimulation experiments in hippocampal slices [212]. A lower long-term potentiation induction threshold was found in p35 knockout mice [212]. It is possible that the down-regulation of Cdk5 activity by the glutamatergic signaling cascade described is a contributing factor to neuronal plasticity.

Cortical neurons, in which it was determined p35 may be degraded as a downstream effect of glutamate binding to post-synaptic NMDA receptors, did not demonstrate neuronal death after 1 hour treatment. Thus, although it is known that prolonged exposure to glutamate may result in cytotoxic events, p35 may be degraded in viable neurons. It was thought exposures of an hour would be necessary to cause degradation of p35, however, shorter exposures were sufficient. This suggested that initial targeting was localized in the post-synaptic region where the ionotropic receptors are concentrated [212].

Extended treatment (6-12 hours) with glutamate produces proteolysis of p35 to p25. Shorter exposures did not generate p25, suggesting proteosomal degradation in some experiments [212]. Others reported p25 appearance in even 30-minute exposures. However, different experimental conditions were cited as a potential explanation [213].

Both experiments used hippocampal neurons, but one treated them with calcium. Treatment with calcium would result in a higher calcium influx and subsequent activation of calpain, which is known to cleave p35 to p25. The glutamate signaling cascade resulting in proteosomal degradation of p35 functions under physiological calcium in contrast to the calpain-directed cleavage [212]. The glutamate signaling cascade resulting in p35 degradation acts via the NMDA or kainate receptors, as mentioned, leading to the inactivation of Cdk5. Inhibition of Cdk5 activity results in decreased levels of phosph-Ser^{67} inhibitor-1, a phenomenon in striatal slices previously attributed to activation of calcineurin (PP2B), the phosphatase responsible for dephosphorylation of this site [206].

Both the glutamate signaling cascade resulting in p35 degradation and PP2B activity are likely responsible in a coordinated fashion. Based on reports in which similar regulation was observed with metabotropic receptors (mGluR), it has been suggested that the expression pattern of glutamate receptors determines the differential regulation of Cdk5 [212].

CONCLUSION

Cdk5, spatially and temporally localized by its activator peptides, p39 and p35, and other peptides is involved in many pathways by virtue of its action as a proline-directed serine-threonine kinase. Typically membrane-bound, Cdk5 exerts influence in the cytosolic domain and in the nuclear compartment. It is involved in the regulation of neuronal synaptic transmission, long-term potentiation and plasticity, neurite outgrowth, neuronal survival and neuronal apoptosis. The involvement of Cdk5 in these pathways poses it in a unique position to be targeted therapeutically in pathological states related to its misregulation: Alzheimer's disease, Parkinson's disease, and Pick's disease for example. Moreover, Cdk5 is critical to neuromuscular junctions and pancreatic islet cells, showing functional relationships between different aspects of these diseases as well as between these neurodegenerative pathologies and diabetes. A unique approach has been taken in selectively inhibiting the aberrant hyperactive p25/Cdk5 complex using the CIP peptide derived from the binding pocket of p35. This appears to return pathological states to normal physiology without disrupting the Cdk5-dependent processes of the cell. This provides a potential avenue for the regulation of the surveillance system of the nervous system: Cdk5.

ACKNOWLEDGMENTS

I thank Dr. Harish Pant of the Laboratory of Neurochemistry of the National Institute of Neurological Disorders and Strokes of the National Institute of Health for critically reviewing this work and for his suggestions. This work was supported by NINDS/NIH intramural summer internship program.

REFERENCES

[1] Kesavapany, S., Li, B.S., Amin, N., Zheng, Y., Grant, P., Pant, H.C. (2004). Neuronal cyclin-dependent kinase 5 : role in nervous system function an its specific inhibition by the Cdk5 inhibitory peptide. *Biochimica et Biophysica Acta* 1697, 143-153.

[2] Shea, B. T., Zheng, Y., Ortiz, D., Pant, H. C. (2004). Cyclin-Dependent Kinase 5 Increases Perikaryal Neurofilament Phosphorylation and Inhibits Neurofilament Axonal Transport in Response to Oxidative Stress. *J. Neurosci. Res.* 76, 795-800.

[3] Pant H. C., Veeranna (1995). Neurofilament phosphorylation. *Biochem. Cell Biol.* 73:575–592.

[4] Shea, T. B., Yabe, J. T., Ortiz, D., Pimenta, A., Loomis, P., Goldman, R. D., Amin, N., Pant, H. C. (2004). CDK5 regulates axonal transport and phosphorylation of neurofilaments in cultured neurons. *J. Cell Sci.* 117:933–941.

[5] Tomkins, J., Usher, P., Slade, J. Y., Ince, P. G., Curtis, A., Bushby, K., Shaw, P. J. (1998). Novel insertion in the KSP region of the neurofilament heavy gene in amyotrophic lateral sclerosis (ALS). *Neuroreport* 9: 3967–3970.

[6] Al-Chalabi A, Andersen PM, Nilsson P, Chioza B, Andersson JL, Russ C, Shaw CE, Powell JF, Leigh PN. (1999). Deletions of the heavy neurofila- ment subunit tail in amyotrophic lateral sclerosis. *Hum. Mol. Genet.* 8:157–164.

[7] Figlewicz, D. A., Krizus, A., Martinoli, M. G., Meininger, V., Dib, M., Rouleau, G. A., Julien, J. P. (1994). Variants of the heavy neurofilament subunit are associated with the development of amyotrophic lateral sclerosis. *Mol. Genet.* 3: 1757–1761.

[8] Julien J. P., Mushynski W.E. (1998). Neurofilaments in health and disease. *Prog. Nucleic Acid Res. Mol. Biol.* 61:1–20.

[9] Cookson M. R., Shaw P. J. (1999). Oxidative stress and motor neurone disease. *Brain Pathol.* 9:165–186.

[10] Nguyen, M. D., Lariviere, R. C., Julien, J. P. (2001). Deregulation of Cdk5 in a mouse model of ALS: toxicity alleviated by perikaryal neurofilament inclusions. *Neuron* 30:135–147.

[11] Strong, M. J., Strong, W. L., Jaffe, H., Traggert, B., Sopper, M. M., Pant, H. C. (2001). Phosphorylation state of the native high-molecular-weight neurofilament subunit protein from cervical spinal cord in sporadic amyotrophic lateral sclerosis. *J. Neurochem.* 76:1315–1325.

[12] Kesavapany, S., Patel, V., Zheng, Y. L., Pareek, T. K., Bjelogrlic, M., Albers, W., Amin, N., Jaffe, H., Gutkind, J. S., Strong, M. J., Grant, P., and Pant, H. C. (2007). Inhibition of Pin1 Reduces Glutamate-induced Perikaryal Accumulation of Phosphorylated Neurofilament-H in Neurons. *Mol. Biol. Cell* 9, 3645–3655

[13] Rudrabhatla, P., Zheng, Y. L., Amin, N. D., Kesavapany, S., Albers, W., Pant, H. C. (2008). Pin1-dependent Prolyl Isomerization Modulates the Stress-induced Phosphorylation of High Molecular Weight Neurofilament Protein. *J. Biol. Chem.* 283 (39), 26737-26747.

[14] Lu, P. J., Wulf, G., Zhou, X. Z., Davies, P., and Lu, K. P. (1999) The prolyl isomerase Pin1 restores the function of Alzheimer-associated phosphorylated tau protein. *Nature* 399, 784 –788.

[15] Xu, Y.-X., Hirose, Y., Zhou, X. Z., Lu, K. P., and Manley, J. L. (2003) Pin1 modulates the structure and function of human RNA polymerase II. *Genes Dev.* 17, 2765–2776.

[16] Chrousos, G. P. (1995). The hypothalamic-pituitary-adrenal axis and immune-mediated inflammation. *N. Engl. J. Med.* 332:1351–1362.

[17] Kino, T., Chrousos, G. P. (2005) Glucocorticoid effect on gene expression. In: Steckler T, Kalin NH, Reul JMHM, eds. Handbook on stress and the brain. Amsterdam: Elsevier BV; 295–312.

[18] Kino, T., Chrousos, G. P. (2004) Glucocorticoid and mineralo- corticoid receptors and associated diseases. *Essays Biochem.* 40:137–155.

[19] Rosenfeld, M. G., Lunyak, V. V., Glass, C. K. (2006) Sensors and signals: a coactivator/corepressor/epigenetic code for integrating signal-dependent programs of transcriptional response. *Genes. Dev.* 20:1405–1428.

[20] Kino, T., Ichijo, T., Amin, N. D., Kesavapany, S., Wang, Y., Kim, N., Rao, S., Player, A., Zheng, Y. L., Garabedian, M. J., Kawasaki, E., Pant, H. C., Chrousos, G. P. (2007). Cyclin-Dependent Kinase 5 Differentially Regulates the Transcriptional Activity of the Glucocorticoid Receptor through Phosphorylation: Clinical Implications for the Nervous System Response to Glucocorticoids and Stress. *Mol. Endocrinology* 21(7):1552-1568.

[21] Kino, T., Chrousos, G., P. (2002). Tissue-specific glucocorticoid resistance-hypersensitivity syndromes: multifactorial states of clinical importance. *J. Allergy Clin. Immunol. 109*: 609–613.

[22] Ismaili, N., Garabedian, M. J. (2004). Modulation of glucocorticoid receptor function via phosphorylation. *Ann. NY Acad Sci.* 1024:86–101.

[23] Itoh, M., Adachi, M., Yasui, H., Takekawa, M., Tanaka, H., Imai, K. (2002). Nuclear export of glucocorticoid receptor is en- hanced by c-Jun N-terminal kinase-mediated phosphor- ylation. *Mol. Endocrinol.* 16:2382–2392.

[24] Szatmary, Z., Garabedian, M., J., Vilcek, J. (2004). Inhibition of glucocorticoid receptor-mediated transcriptional activa- tion by p38 mitogen-activated protein (MAP) kinase. *J. Biol. Chem.* 279:43708–43715.

[25] RogKrstic, M., D., Rogatsky, I., Yamamoto, K., R., Garabedian, M., J., (1997). Mitogen-activated and cyclin-dependent protein kinases selectively and differentially

modulate transcriptional enhancement by the glucocorticoid receptor. *Mol. Cell Biol.* 17:3947–3954.

[26] Kino, T., Jaffe, H., Amin, N. D., Chakrabarti, M., Zheng, Y. L., Chrousos, G. P. (2010). Cyclin-Dependent Kinase 5 Modulates the Transcriptional Activity of the Mieralocorticoid Receptor and Regulates Expression of Brain-Derived Neurotrophic Factor. *Mol. Endocrinology* 24: 941-952.

[27] Nicolaides, N. C., Galata, Z., Kino, T., Chrousos, G. P., Charmandari, E. (2010). The glucocorticoid receptor: molecular basis of biologic function. *Steroids* 75:1–12.

[28] Ortlund, E. A., Bridgham, J. T., Redinbo, M. R., Thornton, J. W. (2007). Crystal structure of an ancient protein: evolution by conformational epistasis. *Science* 317:1544–1548.

[29] Chrousos, G. P. (2001). Glucocorticoid therapy. In: Felig P, Frohman LA, eds. Endocrinology, metabolism, 4th ed. New York: McGraw- Hill; 609–632.

[30] Kovacs, W. J., Orth DN (1998). The adrenal cortex. In: Wilson J, D, Foster DW, eds. Williams textbook of endocrinology. Philadelphia: W.B. Saunders Co.; 517–750.

[31] Kino, T., Ichijo, T., Amin, N., D., Kesavapany, S., Wang, Y., Kim, N., Rao, S., Player, A., Zheng, Y. L., Garabedian, M. J., Kawasaki, E., Pant, H. C., Chrousos, G. P. (2007). Cyclin-dependent kinase 5 differentially regulates the transcriptional activity of the glucocorticoid receptor through phosphorylation: clinical implications for the nervous system response to glucocorticoids and stress. *Mol. Endocrinol.* 21: 1552–1568.

[32] de Kloet, E. R., Derijk, R. H., Meijer, O. C. (2007). Therapy insight: is there an imbalanced response of mineralocorticoid and glucocorticoid receptors in depression? *Nat. Clin. Pract. Endocrinol. Metab.* 3:168 –179.

[33] Shors, T. J., Seib, T. B., Levine, S., Thompson, R. F. (1989). Inescapable versus escapable shock modulates long-term potentiation in the rat hippocampus. *Science* 244:224–226.

[34] Diamond, D. M., Fleshner, M., Rose, G. M. (1994) Psychological stress repeatedly blocks hippocampal primed burst potentiation in behav- ing rats. *Behav. Brain Res.* 62:1–9.

[35] Sousa, N., Almeida, O. F. (2002). Corticosteroids: sculptors of the hippocampal formation. *Rev. Neurosci.* 13:59–84.

[36] Mattson, M. P., Lovell, M. A., Furukawa, K., Markesbery, W. R. (1995). Neurotrophic factors attenuate glutamate-induced accumulation of per- oxides, elevation of intracellular Ca^{2+} concentration, and neurotoxicity and increase antioxidant enzyme activities in hippocampal neurons. *J. Neurochem.* 65:1740–1751.

[37] Brady, S. T. (1993). Axonal dynamics and regeneration. In Neuroregeneration, Gorio A (ed) pp 7–36. New York, NY: Raven Press.

[38] Morfini, G., Szebenyi, G., Brown, H., Pant, H. C., Pigino, G., DeBoer, S., Beffert, U., Brady, S. T. (2004). A novel CDK5-dependent pathway for regulating GSK3 activity and kinesin-driven motility in neurons.

[39] Brady, S.T. (1995). A kinesin medley: biochemical and functional heterogeneity. *Trends Cell Biol* 5: 159–164.

[40] Morfini, G., Szebenyi, G., Richards, B., Brady, S. T. (2001). Regulation of kinesin: implications for neuronal development. *Dev. Neurosci.* 23: 364–376.

[41] Tsai, M-Y., Morfini, G., Szebenyi, G., Brady, S. T. (2000). Modulation of kinesin–vesicle interactions by Hsc70: implications for regulation of fast axonal transport. *Mol. Biol. Cell* 11: 2161–2173.

[42] Stenoien, D. S., Brady, S. T. (1997). Immunochemical analysis of kinesin light chain function. *Mol. Biol. Cell* 8: 675–689.

[43] Morfini, G., Szebenyi, G., Elluru, R., Ratner, N., Brady, S. T. (2002b). Glycogen Synthase kinase 3 phosphorylates kinesin light chains and negatively regulates kinesin-based motility. *EMBO J.* 23: 281–293.

[44] Wang, Q. M., Fiol, C. J., DePaoli-Roach, A. A., Roach, P. J. (1994a). Glycogen synthase kinase-3 beta is a dual specificity kinase differentially regulated by tyrosine and serine/threonine phosphorylation. *J. Biol. Chem.* 269: 14566–14574.

[45] Woodgett, J. R. (1994). Regulation and functions of the glycogen synthase kinase-3 subfamily. *Semin. Cancer Biol.* 5: 269–275.

[46] Ali, A., Hoeflich, K. P., Woodgett, J. R. (2001). Glycogen synthase kinase-3: properties, functions, and regulation. *Chem. Rev.* 101: 2527–2540.

[47] Munoz-Montano, J. R., Lim, F., Moreno, F. J., Avila, J., Diaz-Nido, J. (1999). Glycogen synthase kinase-3 modulates neurite outgrowth in cultured neurons: possible implications for neurite pathology in Alzheimer's disease. *J. Alzheimers Dis.* 1: 361–378.

[48] Lucas, J. J., Hernandez, F., Gomez-Ramos, P., Moran, M. A., Hen, R., Avila, J. (2001). Decreased nuclear beta-catenin, tau hyperphosphorylation and neurodegeneration in GSK-3beta conditional transgenic mice. *EMBO J.* 20: 27–39.

[49] Hoeflich, K. P., Luo, J., Rubie, E. A., Tsao, M. S., Jin, O., Woodgett, J. R. (2000). Requirement for glycogen synthase kinase-3beta in cell survival and NF-kappaB activation. *Nature* 406: 86–90.

[50] Julien, J.P. (1999). Neurofilament functions in health and disease, *Curr. Opin. Neurobiol.* 9 554 – 560.

[51] Miller, C.C.J., Ackerley, S., Brownlees, J., Grierson, A.J., Jacobsen, N.J.O., Thornhill, P. (2002). Axonal transport of neurofilaments in normal and disease states, *Cell. Mol. Life Sci.* 59, 323 – 330.

[52] Pant, H.C., Veeranna, (1995). Neurofilament phosphorylation, *Biochem. Cell. Biol.* 73, 575 – 592.

[53] Pant, H.C., (1988). Dephosphorylation of neurofilament protein enhances their susceptibility to degradation by calpain, *Biochem. J.* 256, 665 – 668.

[54] Giasson, B.I., Mushynski, W.E. (1996). Aberrant stress-induced phosphor- ylation of perikaryal neurofilaments, *J. Biol. Chem.* 271, 30404 – 30409.

[55] Sasaki, T., Taoka, M. Ishiguro, K., Uchida, A., Saito, T., Isobe, T., Hisanaga, S.-I. (2002). In vivo and in vitro phosphorylation at Ser-493 in the glutamate (E)-segment of neurofilament-H subunit by glycogen synthase kinase 3-beta, *J. Biol. Chem.* 277, 36032 – 36039.

[56] Shea, T.B., Jung, C., Pant, H.C. (2003) Does C-terminal phosphorylation regulate neurofilament transport? Re-evaluation of recent data suggests that it does, *Trends Neurosci.* 26, 397 – 400.

[57] Sharma, P., Veeranna, Sharma, M., Amin, N.D., Sihag, R.K., Grant, P., Ahn, N., Kulkarni, A.B., Pant, H.C. (2002). Phosphorylation of MEK1 by cdk5/ p35 down-regulates the mitogen-activated protein kinase pathway, *J. Biol. Chem.* 277, 528 – 534.

[58] W.K.-C. Chan, A. Dickerson, D. Otriz, A. Pimenta, C. Moran, K. Malik, J. Motil, S. Snyder, H.C. Pant, T.B. Shea, Mitogen activated protein kinase regulates neurofilament axonal transport, *J. Cell. Sci.* 117 (2004) 4629 – 4642.

[59] Moran, C.M., Donelly, M., Ortiz, D., Pant, H.C., Mandelkow, E-M., Shea, T.B. (2005). Cdk5 inhibits anterograde axonal transport of neurofilaments but not that of tau by inhibition of mitogen-activated protein kinase activity. *Mol. Brain Res.* 134, 338-344.

[60] E.M. Mandelkow, K. Stamer, R. Vogel, E. Thies, E. Mandelkow, Clogging of axons by tau, inhibition of axonal traffic and starvation of synapses, *Neurobiol. Aging* 24 (2003) 1079 – 1085.

[61] Shea, T.B., Yabe, J.T., Ortiz, D., Pimenta, A., Loomis, P., Goldman, R.D., Amin, N., Pant, H.C. (2004). CDK5 regulates axonal transport and phosphor- ylation of neurofilaments in cultured neurons, *J. Cell. Sci.* 117, 933 – 941.

[62] Shea, T.B., Zheng, Y.-L., Ortiz, D., Pant, H.C. (2004). Cyclin-dependent kinase 5 increases perikaryal neurofilament phosphorylation and inhibits neurofilament axonal transport in response to oxidative stress, *J. Neurosci. Res.* 76, 795 – 800.

[63] Louvi, A. and Artavanis-Tsakonas, S. (2006). Notch signalling in verte- brate neural development. *Nat. Rev. Neurosci.* 7, 93–102.

[64] Saxena, M. T., Schroeter, E. H., Mumm, J. S. and Kopan, R. (2001). Murine notch homologs (N1–4) undergo presenilin-dependent proteolysis. *J. Biol. Chem.* 276, 40268–40273.

[65] Lowell, S., Jones, P., Le Roux, I., Dunne, J. and Watt, F. M. (2000). Stimulation of human epidermal differentiation by delta-notch signalling at the boundaries of stem-cell clusters. *Curr. Biol.* 10, 491–500.

[66] Chitnis, A. and Kintner, C. (1996). Sensitivity of proneural genes to lateral inhibition affects the pattern of primary neurons in Xenopus em- bryos. *Development* 122, 2295–2301.

[67] Selkoe, D., and Kopan, R., (2003). Notch and presenilin: regulated intra- membrane proteolysis links development and degeneration. *Annu. Rev. Neurosci.* 26, 565–597.

[68] Mumm, J. S., Schroeter, E. H., Saxena, M. T., Griesemer, A., Tian, X., Pan, D. J., Ray, W. J. and Kopan, R. (2000). A ligand-induced extracel- lular cleavage regulates gamma-secretase-like proteolytic activa- tion of Notch1. *Mol. Cell.* 5, 197–206.

[69] Kanungo, J., Zheng, Y-L., Amin, N. D., Pant, H. C. (2008). The Notch signaling inhibitor DAPT down-regulates Cdk5 activity and modulates the distribution of neuronal cytoskeletal proteins. *J. Neurochem.* 106, 2236-2248.

[70] Lendahl, U., Zimmerman, L., and McKay, R. D. G. (1990). CNS stem cells express a new class of intermediate filament proteins. *Cell* 60:585–595.

[71] Sejersen, T., and Lendahl, U. (1993). Transient expression of the intermediate filament nestin during skeletal muscle development. *J. Cell Sci.* 106:1291–1300.

[72] Frisen, J., Johansson, C. B., Torok, C., Risling, M., and Lendahl, U. (1995). Rapid, widespread and longlasting induction of nestin contributes to the generation of glial scar tissue after CNS injury. *J. Cell Biol.* 131:453–464.

[73] Vaittinen, S., Lukka, R., Sahlgren, C. M., Hurme, T., Rantanen, J., Lendahl, U., Eriksson, J. E., and Kalimo, H. (2001). The expression of intermediate filament nestin as related to vimentin and desmin in regenerating skeletal muscle. *J. Neuropathol. Exp. Neurol.* 60:588–597.

[74] Chae, T., Kwon, Y. T., Bronson, R., Dikkes, P., Li, E., and Tsai, T. H. (1997). Mice lacking p35, a neuronal specific activator of Cdk5, display cortical lamination defects, seizures, and adult lethality. *Neuron* 18:29–42.

[75] Philpott, A., Porro, E. B., Kirschner, M. W., and Tsai, L. H. (1997). The role of cyclin-dependent kinase 5 and a novel regulatory subunit in regulating muscle differentiation and patterning. *Genes. Dev.* 11:1409–1421.

[76] Sahlgren, C. M., Mihailov, A., Vaittinen, S., Pallari, H-M., Kalimo, H., Pant, H. C., Eriksson, J. E. (2003). Cdk5 Regulates the Organization of Nestin and Its Association with p35. *Mol. and Cell Biol.* 23(14): 5090-5106.

[77] Si, J., Luo, Z,. Mei, L. (1996). Induction of acetylcholine receptor gene ex- pression by ARIA requires activation of mitogen-activated protein kinase, *J. Biol. Chem.* 271 19752 – 19759.

[78] Si, J., Mei, L. (1999). ERK MAP kinase activation is required for acetylcholine receptor inducing activity induced increase in all five acetylcholine receptor subunit mRNAs as well as synapse-specific expression of acetylcholine receptor epsilon-transgene, *Brain Res. Mol. Brain Res.* 67 18 – 27.

[79] Tansey, M.G., Chu, G.C., Merlie, J.P. (1996). ARIA/HRG regulates AChR epsilon subunit gene expression at the neuromuscular synapse via activation of phosphatidylinositol 3-kinase and Ras/MAPK pathway, *J. Cell Biol.* 134 465 – 476.

[80] Fu, A.K., Fu, W.Y., Cheung, J., Tsim, K.W., Ip, F.C., Wang, J.H., Ip, N.Y. (2001). Cdk5 is involved in neuregulin-induced AChR expression at the neuromuscular junction, *Nat. Neurosci.* 4 374 – 381.

[81] Li, B.S., Ma, W., Jaffe, H., Zheng, Y., Takahashi, S., Zhang, L., Kulkarni, A.B., Pant, H.C. Cdk5 is involved in neuregulin-dependent activation of PI-3 kinase and Akt activity mediating neuronal survival, *J. Biol. Chem.* 278 (2003) 35702 – 35709.

[82] Brady, S. T., Witt, A. S., Kirkpatrick, L. L., de Waegh, S. M., Readhead, C., Tu, P. H. and Lee, V. M. (1999). Formation of compact myelin is required for maturation of the axonal cytoskeleton. *J. Neurosci.* 19, 7278–7288.

[83] Kirkpatrick, L. L. and Brady, S. T. (1999). Cytoskeleton of neurons and glia, in (Siegel G. J., Agranoff B. W., Albers R. W., Fisher S. K. and Uhler M. D., eds). pp. 155–174. Basic Neurochemistry: Molecular Cellular and Medical Aspects. Philadelphia: Lippincott- Raven.

[84] Kirkpatrick, L. L. and Brady, S. T. (1994) Modulation of the axonal microtubule cytoskeleton by myelinating schwann cells. *J. Neurosci.* 14, 7440–7450.

[85] Schachner, M., and Bartsch, U. (2000). Multiple functions of the myelin- associated glycoprotein MAG (siglec-4a) in formation and main- tenance of myelin. *Glia* 29, 154–165.

[86] Li, C., Trapp, B., Ludwin, S., Peterson, A., and Roder, J. (1998) Myelin associated glycoprotein modulates glia-axon contact in vivo. *J. Neurosci. Res.* 51, 210–217.

[87] Dashiell, S. M., Tanner, S. L., Pant, H. C., Quarles, R. H. (2002). Myelin-associated glycoprotein modulates expression and phosphorylation of neuronal cytoskeletal elements and their associated kinases. *J. Neurochem.* 81, 1263-1272.

[88] Sharma, P., Veeranna, Sharma, M., Amin, N. D., Sihag, R. K., Grant, P., Ahn, N., Kulkarni, A. B., Pant, H. C. (2002). Phosphorylation of MEK1 by cdk5/p35 Down-regulates the Mitogen-activated Protein Kinase Pathway. *J. Biol. Chem.* 277(1): 528-534.

[89] Cowley, S., Paterson, H., Kemp, P., and Marshall, C. J. (1994). Activation of MAP kinase kinase is necessary and sufficient for PC12 differentiation and for transformation of NIH 3T3 cells. *Cell* 77, 841–852.

[90] Pearson, G., Robinson, F., Beers Gibson, T., Xu, B., Karandikar, M., Berman, K., and Cobb, M. H. (2001). Mitogen-activated protein (MAP) kinase pathways: regulation and physiological functions. *Endocr. Rev.* 22, 153–183

[91] Kesavapany, S., Amin, N. D., Zheng, Y-L., Nijhara, R., Jaffe, H., Sihag, R., Gutkind, J. S., Takahashi, S., Kulkarni, A., Grant, P., Pant, H. C. (2004). P35/Cyclin-dependent Kinase 5 Phosphorylation of Ras Guanine Nucleotide Releasing Factor 2 (RasGRF2) Mediates Rac-Dependent Extracellular Signal-Regulated Kinase 1/2 Activity, Altering RasGRF2 and Microtubule-Associated Protein 1b Distribution in Neurons. *J. Neuroscience* 24(18): 4421-4431.

[92] Rashid T, Banerjee M, Nikolic M (2001) Phosphorylation of Pak1 by the p35/ Cdk5 kinase affects neuronal morphology. *J Biol Chem* 276:49043– 49052.

[93] Nikolic, M., Chou, M. M., Lu, W., Mayer, B. J., Tsai, L.H. (1998). The p35/Cdk5 kinase is a neuron-specific Rac effector that inhibits Pak1 activity. *Nature* 395: 194 –198.

[94] Fan, W. T., ,Koch, C. A., de Hoog, C. L., Fam, N. P., Moran, M. F. (1998). The exchange factor Ras-GRF2 activates Ras-dependent and Rac-dependent mitogen-activated protein kinase pathways. *Curr. Biol.* 8:935–938.

[95] Yang, H., Cooley, D., Legakis, J. E., Ge, Q., Andrade, R., Mattingly, R. R. (2003). Phosphorylation of the Ras-GRF1 exchange factor at Ser916/898 reveals activation of Ras signaling in the cerebral cortex. *J. Biol. Chem.* 278:13278– 13285.

[96] Kiyono, M., Kaziro, Y., Satoh, T. (2000a). Induction of rac-guanine nucleotide exchange activity of Ras-GRF1/CDC25(Mm) following phosphorylation by the nonreceptor tyrosine kinase Src. *J. Biol. Chem.* 275:5441–5446.

[97] Kiyono, M., Kato, J., Kataoka, T., Kaziro, Y., Satoh, T. (2000b). Stimulation of Ras guanine nucleotide exchange activity of Ras-GRF1/CDC25(Mm) after tyrosine phosphorylation by the Cdc42-regulated kinase ACK1. *J. Biol. Chem.* 275:29788 – 29793.

[98] Kesavapany, S., Pareek, T. K., Zheng, Y-L., Amin, N. D., Gutkind, J. S., Ma, W., Kulkarni, A. B., Grant, P., Pant, H. C. (2006). Neuronal Nuclear Organization is Controlled by Cyclin-Dependent Kinase 5 Phosphorylation of Ras Guanine Nucleotide Releasing Factor-1. *Neurosignals* 07(15): 157-173.

[99] Kiyono, M., Kato, J., Kataoka, T., Kaziro, Y., Satoh, T. (2000). Stimulation of Ras guanine nucleotide exchange activity of Ras-GRF1/CDC25(Mm) upon tyrosine phosphorylation by the Cdc42-regulated kinase ACK1. *J. Biol. Chem.* 275:29788–29793.

[100] Kiyono, M., Kaziro, Y., Satoh, T. (2000). Induction of rac-guanine nucleotide exchange activity of Ras-GRF1/CDC25(Mm) following phosphorylation by the nonreceptor tyrosine kinase Src. *J. Biol. Chem.* 275:5441–5446.

[101] Yang, H., Cooley, D., Legakis, J. E., Ge, Q., Andrade, R., Mattingly, R. R. (2003). Phosphorylation of the Ras-GRF1 exchange factor at Ser916/898 reveals activation of Ras signaling in the ce- rebral cortex. *J. Biol. Chem.* 278:13278– 13285.

[102] Buchsbaum, R., Telliez, J. B., Goonesekera, S., Feig, L. A. (1996). The N-terminal pleckstrin, coiled- coil, and IQ domains of the exchange factor Ras-GRF act cooperatively to facilitate acti- vation by calcium. *Mol. Cell Biol.* 16: 4888–4896.

[103] Patrick, G. N., Zukerberg, L., Nikolic, M., de la Monte, S., Dikkes, P., Tsai, L. H. (1999). Conversion of p35 to p25 deregulates Cdk5 activity and promotes neurodegeneration. *Nature* 402:615–622.

[104] Tian, X., Feig, L. A. (2001). Basis for signaling specific- ity difference between Sos and Ras-GRF guanine nucleotide exchange factors. *J. Biol. Chem.* 276:47248–47256.

[105] Tian, X., Gotoh, T., Tsuji, K., Lo, E. H., Huang, S., Feig, L. A. (2004). Developmentally regulated role for Ras-GRFs in coupling NMDA glutamate receptors to Ras, Erk and CREB. *EMBO J.* 23:1567–1575.

[106] Li, B. S., Sun, M. K., Zhang, L., et al (2001). Regulation of NMDA receptors by cyclin-dependent ki- nase-5. *Proc. Natl. Acad. Sci. USA* 98: 12742–12747.

[107] Font, de Mora, J., Esteban, L. M., Burks, D. J., et al (2003). Ras-GRF1 signaling is required for normal beta-cell development and glucose homeo- stasis. *EMBO J.* 22:3039–3049.

[108] Ubeda, M., Kemp, D. M., Habener, J. F. (2004). Glucose- induced expression of the cyclin-dependent protein kinase 5 activator p35 involved in Alzheimer's disease regulates insulin gene transcription in pancreatic beta-cells. *Endo- crinology* 145:3023–3031.

[109] Li, B.S., Zhang, L., Takahashi, S., Ma, W., Jaffe, H., Kulkarni, A.B., Pant, H.C. (2002). Cyclin-dependent kinase 5 prevents neuronal apoptosis by negative regulation of c-Jun N-terminal kinase 3, *EMBO J.* 21 324 – 333.

[110] Nishitani, H., Lygerou, Z. (2002). Control of DNA replication licensing in a cell cycle. *Genes Cells* 7: 523-34.

[111] Hook, S. S., Lin, J. J., Dutta, A. (2007). Mechanisms to control rereplication and implications for cancer. *Curr Opin Cell Biol* 19: 663-71.

[112] Nigg, E. A. (2001). Mitotic kinases as regulators of cell division and its checkpoints. *Nat. Rev. Mol. Cell Biol* 2:21-32.

[113] Gong, X., Tang, X., Wiedmann, M., Wang, X., Peng, J., Zheng, D., et al. (2003). Cdk5-mediated inhibition of the protective effects of transcription factor MEF2 in neu-rotoxicity-induced apoptosis. *Neuron* 38:33-46.

[114] Hamdane, M., Bretteville, A., Sambo, A. V., Schindowski, K., Bégard, S., Delacourte, A., et al. (2005). p25/Cdk5-mediated retinoblastoma phosphorylation is an early event in neuronal cell death. *J. Cell Sci.* 118:1291- 8.

[115] Tian, B., Yang, Q., Mao, Z. (2009). Phosphorylation of ATM by Cdk5 mediates DNA damage signalling and regu- lates neuronal death. *Nat. Cell Biol.* 11:211- 34. 8.

[116] Lee, K. Y., Mummery, A., Park, J., Tariq, H., Rosales, J. L. (2010). Localization of CDK5 in the midbody and increased aneuploidy in CDK5-/- cells. *Cell Cycle* 9:3629-30.

[117] Kim, T., Law, V., Rosales, J. L., Lee, K. Y. (2010). Cdk5 variant 1 (cdk5-v1), but not full-length cdk5, is a centrosomal protein. *Cell Cycle* 9:2251-3.

[118] Patrick, G. N., Zukerberg, L., Nikolic, M., de la Monte, S., Dikkes, P., Tsai, L. H. (1999). Conversion of p35 to p25 deregu- lates Cdk5 activity and promotes neurodegeneration. *Nature* 402:615-22.

[119] Utreas, E., Maccioni, R., Gonzlez-Billault, C. (2009). Cyclin-dependent kinase 5 activator p35 overexpression and amyloid beta synergism increase apoptosis in cultured neuronal cells. *Neuroscience* 161: 978-987.

[120] Futatsugi, A., Utreras, E., Rudrabhatla, P., Jaffe, H., Pant, H. C., Kulkarni, A. B. (2012). Cyclin-dependent kinase 5 regulates E2F transcription factor through phosphotylation of Rb proteins in neurons. *Cell Cycle* 11(8): 1603-1610.

[121] Ubeda, M., Kemp, D. M., Habener, J. F. (2004). Glucose-induced expression of the cyclin-dependent protein kinase 5 activator p35 involved in Alzheimer's disease regulates insulin gene transcription in pancreatic beta-cells. E*ndocrinology* 145: 3023–31.

[122] Lilja, L., Yang, S. N., Webb, D. L., Juntti-Berggren, L., Berggren, P. O., Bark, C. (2001). Cyclin-dependent kinase 5 promotes insulin exocytosis. *J. Biol. Chem.* 276: 34199–205.

[123] Wei, F. Y., Nagashima, K., Ohshima, T., Saheki, Y., Lu, Y. F., Matsushita, M., et al. (2005). Cdk5-dependent regulation of glucose-stimulated insulin secretion. *Nat. Med.* 11: 1104–8.

[124] Zheng, Y-L., Hu, Y-F., Zhang, A., Wang, W., Li,, B., Amin, N. D., Grant, P., Pant, H. C. (2010). Overexpression of p35 in Min6 pancreatic beta cells induces a stressed neuron-like apoptosis. *Journal of Neurological Sciences* 299: 101-107.

[125] Lilja L., Johansson, J. U., Gromada, J., Mandic, S. A., Fried, G., Berggren, P. O., et al. (2004). Cyclin- dependent kinase 5 associated with p39 promotes Munc18-1 phosphorylation and Ca(2+)-dependent exocytosis. *J. Biol. Chem.* 279: 29534–41.

[126] Ubeda, M., Rukstalis, J. M., Habener, J. F. (2006). Inhibition of cyclin-dependent kinase 5 activity protects pancreatic beta cells from glucotoxicity. *J. Biol. Chem.* 281:28858–64.

[127] Macfarlane, W. M., Frayling, T. M., Ellard, S., Evans, J. C., Allen, L. I., Bulman, M. P., et al. (1999). Missense mutations in the insulin promoter factor-1 gene predispose to type 2 diabetes. *J. Clin. Invest.* 104: R33–9.

[128] Janson, J., Laedtke, T., Parisi, J. E., O'Brien, P., Petersen, R. C., Butler, P. C. (2004). Increased risk of type 2 diabetes in Alzheimer disease. *Diabetes* 53: 474–81.

[129] Hope, B., Kosofsky, B., Hyman, S. E. and Nestler, E. J. (1992) Regulation of immediate early gene expression and AP-1 binding in the rat nucleus accumbens by chronic cocaine. *Proc. Natl. Acad. Sci. USA* 89, 5764 –5768.

[130] Nestler, E. J., Hope, B. T. and Widnell, K. L. (1993). Drug addiction: a model for the molecular basis of neural plasticity. Neuron 11, 995–1006.

[131] Hope, B. T., Nye, H. E., Kelz, M. B., Self, D. W., Iadarola, M. J., Nakabeppu, Chergui, K., Svenningsson, P. and Greengard, P. (2004). Induction of a long-lasting AP-1 complex composed of altered Fos-like proteins in brain by chronic cocaine and other chronic treatments. *Proc. Natl. Acad. Sci.* USA 101, 2191–2196.

[132] Bibb, J. A., Snyder, G. L., Nishi, A., Yan, Z., Meijer, L., Fienberg, A. A., Tsai, L. H., Kwon, Y. T., Girault, J. A., Czernik, A. J., et al. (1999). Phosphorylation of DARPP-32 by Cdk5 modulates dopamine signalling in neurons. *Nature* 402, 669 – 671.

[133] Takahashi, S., Oshima, T., Cho, A., Sreenath, T., Iadarola, M. J., Pant, H. C., Kim, Y., Nairn, A. C., Brady, R. O., Greengard, P., Kulkarni, A. B. (2004). Increased activity of

cyclin-dependent kinase 5 leads to attenuation of cocaine-mediated dopamine signaling. *PNAS* 102(5): 1737-1742.

[134] Pareek, T. K., Keller, J., Kesavapany, S., Pant, H. C., Iadarola, M. J., Brady, R. O., Kulkarni, A. B. (2005). Cyclin-dependent kinase 5 activity regulates pain signaling. *PNAS* 103(3): 791-796.

[135] Hunt, S. P. and Mantyh, P. W. (2001). The molecular dynamics of pain control. Nat. Rev. Neurosci. 2, 83–91.

[136] Malik-Hall, M., Dina, O. A. and Levine, J. D. (2005). Primary afferent nociceptor mechanisms mediating NGF-induced mechanical hyperalgesia. *Eur. J. Neurosci.* 21, 3387–3394.

[137] Obata, K. and Noguchi, K. (2004). MAPK activation in nociceptive neurons and pain hypersensitivity. *Life Sci.* 74, 2643–2653.

[138] Chen, J., Zhang, Y., Kelz, M. B., Steffen, C., Ang, E. S., Zeng, L. and Nestler, E. J. (2000). Induction of cyclin-dependent kinase 5 in the hippocampus by chronic electroconvulsive seizures: role of [Delta]FosB. *J. Neurosci.* 20, 8965–8971.

[139] Bullitt, E. (1990). Expression of c-fos-like protein as a marker for neuronal activity following noxious stimulation in the rat. *J. Comp. Neurol.* 296, 517–530.

[140] Montell, C., Birnbaumer, L., Flockerzi, V. (2002). The TRP channels, a remarkably functional family. *Cell* 108:595–598.

[141] Montell, C., Birnbaumer, L., Flockerzi, V., Bindels, R. J., Bruford, E. A., Caterina, M. J., Clapham, D. E., Harteneck, C., Heller, S., Julius, D., et al. (2002). A unified nomenclature for the superfamily of TRP cation channels. *Mol. Cell* 9:229–231.

[142] Hwang, S. W., Cho, H., Kwak, J., Lee, S. Y., Kang, C. J., Jung, J., Cho, S., Min, K. H., Suh, Y. G., Kim, D., et al. (2000). Direct activation of capsaicin receptors by products of lipoxygenases: endogenous capsaicin-like substances. *Proc. Natl. Acad. Sci. USA* 97:6155–6160.

[143] Caterina, M. J., Leffler, A., Malmberg, A. B., Martin, W. J., Trafton, J., Petersen-Zeitz, K. R., Koltzenburg, M., Basbaum, A. I., Julius, D. (2000). Impaired nociception and pain sensation in mice lacking the capsaicin receptor. *Science* 288:306 –313.

[144] Ghilardi, J. R., Rohrich, H., Lindsay, T. H., Sevcik, M. A., Schwei, M. J., Kubota, K., Halvorson, K. G., Poblete, J., Chaplan, S. R., Dubin, A. E., et al. (2005). Selective blockade of the capsaicin receptor TRPV1 attenuates bone cancer pain. *J. Neurosci.* 25:3126 –3131.

[145] Nilius, B., Voets, T., Peters, J. (2005). TRP Channels in Disease. *Sci. STKE* 2005:re8.

[146] Bhave, G., Zhu, W., Wang, H., Brasier, D. J., Oxford, G. S., Gereau, R. W. T., IV (2002). cAMP-dependent protein kinase regulates desensitization of the capsaicin receptor (VR1) by direct phosphorylation. *Neuron* 35:721–731.

[147] Jung, J., Shin, J. S., Lee, S. Y., Hwang, S. W., Koo, J., Cho, H., Oh, U. (2004). Phosphorylation of vanniloid receptor 1 by Ca^{2++}/calmodulin-dependent kinase II regulates its vanilloid binding. *J. Biol. Chem.* 279:7048 –7054.

[148] Pareek, T. K., Keller, J., Kesavapany, S., Agarwal, N., Kuner, R., Pant, H. C., Iadarola, M. J., Brady, R. O., Kulkarni, A. B. (2006). Cyclin-dependent kinase 5 modulates nociceptive signaling through direct phosphorylation of transient receptor vanilloid 1. *PNAS* 104(2): 660-665.

[149] Utreras, E., Futatsugi, A., Rudrabhatla, P., Keller, J., Iadarola, M. J., Pant, H.C., Kulkarni, A. B., (2008). Tumor Necrosis Factor- α Regulates Cyclin-dependent Kinase 5 Activity during Pain Signaling through Transcriptional Activation of p35. *J. Biol. Chem.* 284 (4), 2275-2284.

[150] Cheng, B., Chistakos, S., and Mattson, M. P. (1994). Tumor necrosis factors protect neurons against metabolic-excitotoxic insults and promote maintenance of calcium homeostasis. *Neuron* 12, 139–153.

[151] Arruda, J. L., Colburn, R. W., Rickman, A. J., Rutkowski, M. D., and DeLeo, J. A. (1998). Increase of interleukin-6 mRNA in the spinal cord following peripheral nerve injury in the rat: potential role of IL-6 in neuropathic pain. *Brain Res. Mol. Brain Res.* 62, 228 –235.

[152] Barbin, G., Roisin, M.P., and Zalc, B. (2001). Tumor Necrosis Factor-α Regulates Cyclin-dependent Kinase 5 Activity during Pain Signaling through Transcriptional Activation of p35. *Neurochem. Res.* 26, 107–112.

[153] Safieh-Garabedian, B., Poole, S., Allchorne, A., Winter, J., and Woolf, C.J. (1995). Contribution of interleukin-1 beta to the inflammation-induced increase in nerve growth factor levels and inflammatory hyperalgesia. *Br. J. Pharmacol.* 115, 1265–1275.

[154] Myers, R. R., Campana, W. M., and Shubayev, V. I. (2006). The role of neuroinflammation in neuropathic pain: mechanisms and therapeutic targets. *Drug Discov. Today* 11, 8 –20.

[155] Ahlijanian MK, Barrezueta NX, Williams RD, Jakowski A, Kowsz KP, McCarthy S, Coskran T, Carlo A, Seymour PA, Burkhardt JE, Nelson RB, McNeish JD (2000). Hyperphosphorylated tau and neurofilament and cytoskeletal disruptions in mice overexpres- sing human p25, an activator of cdk5. *Proc. Natl. Acad. Sci. USA* 97: 2910–2915.

[156] Cruz, J. C., Tseng, H. C., Goldman, J. A., Shih, H., Tsai, L. H. (2003). Aberrant Cdk5 activation by p25 triggers pathological events leading to neurodegeneration and neurofibrillary tangles. *Neuron* 40: 471–483.

[157] Julien, J. P., Mushynski, W. E. (1998). Neurofilaments in health and disease. *Prog Nucleic Acid Res Mol Biol* 61: 1–23.

[158] Kusakawa, G., Saito, T., Onuki, R., Ishiguro, K., Kishimoto, T., Hisanaga, S. (2000). Calpain-dependent proteolytic cleavage of the p35 cyclin-dependent kinase 5 activator to p25. *J. Biol. Chem.* 275: 17166–17172.

[159] Alvarez, A., Toro, R., Caceres, A., Maccioni, R. B. (1999). Inhibition of tau phosphorylating protein kinase cdk5 prevents beta-amyloid-in- duced neuronal death. *FEBS Lett.* 459: 421–426.

[160] Town, T., Zolton, J., Shaffner, R., Schnell, B., Crescentini, R., Wu, Y., Zeng, J., DelleDonne, A., Obregon, D., Tan, J., Mullan, M. (2002). p35/Cdk5 pathway mediates soluble amyloid-beta peptide-induced tau phosphorylation in vitro. *J. Neurosci. Res.* 69: 362–372.

[161] Michaelis, M. L., Dobrowsky, R. T., Li, G. (2002). Tau neurofibrillary pathology and microtubule stability. *J. Mol. Neurosci.* 19: 289–293.

[162] Zheng, Y-L., Kesavapany, S., Gravell, M., Hamilton, R. S., Schubert, M., Amin, N. D., Albers, W., Grant, P., Pant, H. C. (2005). A cdk5 inhibitor peptide reduces tau hyperphosphorylation and apoptosis in neurons. *EMBO* 24: 209-220.

[163] Gengyo-Ando, K., Kamiya, Y., Yamakawa, A., Kodaira, K., Nishiwaki, K., Miwa, J., Hori, I., Hosono, R., (1993). The C. elegans unc-18 gene encodes a protein expressed in motor neurons. *Neuron* 11, 703–711.

[164] Graham, T.R., Emr, S.D., (1991). Compartmental organization of Golgi- specific protein modification and vacuolar protein sorting events defined in a yeast sec18 (NSF) mutant. *J. Cell Biol.* 114, 207–218.

[165] Hata, Y., Slaughter, C.A., Sudhof, T.C., 1993. Synaptic vesicle fusion complex contains unc-18 homologue bound to syntaxin. *Nature* 366, 347–351.

[166] Sharma, V. M., Shareef, M. M., Bhaskar, K., Kalidas, S., Shetty, P. A., Christopher, R., Ramamohan, Y., Pant, H. C., Shetty, T. K. (2005). Nuclear localization of Munc18-1 (p67) in the adult rat brain and PC12 cells. *Neurochemistry International* 47: 225-234.

[167] Paglini, G., Peris, L., Diez-Guerra, J., Quiroga, S., Caceres, A. (2001). The Cdk5– p35 kinase associates with the Golgi apparatus and regulates membrane traffic. *EMBO Rep.* 2:1139 –1144.

[168] Shuang, R., Zhang, L., Fletcher, A., Groblewski, G. E., Pevsner, J., Stuenkel, E. L. (1998). Regulation of Munc-18/syntaxin 1A interaction by cyclin- dependent kinase 5 in nerve endings. *J. Biol. Chem.* 273:4957– 4966.

[169] Smillie, K. J., Cousin, M. A. (2005). Dynamin I phosphorylation and the control of synaptic vesicle endocytosis. *Biochem. Soc. Symp.* 87–97.

[170] Amin, N. D., Zheng, Y-L., Kesavapany, S., Kanungo, J., Guszczynski, T., Sihag, R. K., Rudrabhatla, P., Albers, W., Grant, P., Pant, H. C. (2008). Cyclin-dependent Kinase 5 Phosphorylation of Human Septin SEPT5 (hCDCrel-1) Modulates Exocytosis. *J. Neuroscience* 28(14): 3631-3643.

[171] Nagy, G., Matti, U., Nehring, R. B., Binz, T., Rettig, J., Neher, E., Sorensen, J. B. (2002). Protein kinase C-dependent phosphorylation of synaptosome-associated protein of 25 kDa at Ser187 potentiates vesicle recruitment. *J. Neurosci.* 22:9278 –9286.

[172] Marsh, M., McMahon, H.T. (1999). The structural era of endocytosis, *Science* 285: 215 – 220.

[173] Wigge, P., McMahon, H.T. (1998). The amphiphysin family of proteins and their role in endocytosis at the synapse, *Trends Neurosci.* 21: 339 – 344.

[174] Rosales, J.L., Nodwell, M.J., Johnston, R.N., Lee, K.Y. (2000). Cdk5/ p25(nck5a) interaction with synaptic proteins in bovine brain, *J. Cell. Biochem.* 78: 151–159.

[175] Floyd, S.R., Porro, E.B., Slepnev, V.I., Ochoa, G.C., Tsai, L.H., De Camilli, P. (2001). Amphiphysin 1 binds the cyclin-dependent kinase (cdk) 5 regulatory subunit p35 and is phosphorylated by cdk5 and cdc2, *J. Biol. Chem.* 276: 8104 – 8110.

[176] Cousin, M.A. (2000). Synaptic vesicle endocytosis: calcium works overtime in the nerve terminal, *Mol. Neurobiol.* 22: 115 – 128.

[177] Cousin, M.A., Robinson, P.J. (2000). Ca(2+) influx inhibits dynamin and arrests synaptic vesicle endocytosis at the active zone, *J. Neurosci.* 20: 949 – 957.

[178] Cousin, M.A., Robinson, P.J. (2001). The dephosphins: dephosphorylation by calcineurin triggers synaptic vesicle endocytosis, *Trends Neuro- sci.* 24: 659 – 665.

[179] Westbrook, G.L. (1994). Glutamate receptor update, *Curr. Opin. Neurobiol.* 4: 337 – 346.

[180] Mori, H., Mishina, M. (1995). Structure and function of the NMDA receptor channel, *Neuropharmacology* 34: 1219 – 1237.

[181] Ryan, T. A., Ziv, N. E., Smith, S. J. (1996). Potentiation of evoked vesicle turn- over at individually resolved synaptic boutons, *Neuron* 17: 125 – 134.

[182] Li, B. S., Sun, M. K., Zhang, L., Takahashi, S., Ma, W., Vinade, L., Kulkarni, A.B., Brady, R.O., Pant, H.C. (2001). Regulation of NMDA receptors by cyclin-dependent kinase-5, *Proc. Natl. Acad. Sci. U S A*. 98: 12742 – 12747.

[183] Tomizawa, K., Ohta, J., Matsushita, M., Moriwaki, A., Li, S. T., Takei, K., Matsui, H. (2002). Cdk5/p35 regulates neurotransmitter release through phosphorylation and downregulation of P/Q-type volt- age-dependent calcium channel activity, *J. Neurosci.* 22: 2590 – 2597.

[184] Lee, S. Y., Wenk, M. R., Kim, Y., Nairn, A. C., De Camilli, P. (2004). Regulation of synaptojanin 1 by cyclin-dependent kinase 5 at synapses. *Proc. Natl. Acad. Sci. U S A* 101:546 –551.

[185] Fu, A. K., Fu, W. Y., Cheung, J., Tsim, K. W., Ip, F. C., Wang, J. H., Ip, N. Y. (2001). Cdk5 is involved in neuregulin-induced AChR expression at the neuromuscular junction. *Nat. Neurosci.* 4:374 –381.

[186] Morabito, M. A., Sheng, M., Tsai, L. H. (2004). Cyclin-dependent kinase 5 phosphorylates the N-terminal domain of the postsynaptic density protein PSD-95 in neurons. *J. Neurosci.* 24:865– 876.

[187] Poore, C. P., Sundaram, J. R., Pareek, T. K., Fu, A., Amin, N. D., Modamed, N. E., Zheng, Y-L., Goh, A. X. H., Lai, M. K., Ip, N. Y., Pant, H. C., Kesavapany, S. (2010). Cdk5-Mediated Phosphorylation of ∂-Catenin Regulates Its Localization and GluR2-Mediated Synaptic Activity. *J. Neuroscience* 30(25): 8457-8467.

[188] Tanahashi, H., Tabira, T. (1999). Isolation of human delta-catenin and its binding specificity with presenilin 1. *Neuroreport* 10:563–568.

[189] Medina, M., Marinescu, R. C., Overhauser, J., Kosik, K. S. (2000). Hemizygosity of delta-catenin (CTNND2) is associated with severe mental retardation in cri-du-chat syndrome. *Genomics* 63:157–164.

[190] Lu, Q., Paredes, M., Medina, M., Zhou, J., Cavallo, R., Peifer, M., Orecchio, L., Kosik, K. S. (1999). Delta-catenin, an adhesive junction-associated protein which promotes cell scattering. *J. Cell Biol.* 144:519 –532.

[191] Silverman, J. B., Restituito, S., Lu, W., Lee-Edwards, L., Khatri, L., Ziff, E. B. (2007). Synaptic anchorage of AMPA receptors by cadherins through neural plakophilin-related arm protein AMPA receptor-binding protein com- plexes. *J. Neurosci.* 27:8505– 8516.

[192] Misra, C., Restituito, S., Ferreira, J., Rameau, G. A., Fu, J., Ziff, E. B. (2010). Regulation of synaptic structure and function by palmitoylated AMPA receptor binding protein. *Mol. Cell Neurosci.* 43:341–352.

[193] Ohshima, T., Ward, J. M., Huh, C. G., Longenecker, G., Veeranna, Pant, H. C., Brady, R. O., Martin, L. J., and Kulkarni, A. B. (1996). Targeted disruption of the cyclin-dependent kinase 5 gene results in abnormal corticogenesis, neuronal pathology and perinatal death. *Proc. Natl. Acad. Sci. USA* 93, 11173–11178.

[194] Philpott, A., Tsai, L., and Kirschner, M. W. (1999). Neuronal differentiation and patterning in Xenopus: the role of cdk5 and a novel activator xp35.2. *Dev. Biol.* 207, 119–132.

[195] Zhang, J., and Herrup, K. (2008). Cdk5 and the non-catalytic arrest of the neuronal cell cycle. *Cell Cycle* 7, 3487–3490.

[196] Lagace, D. C., Benavides, D. R., Kansy, J. W., Mapelli, M., Greengard, P., Bibb, J. A., and Eisch, A. J. (2008). Cdk5 is essential for adult hippocampal neuro- genesis. *Proc. Natl. Acad. Sci. USA* 105, 18567–18571.

[197] Zheng, Y-L., Li, B-S., Rudrabhatla, P., Shukla, V., Amin, N. D., Maric, D., Kesavapany, S., Kanungo, J., Pareek, T. K., Takahashi, S., Grant, P., Kulkarni, A. B., Pant, H. C. (2010). Phosphorylation of p27Kip1 at Thr187 by Cyclin-dependent Kinase 5 modulates Neural Stem Cell Differentiation. *Mol. Biol. Cell* 21: 3601-3614.

[198] Sherr, C. J., and Roberts, J. M. (1999). CDK inhibitors: positive and negative regulators of G1-phase progression. *Genes Dev.* 13, 1501–1512.

[199] Nguyen, L., Besson, A., Roberts, J. M., and Guillemot, F. (2006b). Coupling cell cycle exit, neuronal differentiation and migration in cortical neurogenesis. *Cell Cycle* 5, 2314–2318.

[200] Nguyen, L., Besson, A., Heng, J. I., Schuurmans, C., Teboul, L., Parras, C., Philpott, A., Roberts, J. M., and Guillemot, F. (2006a). p27kip1 independently promotes neuronal differentiation and migration in the cerebral cortex. *Genes Dev.* 20, 1511–1524.

[201] Montagnoli, A., Fiore, F., Eytan, E., Carrano, A.C., Draetta, G.F., Hershko, A., and Pagano, M. (1999). Ubiquitination of p27 is regulated by Cdk-dependent phosphorylation and trimeric complex formation. *Genes Dev* 13, 1181–1189.

[202] Lee, M.H., Nikolic, M., Baptista, C.A., Lai, E., Tsai, L.H., and Massague, J. (1996). The brain-specific activator p35 allows Cdk5 to escape inhibition by p27Kip1 in neurons. *Proc. Natl. Acad. Sci. USA* 93, 3259–3263.

[203] Matsubara, M., Kusubata, M., Ishiguro, K., Uchida, T., Titani, K., and Taniguchi, H. (1996). Site-specific phosphorylation of synapsin I by mitogen-activated protein kinase and Cdk5 and its effects on physiological function. *J. Biol. Chem.* 271, 21 108–21 1113.

[204] Tomizawa, K., Ohta, J., Matsushita, M., Moriwaki, A., Li, S.-T., Takei, K., and Matsui, H. (2002). Cdk5/p35 regulates neurotransmitter release through phosphorylation and downregulation of P/Q-type voltage-dependent calcium channel activity. *J. Neurosci.* 22, 2590–2597.

[205] Bibb, J. A., Nishi, A., O'Callaghan, J. P. et al. (2001). Phosphorylation of protein phosphatase inhibitor-1 by Cdk5. *J. Biol. Chem.* 276, 1 44 990–1 14 497.

[206] Bibb, J. A., Chen, J., Taylor, J. R. et al. (1999) Phosphorylation of DARPP-32 by Cdk5 modulates dopamine signalling in neurons. *Nature* 402, 669–671.

[207] Dhavan, R., Greer, P. L., Morabito, M. A., Orlando, L. R., and Tsai, L.-H. (2002). The cyclin-dependent kinase 5 activators p35 and p39 ineract with the alpha-subunit of Ca2+-calmodulin-dependent pro- tein kinase II and alph-actinin-1 in a calcium-dependent manner. *J. Neurosci.* 22, 7879–7891.

[208] Amin, N. D., Albers, W., and Pant, H. C. (2002). Cyclin-dependent kinase 5 (Cdk5) activation requires interaction with three domains of p35. *J. Neurosci. Res.* 67, 354–362.

[209] Tokuoka, H., Saito, T., Yorifuji, H., Wei, F.-Y., Kishimoto, T., and Hisanaga, S. (2000). BDNF-induced phosphorylation of neurofilament-H subunit in primary culture of rat cortical neurons. *J. Cell Sci.* 113, 1059–1068.

[210] Paglini, G., Pigino, G., Kunda, P., Morfini, G., Maccioni, R., Quiroga, S., Fe, A., and Caceres, A. (1998). Evidence for the participation of the neuron-specific CDK5 activator P35 during laminin enhanced axonal growth. *J. Neurosci.* 18, 9858–9869.

[211] Wei, F-Y., Tomizawa, K., Ohshima, T., Asada, A., Saito, T., Nguyen, C., Bibb, J. A., Ishiguro, K., Kulkarni, A. B., Pant, H. C., Mikoshiba, K., Matsui, H., Hisanaga, S. (2005). Control of cyclin-dependent kinase 5 (Cdk5) activity by glutamergic regulation of p35 stability. *J. Neurochem.* 93: 502-512.

[212] Kerokoski, P., Suuronen, T., Salminen, A., Soininen, H., and Pirttila, T. (2002). Cleavage of the cyclin-dependnet kinase 5 activator p35 to p25 does not induce tau hyperphosphorylation. *Biochem. Biophys. Res. Commun.* 298, 693–698.

In: New Developments on Signal Transduction Research
Editor: Masayoshi Yamaguchi
ISBN: 978-1-62618-230-1

Chapter 5

The Role of Signal Transduction in Abnormal Embryonic Development

Carolyn M. Kapron*
Department of Biology, Trent University,
Peterborough, Ontario, Canada

Abstract

A wide variety of environmental chemicals and drugs can interfere with normal embryonic development, leading to malformations and other anomalies. While the underlying cellular and molecular mechanisms are not always clear, there have been numerous studies implicating disruptions in signal transduction pathways in the induction of abnormalities by embryotoxic compounds. The types of intracellular signalling pathways influenced by teratogens are quite diverse and, in many cases, the pathway is activated by treatment. The Mitogen-Activated Protein Kinase family (MAPK) includes members that influence both cell proliferation and cell death. Stress-induced members of this family (c-Jun N-terminal kinase, (JNK) and p38) have been found to be induced by exposure of vertebrate embryos to toxic metals, cyclophosphamide, hydroxyurea, perfluorooctane sulfonate (PFOS) and staurosporine. The activation of these stress proteins is often linked to reactive oxygen species and oxidative stress, and typically culminates in apoptosis. Activation of Ras, which typically leads to cell growth and differentiation, is involved in phenytoin embryopathy. Rather than being activated, other pathways are inhibited. The Protein Kinase B/Akt pathway, which is linked to cell survival, has been found to be down-regulated after limb cells are exposed to the well-known teratogen, thalidomide. Developmentally important signalling pathways, such as Sonic Hedgehog (Shh), are disrupted by exposure to ethanol and retinoic acid. The primary consequences of the induction of many of these pathways include alterations in cell proliferation, differentiation, apoptosis and cell migration. These changes, in turn, disrupt the normal processes of embryonic development, leading to malformations or embryonic death.

This chapter will discuss the major disturbances in cell signalling by treatments that interfere with embryogenesis, with the aim of determining whether commonalities exist

* Phone: 1-705-748-1011 ext. 7641; Fax: 1-705-748-1139; Email: ckapron@trentu.ca

among the toxins, outcomes and pathways affected that would help to explain the mechanisms of toxin-induced abnormal development.

Keywords: embryo, teratogen, signal transduction, reactive oxygen species, abnormal development

Introduction

The occurrence of birth defects in the human population remains a significant health issue, as over 6% of babies worldwide are born with a structural abnormality as a result of disruption of normal development (Christianson et al., 2006). While some developmental abnormalities can be traced to specific genetic or environmental causes, in the vast majority of cases the cause is unknown and is possibly the result of a combination of genetic and environmental factors (see review by Finnell et al., 2002). Even when the environmental trigger of abnormal development is known, the cellular and molecular mechanisms that lead to the defect have often not been determined. Many factors will influence the type and severity of the malformation produced. These factors include the specific teratogen being studied, the species that is exposed, the time of development at which the exposure occurs, and the concentration, route and duration of exposure.

While direct damage to cellular components—such as enzyme disruption or membrane damage—may lead to abnormal development, recent evidence points to mechanisms related to changes in normal developmental programming through disruptions to gene expression brought about by interference with cell signalling pathways. The types of intracellular signalling pathways influenced by teratogens are quite diverse and there are numerous examples of pathways being activated or inhibited by treatment. In fact, different teratogens that cause the same developmental disorder may do so by activating different cellular responses (Dufour-Rainfray et al., 2011).

During development, the cells of the embryo must communicate with each other and with their surrounding environment in order to survive, migrate and differentiate, the result of which will be a normal embryo. Development depends on the expression of specific proteins and genes to proceed properly and the developmental changes rely on cellular interactions and signal transduction pathways. These behaviours are often mediated by the activation of surface or cytoplasmic receptors and the transmission of the stimulus through common intracellular signalling pathways. Although some of these pathways are uncommon in adults, most of the time the signalling pathways are not unique to embryos and the intracellular outcome of their activation (e.g. apoptosis or cell proliferation) is similar in adults and embryos. However, in the embryo that is changing very rapidly as it develops, the functional consequence of the stimulation of a particular pathway and the ensuing changes in gene expression may have a profound functional consequence if the normal program of development is disrupted.

The primary consequences of the induction or disruption of many of the signal transduction pathways in the embryo include alterations in cell proliferation, differentiation, apoptosis, and cell migration. These changes, in turn, disrupt the normal processes of embryonic development, leading to malformations or embryonic death. This review will discuss some of the major disturbances in cell signalling by harmful environmental or endogenous conditions that interfere with embryogenesis, with the aim of determining whether commonalities exist among the toxins, outcomes and pathways affected that would help to explain mechanisms of toxin-induced development.

SONIC HEDGEHOG PATHWAY

Sonic hedgehog (Shh) is a signalling molecule that is involved in a number of developmental processes, including patterning of both the neural tube and the limb buds in many vertebrates. Shh is part of the highly conserved Hedgehog family of proteins that influence morphogenesis in a broad variety of organisms. After synthesis, Shh is modified and secreted as a cholesterol-bound ligand. It binds to its receptor, Patched 1 (Ptc1), which when unbound inhibits the associated protein, Smoothened (Smo). When Shh binds Ptc1, the inhibition of Smo is released, initiating a downstream signalling cascade that leads to the activation of Gli transcription factors and the expression of target genes (reviewed in Ingham and McMahon, 2001).

Genetic deficiency in Shh causes various malformations, including cyclopia and holoprosencephaly (Chiang et al., 1996). Some teratogenic compounds that cause similar abnormalities have been found to interfere with Shh signalling (Cohen and Shiota, 2002). The archetype molecules that interfere with the functions of Shh in cranial development are cylcopamine and jervine, components of the plant *Veratrum californicum*. Consumption of this plant is teratogenic in animals such as sheep, in which malformations like cyclopia and holoprosencephaly were observed under both natural and experimental conditions (Keeler and Binns, 1968). Cyclopamine blocks Shh signalling in tissues that are normally responsive to this ligand. Neural explants cultured with Shh take on ventral neural tube characteristics but this response is blocked in the presence of cyclopamine, which has been shown to interfere with Shh signalling by binding to the Smoothened receptor (Chen et al., 2002) and blocking Smoothened activation (Cooper et al., 1998, Incardona et al., 1998).

The Shh pathway is also disrupted by exposure to ethanol, suggesting that it may play a role in human fetal alcohol spectrum disorders. Experimental animals that are exposed to ethanol develop abnormalities that are consistent with disruption of Shh signalling. Studies have shown a reduction in Shh signalling in chick and zebrafish embryos that are exposed to an ethanol treatment that stimulates axial defects (Ahlgren et al., 2002; Loucks et al., 2007).The abnormalities observed included dorsal curvature of the tail, somite abnormalities and cyclopia. These were somewhat distinct from those induced by cyclopamine and forskolin, another Shh pathway disruptor, which suggests that the modes of action are different (Loucks and Ahlgren, 2009). However, Shh does appear to play a direct role in ethanol teratogenicity, as addition of exogenous Shh to ethanol-treated embryos prevents the associated abnormalities (Loucks and Ahlgren, 2009). The varying outcomes suggest that Shh may interact with other cell signalling pathways in ethanol teratogenesis. Indeed, in mouse embryos exposed to ethanol in concentrations that induce limb abnormalities, decreases in Fibroblast growth factor 8 (Fgf8) were found in the apical ectodermal ridge of the limb bud and in Shh in the zone of polarizing activity (Chrisman et al., 2004), two regions that are essential for patterning the growing limb. The expression domain of fibroblast growth factors 8 and 4 (Fgf8/4) was also reduced in the apical ectodermal ridge of the developing limb that was exposed to cadmium on day 9 of gestation and there was also a delayed onset of expression and a subsequent reduction in the domain of expression of Shh in the posterior mesenchyme (Elsaid et al., 2007). These two pathways contribute to patterning in the limb and their disruption may result in abnormal limb morphogenesis, such as that which occurs with cadmium exposure (Chernoff, 1973). However, the specific effect of cadmium appears to be dependent on the stage of development at the time of exposure, as embryos that are exposed just slightly later in development lose Shh activity in the zone of polarizing activity but show no change in the pattern of expression (Scott et al., 2005). Another teratogen that targets Shh is retinoic acid. Excess retinoic acid can have profound effects on embryonic development, including the induction of limb abnormalities (Collins and Mao, 1999). In zebrafish embryos that are exposed to retinoic acid at the time the pectoral fin is starting to develop, there is an expansion of the domain of expression of Shh and Ptc1 in the limb, and ultimately ectopic expression of these genes in the anterior region (Hoffman et al., 2002). The retinoic acid and Shh pathways may interact within the embryo. Concentrations of glyphosate based herbicides that induced a variety of malformation in *Xenopus* and chick embryos (shortening of the trunk, cyclopia, and reduction of neural crest) were found to increase the activity of retinoic acid signalling and decrease the expression of Shh (Paganelli et al. 2010).

The Shh pathway may be one that is particularly sensitive to a wide variety of teratogens. In a screen of a relatively small library of 4240 chemicals that included pesticides, drugs, and natural products, three (a dietary supplement, an anti-fungal agent and a hormone) were found to disrupt Shh signalling, although to a lesser extent than does cyclopamine (Lipinski and Bushman, 2010). The potential role of such compounds in human teratogenesis is unclear, although it is possible that they would interact with genes to increase the risk of holoprosencephaly or other malformations (Lipinski and Bushman, 2010).

Cholesterol has been found to regulate the Shh pathway at several points. It modifies Shh and may serve to anchor it in the membrane of cells that produce it, while also appearing to regulate Shh signalling at the level of the Ptc-1 receptor (reviewed in Incardona and Eaton, 2000). Modification of Shh by cholesterol supports the cell-surface retention of Shh, enhances the signal potency of the Shh ligand, and increases the amount of Shh associated with receiving cells (Grover et al. 2011). Cholesterol homeostasis may be disrupted by ethanol in the developing brain (Guizzetti and Costa, 2007). Further evidence for the involvement of cholesterol in cranial development is seen in Smith-Lemli-Opitz syndrome. This human disorder that results from a genetic deficiency in dehydrocholesterol reductase activity, an enzyme essential for cholesterol biosynthesis, is associated with less severe forms of holoprosencephaly (Tint et al., 1994).

WNT AND TGFβ PATHWAYS

Another pathway that plays an important role in embryonic development is the Wnt pathway. The Wnt family comprises a large group of secreted glycoproteins that bind to the Frizzled receptor, typically activating Dishevelled. This inhibits glycogen synthase kinase 3 (GSK3), which prevents the degradation of β-catenin, a component of a transcription factor complex (Logan and Nusse 2004). Perhaps not surprisingly, members of the family play a role in teratogenesis, sometimes in combination with members of the Transforming growth factor β (TGF-β) family, another large family of secreted proteins that stimulate intracellular pathways upon binding to receptors. The Wnt pathway has been found to play a role in thalidomide-induced limb truncation, microphthalmia and programmed cell death in chick embryos (Knobloch et al. 2007). Abnormalities induced by thalidomide exposure were the result of the generation of reactive oxygen species that induced increased signalling by Bone Morphogenetic Proteins (BMPs), which led to an increase in Dickkopf1 (Dkk1), inhibition of Wnt/β-catenin signalling and increased programmed cell death (Knobloch et al. 2007). In human neural stem cells that were isolated from fetal brains and grown in culture, ethanol caused changes in several developmentally important pathways, including Shh, transforming growth factor β, and Wnt. Both Wnt 3a and Wnt 5a were decreased in neural cells exposed to ethanol and proteins in the Wnt pathway, such as β-catenin and GSK3β, were also less abundant (Vangipuram and Lyman, 2012).

MAPK/RAS PATHWAYS

The Mitogen-activated protein kinase family (MAPK) is a family of signalling proteins that are activated in response to a wide variety of environmental stimuli (Cobb, 1999; Kyriakis et al., 1994). In addition to influencing morphogenetic events, these signalling proteins also influence both cell proliferation and cell death. This highly conserved family has several members, including extracellular signal-regulated kinases 1/2 (ERK 1/2), c-Jun N-terminal kinase (JNK, also known as stress-activated protein kinase (SAPK)), and p38 (English et al., 1999). While ERKs 1 and 2 are activated by growth factors, such as the Fibroblast growth factor family, and typically influence cell proliferative responses (Cobb, 1999), JNK and p38 are stimulated in response to a variety of cellular stresses, for example, inflammatory cytokines and environmental stresses, and often promote an apoptotic response (Kyriakis et al., 1994). All members of the family are phosphorylated and activated by a series of upstream kinases. When MAPK family proteins are activated, they can translocate to the nucleus and phosphorylate and activate components of the transcriptional complex AP-1, c-Jun, ATF-2 and other proteins.

Many of the members of the MAPK family have been implicated in the teratogenicity of various compounds and they tend to be differentially induced in a tissue- and teratogen-specific manner. In mouse embryos in whole mouse embryo culture, a heat shock of 43°C for 15 min on day 9 of development led to a rapid phosphorylation of ERK, JNK and p38, with a maximum activation observed at 30 minutes. Cyclophosphamide induced p38 activation after 5 hours and staurosporine after 3 hours, but neither one induced ERK or JNK activation, even after 7.5 hours (Mirkes et al. 2000). Although all three signalling proteins were induced in the three portions of the embryo that were examined after heat shock exposure, heads, hearts and trunks, apoptosis was only observed in heads and trunks (Mirkes and Little, 1998). This suggests that activation of these pathways does not necessarily lead to the induction of apoptosis (Mirkes et al., 2000). In contrast, hydroxyurea exposure on day 9 of gestation in CD-1 mice, which causes limb and skeletal abnormalities, as well as increased mortality, was found to increase the activation of p38 and JNK but not ERK 1/2 (Yan and Hales, 2008). The function of these signalling pathways in the embryotoxic response appears to be different, as blocking p38 caused an increase in mortality but not malformations and blocking JNK caused an increase in hindlimb malformations but not mortality. A third pattern of MAPK-family activation was observed in day 11 BALB/c mouse embryo limb bud cells that were exposed to cadmium, an embryotoxic metal, in micromass culture. While JNK was phosphorylated very rapidly after cadmium exposure, with a peak at 60 minutes, neither p38 nor ERK 1/2 showed a response up to 120 minutes (Liu and Kapron, 2010).

ERK 1/2 is often observed to be activated in response to growth-promoting treatments and it may also be inhibited in response to teratogens. The detrimental effects of valproic acid (VPA), and its teratogenic derivatives, on cell growth and motility in L929 cells in culture were correlated with reductions in ERK 1/2 activation, as indicated by phosphorylation (Gotfryd et al., 2011). Similarly, treatment of mouse embryos with cadmium was found to reduce the amount of activate ERK1/2 throughout the mesenchyme of the forelimb buds of embryos that are sensitive and develop limb abnormalities after cadmium treatment (C57BL/6N) but not in embryos that are resistant to the effects (SWV) (Elsaid et al., 2007).

Maternal metabolic disorders that can lead to abnormal embryonic development have also been shown to interfere with MAPK signalling in developing embryos, as the offspring of diabetic rat mothers showed decreased phosphorylation of ERK 1/2 and increased phosphorylation of JNK and p38, linked to down-regulation and decreased DNA-binding of retinoic acid receptors (RARs) and retinoid X receptors (RXRs) (Singh et al., 2012).

NFκB

The Nuclear factor kappa B (NFκB) family of highly-regulated transcription factors are stimulated by a wide variety of extracellular signals to lose inhibition, translocate to the nucleus and bind to the κB site on a variety of genes that control cellular processes , including those important for embryonic development (reviewed in Gilmore, 2006). Because NFκB has been found to regulate apoptosis in both a positive and negative manner, its potential role in teratogenesis has been addressed. Cyclophosphamide treatment of ICR mice on the 12th day of gestation caused significant malformations, with craniofacial, limb and trunk abnormalities being observed (Torchinsky et al., 2002). Brain tissue, which has elevated apoptosis after cyclophosphamide treatment, also showed a cyclophosphamide-induced suppression of NFκB binding to DNA, whereas liver tissue, which quickly recovers from apoptosis induction, did not exhibit a suppression of DNA binding, suggesting that NFκB binding to DNA might have a protective effect (Torchinsky et al., 2002). Reduction of NFκB binding to DNA was also observed after thalidomide exposure in rabbit embryos, which are sensitive to the teratogenic effects of the drug, but not in rat embryos, which are resistant (Hansen et al., 2002). Suppression of NFκB binding to appropriate regions of DNA would prevent the expression of genes that it normally upregulates. It is proposed that the subsequent lack of expression of genes such as *Twist*, *Fgf8* and *Fgf10*, all essential for normal limb outgrowth, leads to the observed limb abnormalities (Hansen and Harris, 2004). NFκB may also influence the development of extraembryonic tissues. Its expression was significantly higher in the placenta of rats exposed to elevated levels of lead compared to controls and the amount of NFκB in human placenta was positively correlated with the concentration of lead in maternal blood, with both showing placental abnormalities (Wang et al., 2009).

The tumour suppressor, p53, has been shown to play a role in the response of embryonic tissues to various teratogens and one of its mechanisms of action may be to regulate the NFκB pathway. In response to exposure to cyclophosphamide, increased malformations and apoptosis, decreased cell proliferation, increased caspase activity and decreased DNA binding of NFκB were observed in $p53^{+/+}$ but not in $p53^{-/-}$ mouse embryos, which suggested that p53 is necessary for the NFκB activity (Pekar et al., 2007). In some instances, p53 may not play an essential role in the teratogenic outcome. While $p53^{-/-}$ mouse embryos showed a different pattern of limb bud apoptosis after cadmium exposure, compared to wildtype or heterozygous embryos, there was no difference in the frequency or type of limb abnormality that was induced (Elsaid et al., 2010).

OXIDATIVE STRESS

The generation of oxidative stress and/or reactive oxygen species (ROS) by exposure of embryos to a wide variety of compounds has been linked to teratogenic outcomes (for example see Brocardo et al., 2011). In fact, the production of ROS appears to be a central element in the mechanism of action of many teratogens (Kovacic and Somanathan, 2006, Alexander and Tuan, 2010). Because of their reactive nature, it would be expected that ROS could interfere with multiple cellular processes, inhibit or activate signal transduction pathways and damage cellular macromolecules. The risk of malformations or death of the embryo would depend on the number of processes affected and the magnitude of the effect (Wells et al., 2005). However, if multiple pathways ultimately converge on one process, it might also be possible that after a certain point, no additional damage would be done by the stimulation of additional pathways.

As reviewed by Brocardo et al. (2011), exposure of animals that are a model for fetal alcohol spectrum disorder (FASD) to ethanol at varying gestational periods, and for varying lengths of time, caused changes in markers of oxidative stress, including reduced glutathione (GSH) levels, increased DNA, protein and lipid damage and alterations in endogenous antioxidant enzymes. Bromodeoxyuridine causes skeletal and digit abnormalities when mouse embryos are exposed on gestational day 9 (Sahambi and Hales, 2006). The occurrence of abnormalities with BRDU was linked to the generation of oxidative stress, which led to the induction of c-Fos dependent AP-1 binding activity in the embryo (Sahambi and Hales, 2006).

Activation of Ras, which typically triggers cell growth, survival and differentiation, is involved in adverse developmental outcomes caused by exposure to phenytoin, a ROS-inducing teratogenic anti-convulsant (Winn and Wells, 2002). It appears that the drug is not acting through mutation of Ras but rather by increasing the amount of active GTP-bound Ras (Winn and Wells, 2002). Perfluorooctane sulfonate (PFOS), a developmental toxicant that causes oxidative stress and apoptosis in zebrafish embryos, did not cause an observable change in ERK expression but was linked to increased expression of JNK and p38, as well as upregulation of Nuclear factor erythroid 2–related factor 2 (Nrf2), which regulates the expression of many antioxidant enzymes (Shi and Zhou, 2010).

In some studies, the generation of oxidative stress by a teratogenic environment is inferred because of the inhibition of malformations and suppression of the signalling response by the pre- or co-exposure of the embryos to an antioxidant. For example, the reduction in oxidative stress and abnormal embryo development observed by treatment with resveratrol was accompanied by a reduction in the activation of JNK and p38 in the offspring of diabetic mothers (Singh et al. 2012).

Under whole embryo culture conditions, an increase in oxidative stress was observed in untreated cultures, as indicated by a rapid disruption in glutathione homeostasis that could be prevented by treatment with antioxidants (Ozolinš and Hales, 1997).

At the same time, an increase in the expression *c-fos, c-jun, junB and junD* mRNAs was observed, along with enhanced DNA-binding of the Activator Protein 1 (AP-1) transcription factor that could be prevented by treatment with antioxidants (Ozolinš and Hales, 1997). Exposure of mouse embryos to BrdU in vivo also causes an increase in c-fos expression, predominantly in the neuroepithelium where most BrdU was incorporated and most oxidative damage was observed (Gnanabakthan and Hales, 2009). However, some well-known antioxidants may function in multiple ways, as seen in mouse embryo limb bud cells exposed to cadmium in vitro. The cytotoxicity of cadmium and the induction of JNK phosphorylation were both decreased by the addition of glutathione (GSH) to the culture medium; however, it is likely that GSH was functioning extracellularly, perhaps by binding to cadmium through its sulfhydryl groups (MacKinnon and Kapron, 2010).

Akt/PKB

The phosphoinositide 3-kinase/Akt (PI3K/Akt) pathway may also be involved in cadmium-induced limb deformities (Elsaid et al., 2007) and preliminary evidence suggests that it interacts with the JNK pathway in causing reduced cell survival and differentiation in limb bud cells in vitro (Kapron et al., 2012). Akt, also known as protein kinase B (PKB), plays a central role in mediating cell signalling, intracellular trafficking/targeting, protein interactions, and transcription. When it is activated by phosphorylation on both Ser473 and Thr308, it induces cell growth, differentiation, survival, and proliferation by activating or inhibiting apoptotic pathways and stimulating changes in gene expression (Rokutanda et al., 2009). The PI3K/Akt pathway is also linked to activation of Fgf8/4 by cadmium in the developing limb (Elsaid et al., 2007). In contrast, its activity was found to be decreased in chick embryos that are exposed to thalidomide, along with an increase in one of its primary negative regulators, PTEN, which is protected from degradation by BMPs (Knobloch et al., 2008). Blocking PTEN activity with an inhibitor, or enhancing Akt activity with insulin both prevented the limb and eye malformations induced by thalidomide (Knobloch et al., 2008), suggesting that the Akt pathway is directly involved in the teratogenesis of thalidomide.

Conclusion

A wide variety of environmental chemicals and drugs can interfere with normal embryonic development, leading to malformations and other anomalies. While the underlying cellular and molecular mechanisms are not always clear, there have been numerous studies implicating disruptions in signal transduction pathways in the induction of abnormalities by embryotoxic compounds. The activation of these stress proteins is often linked to reactive oxygen species and oxidative stress, and typically culminates in apoptosis. However, it is apparent that exposure of cells, including embryonic cells, to toxic compounds can often generate a simultaneous inhibition and activation of numerous cellular responses that are related to signalling pathways. For example, exposure of various cell types to cadmium can induce signalling pathways that may be interlinked, such as ROS, MAPK, Akt/PKB, NFκB, Shh and Wnt pathways (Thévenod, 2009). Understanding how these pathways are linked and how their activation influences embryonic development is essential to our understanding of teratogenesis. Recent efforts to profile larger-scale changes in gene expression in embryos exposed to teratogens have the potential to uncover common teratogenic cell signalling pathways and how they interact. In one such study, exposure of organogenesis-stage mouse limb buds to a pre-activated form of cyclophosphamide changed the expression of a number of genes that were thought to fall into general categories of developmentally-regulated signalling pathways and genes that are a more general response to cellular damage (Huang and Hales, 2009). SB-236057, a serotonin receptor inverse agonist that is teratogenic in rodents, disrupts a number of genes that are involved in axial formation, suggesting an interaction between a nuclear factor r-espl and the Notch and Shh pathways (Augustine-Rauch et al., 2004). Future studies that attempt to understand and integrate the various signalling pathways that are induced by a teratogen will be helpful in elucidating the molecular mechanisms of environmentally-induced birth defects.

REFERENCES

Ahlgren, S.C., Thakur, V. and Bronner-Fraser, M. (2002) Sonic hedgehog rescues cranial neural crest from cell death induced by ethanol exposure. *Proceedings of the National Academy of Sciences* 99:10476–10481.

Augustine-Rauch, K.A., Zhang, Q.J., Leonard, J.L., Chadderton, A., Welsh, M.J., Rami, H.K., Thompson, M., Gaster, L., and Wier, P.J. (2004) Evidence for a molecular mechanism of teratogenicity of SB-236057, a 5-HT1B receptor inverse agonist that alters axial formation. *Birth Defects Research* (Part A) 70:789–807.

Alexander, P.G. and Tuan, R.S. (2010) Role of environmental factors in axial skeletal dysmorphogenesis. *Birth Defects Research (Part C)* 90:118–132.

Brocardoa, P.S., Gil-Mohapela, J., Christieam, B.R. (2011) The role of oxidative stress in fetal alcohol spectrum disorders. *Brain Research Reviews* 67:209–225.

Chen, J.K., Taipale, J., Cooper, M.K., Beachy, P.A. (2002) Inhibition of Hedgehog signaling by direct binding of cyclopamine to smoothened. *Genes and Development 16*: 2743-2748.

Chernoff, N. (1973) Teratogenic effects of cadmium in rats. *Teratolog*y 8:29–32.

Chiang, C., Litingung, Y., Lee, E., Young, K.E., Corden, J.E., Westphal, H., Beachy, P.A. (1996) Cyclopia and defective axial patterning in mice lacking sonic hedgehog gene function. *Nature* 383:407–413.

Chrisman, K., Kenney, R., Comin, J., Thal, T., Suchocki, L., Yueh, Y.G. and Gardner, D.P. (2004) Gestational ethanol exposure disrupts the expression of FGF8 and Sonic hedgehog during limb patterning. *Birth Defects Research (Part A)* 70:163–171.

Christianson, A., Howson, C.P. and Modell, B. (2006) Global report on birth defects: The hidden toll of dying and disabled children. March of Dimes Birth Defects Foundation. White Plains, NY.

Cobb, M.H. (1999) MAP kinase pathways. *Progress in Biophysics and Molecular Biology* 71:479–500.

Cohen, M.M., Shiota, K. (2002) Teratogenesis of holoprosencephaly. *American Journal of Medical Genetics* 109:1 – 15.

Collins, M.D. and Mao, G.E. (1999) Teratology of retinoids. *Annual Review of Pharmacology and Toxicology 39*:399–430.

Cooper, M.K., Porter, J.A., Young, K.E. and Beachy, P.A. (1998) Teratogen-mediated inhibition of target tissue response to Shh signaling. *Science* 280:1603-1607.

Dufour-Rainfray, D., Vourc'h, P., Tourleta, S., Guilloteau, D., Chalon, S. and Andres, C.R. (2011) Fetal exposure to teratogens: Evidence of genes involved in autism. *Neuroscience and Biobehavioral Reviews* 35:1254–1265.

English, J., Pearson, G., Wilsbacher, J., Swantek, J., Karandikar, M., Xu, S. and Cobb, M.H. (1999) New insights into the control of MAP kinase pathways. *Experimental Cell Reseach* 253:255–270.

Elsaid, A.F., Delot, E.C. and Collins, M.D. (2007) Differential perturbation of the Fgf/Erk1/2 and Shh pathways in the C57BL/6N and SWV embryonic limb buds after mid-gestational cadmium chloride administration. *Molecular Genetics and Metabolism* 92:258–270.

Elsaid, A.F., Koriem, K.M.M. and Collins, M.D. (2010) Sensitivity to cadmium chloride-induced forelimb ectrodactyly is independent of the p53 gene-dosage in the C57BL/6J mouse. *Birth Defects Research (Part A)* 88:223–227.

Finnell, R.H., Gelineau-van Waes, J., Eudy, J.D. and Rosenquist, T.H. (2002) Molecular basis of environmentally induced birth defects. *Annual Review of Pharmacology and Toxicology* 42:181–208.

Gilmore, T.D. (2006) Introduction to NF-κB: players, pathways, perspectives. *Oncogene* 25:6680–6684.

Gnanabakthan, N. and Hales, B.F. (2009) The oxidative stress response is region specific in organogenesis stage mouse embryos exposed to 5-bromo-2'-deoxyuridine. *Birth Defects Research (Part A)* 85:202-210.

Gotfryd, K., Hansen, M., Kawa, A., Ellerbeck, U., Nau, H., Berezin, V., Bock, E, and Walmod, P.S. (2011) The teratogenic potencies of valproic acid derivatives and their effects on biological end-points are related to changes in histone deacetylase and Erk1/2 activities. *Basic and Clinical Pharmacology and Toxicology* 109:164–174.

Grover, V.K., Valadez, J.G., Bowman, A.B. and Cooper, M.K. (2011) Lipid modifications of Sonic hedgehog ligand dictate cellular reception and signal response. PLoS ONE 6(7): e21353. doi:10.1371/journal.pone.0021353.

Guizzetti, M. and Costa, L.G. (2007) Cholesterol homeostasis in the developing brain: a possible new target for ethanol. *Human and Experimental Toxicology* 26:355–360.

Hansen, J.M. and Harris, C. (2004) A novel hypothesis for thalidomide-induced limb teratogenesis: Redox misregulation of the NF-kB pathway. *Antioxid. Redox Signal.* 6: 1–14.

Hansen, J. M., Gong, S.-G., Philbert, M. and Harris, C. (2002) Misregulation of gene expression in the redox-sensitive NF-κB-dependent limb outgrowth pathway by thalidomide. *Developmental Dynamics* 255:186–194.

Hoffman, L., Miles, J., Avaron, F., Laforest, L. and Akimenko, M-A. (2002) Exogenous retinoic acid induces a stage-specific, transient and progressive extension of Sonic hedgehog expression across the pectoral fin bud of Zebrafish. *International Journal of Developmental Biology* 46: 949-956.

Huang, C. and Hales, B.F. (2009) Teratogen responsive signaling pathways in organogenesis stage mouse limbs. *Reproductive Toxicology* 27:103–110.

Incardona, J.P. and Eaton, S. (2000) Cholesterol in signal transduction. *Current Opinion in Cell Biology* 12:193–203.

Incardona, J.P., Gaffield, W., Kapur, R.P. and Roelink, H. (1998) The teratogenic Veratrum alkaloid cyclopamine inhibits Sonic hedgehog signal transduction. *Development* 125:3553-3562.

Ingham, P.W. and McMahon, A.P. (2001) Hedgehog signaling in animal development: Paradigms and principles. *Genes and Development* 15:3059-3087.

Kapron, C.M., Backman, K., Ostler, C. and Wagner, K. (2012) Activation of Akt by cadmium in micromass cultures of mouse embryo limb bud cells. *Birth Defects Research Part A.* 95: 318.

Keeler, R.F. and Binns, W. (1968) Teratogenic compounds of *Veratrum californicum. Teratology* 1:5-10.

Knobloch, J., Shaughnessy, Jr., J.D. and Rüther, U. (2007) Thalidomide induces limb deformities by perturbing the Bmp/Dkk1/Wnt signaling pathway. *FASEB Journal* 21:1410–1421.

Knobloch, J., Schmitz, I., Götz, K., Schulze-Osthoff, K. and Rüther, U. (2008) Thalidomide induces limb anomalies by PTEN stabilization, Akt suppression, and stimulation of caspase-dependent cell death. *Molecular and Cellular Biology* 28:529-538.

Kovacic, P. and Somanathan, R. (2006) Mechanism of teratogenesis: Electron transfer, reactive oxygen species, and antioxidants. *Birth Defects Research (Part C)* 78:308–325.

Kyriakis, J.M., Banerjee, P., Nikolakaki, E., Dai, T., Rubie, E.A., Ahmad, M.F., Avruch, J. and Woodgett, J.R. (1994) The stress-activated protein kinase subfamily of c-Jun kinases. *Nature* 369:156–160.

Lipinski, R.J. and Bushman, W. (2010) Identification of Hedgehog signaling inhibitors with relevant human exposure by small molecule screening. *Toxicology in Vitro* 24:1404–1409.

Liu, J. and Kapron, C.M. (2010) Differential induction of MAP kinase signalling pathways by cadmium in primary cultures of mouse embryo limb bud cells. *Reproductive Toxicology* 29:286–291.

Logan, C.Y. and Nusse, R. (2004) The Wnt signaling pathway in development and disease. *Annual Review of Cell and Developmental Biology* 20:781–810.

Loucks, E. J. and Ahlgren, S.C. (2009) Deciphering the role of Shh signalling in axial defects produced by ethanol exposure. *Birth Defects Research (Part A)* 85:556-567.

Loucks, E. J., Schwend, T. and Ahlgren, S.C. (2007) Molecular changes associated with teratogen-induced cyclopia. *Birth Defects Research (Part A)* 79:642651.

MacKinnon, Y. and Kapron, C.M. (2010) Reduction in cadmium-induced toxicity and c-Jun N-terminal kinase activation by glutathione in cultured mouse embryonic cells. *Birth Defects Research (Part A)* 88:707–714.

Mirkes, P.E. and Little, S.A. (1998) Teratogen-induced cell death in postimplantation mouse embryos: differential tissue sensitivity and hallmarks of apoptosis. *Cell Death and Differentiation* 5:592–600.

Mirkes, P.E., Wilson, K.L. and Cornel, L.M. (2000) Teratogen-induced activation of ERK, JNK, and p38 MAP kinases in early postimplantation murine embryos. *Teratology* 62:14–25.

Ozolinš, T.R.S. and Hales, B.F. (1997) Oxidative stress regulates the expression and activity of transcription factor activator protein-1 in rat conceptus. The Journal of Pharmacology and Experimental *Therapeutics* 280:1085–1093.

Paganelli, A., Gnazzo, V., Acosta, H., López, S.L. and Carrasco, A.E. (2010) Glyphosate-based herbicides produce teratogenic effects on vertebrates by impairing retinoic acid signaling. *Chemical Research in Toxicology* 23:1586-1595.

Pekar, O., Molotski, N., Savion, S., Fein, A., Toder, V. and Torchinsky, A. (2007) p53 regulates cyclophosphamide teratogenesis by controlling caspases 3, 8, 9 activation and NF-κB DNA binding. *Reproduction* 134:379–388.

Rokutanda, S., Jujita, T., Kanatani, N., Yoshida, C.A., Komori, H., Liu, W., Mizuno, A. and Komori T. (2009) Akt regulates skeletal development through GSK3, mTOR, and FoxOs. *Developmental Biology*, 328:78-93.

Sahambi, S.K. and Hales, B.F. (2006) Exposure to 5-bromo-20-deoxyuridine induces oxidative stress and activator protein-1 DNA binding activity in the embryo. *Birth Defects Research (Part A)* 76:580–591.

Scott Jr., W.J., Schreiner, C.M., Goetz, J.A., Robbins, D. and Bell, S.M. (2005) Cadmium-induced postaxial forelimb ectrodactyly: association with altered sonic hedgehog signaling. *Reproductive Toxicology* 19:479–485.

Shi, X. and Zhou, B. (2010) The role of Nrf2 and MAPK pathways in PFOS-induced oxidative stress in zebrafish embryos. *Toxicological Sciences* 115:391–400.

Singh, C.K., Kumar, A., LaVoie, H.A. DiPette, D.J. and Singh, U.S. (2012) Resveratrol prevents impairment in activation of retinoic acid receptors and MAP kinases in the embryos of a rodent model of diabetic embryopathy. *Reproductive Sciences* 19 949-961.

Thévenod, F. (2009) Cadmium and cellular signaling cascades: To be or not to be? *Toxicology and Applied Pharmacology* 238:221–239.

Tint, G.S., Irons, M., Elias, E.R., Batta, A.K., Frieden, R., Chen, T,S. and Salen, G. (1994) Defective cholesterol biosynthesis associated with the Smith-Lemli-Opitz syndrome. *The New England Journal of Medicine* 330:107-13.

Torchinsky, A., Lishanski, L., Wolstein, O., Shepshelovich, J., Orenstein, H., Savion, S., Zaslavsky, Z., Carp, H., Brill, A., Dikstein, R., Toder, V. and Fein, A. (2002) NF-κB DNA-binding activity in embryos responding to a teratogen, cyclophosphamide. *BMC Developmental Biology* 2:2.

Vangipuram, S.D. and Lyman, W.D. (2012) Ethanol affects differentiation-related pathways and suppresses Wnt signalling protein expression in human neural stem cells. Alcoholism: *Clinical and Experimental Research* 36: 788-797.

Wang, Y.Y., Sui, K.X., Li, H. and Ma, H.Y. (2009) The effects of lead exposure on placental NF-κB expression and the consequences for gestation. *Reproductive Toxicology 27*:190–195.

Wells, P.G., Bhuller, Y., Chen, C.S., Jeng, W., Kasapinovic, S., Kennedy, J.C., Kim, P.M., Laposa, R.R., McCallum, G.P., Nicol, C.J., Parman, T., Wiley, M.J. and Wong, A.W. (2005) Molecular and biochemical mechanisms in teratogenesis involving reactive oxygen species. *Toxicology and Applied Pharmacology* 207:S354 – S366.

Winn, L.M. and Wells, P.G. (2002) Evidence for ras-dependent signal transduction in phenytoin teratogenicity. *Toxicology and Applied Pharmacology* 184:144 –152.

Yan, J. and Hales, B.F. (2008) p38 and c-Jun N-terminal kinase mitogen-activated protein kinase signaling pathways play distinct roles in the response of organogenesis-stage embryos to a teratogen. *The Journal of Pharmacology and Experimental Therapeutics* 326:764–772.

In: New Developments on Signal Transduction Research
Editor: Masayoshi Yamaguchi
ISBN: 978-1-62618-230-1

Chapter 6

REGUCALCIN, A SUPPRESSOR PROTEIN IN SIGNAL TRANSDUCTION

Masayoshi Yamaguchi*
Division of Diabetes, Endocrinology and Metabolism,
Department of Medicine,
Baylor College of Medicine, Houston, TX, US

ABSTRACT

Regucalcin was discovered in 1978 as a novel calcium-binding protein that does not contain EF-hand motif of calcium-binding domain, which differs from calmodulin and other calcium-related proteins. The name of regucalcin was proposed for this calcium-binding protein, which regulates various Ca^{2+}- or Ca^{2+}/calmodulin-dependent enzyme activations. Regucalcin and its gene (RGN) are identified in over 15 species consisting of regucalcin family. The regucalcin gene is localized on the chromosome X, and the organization of the regucalcin gene consists of seven exons and six introns. AP-1, NF1-A1, and RGPR-p117, which is a transcription factor, bind to the promoter region of the regucalcin gene and enhance transcription activity of the regucalcin gene expression that is mediated through calcium and other signalings. Regucalcin mRNA expression and its protein content are pronounced in the liver, kidney cortex, and other tissues. Regucalcin has been demonstrated to play a multifunctional role in cell regulation: regulation of intracellular Ca^{2+} homeostasis, inhibition of Ca^{2+}-dependent and independent enzyme activations including protein kinases and protein phosphatases, suppression of nuclear signaling-related deoxyribonucleic acid (DNA) and ribonucleic acid (RNA) synthesis, and depression of apoptotic cell death and cell proliferation induced by various signaling factors in many cell types *in vitro*. Morever, regucalcin has been shown to have a suppressive effect on protein synthesis and an activating effect on protein degradation. Regucalcin suppresses the over-proliferation in regenerating liver cells with partial hepatoectomy and the development of carcinogenesis in liver tissues *in vivo*. Regucalcin is proposed to play a pivotal role in maintaining of cell homeostasis and function. The revelation of these effects is related to role of regucalcin as a suppressor in various cell-

* Corresponding Author address: Division of Diabetes, Endocrinology and Metabolism, Department of Medicine, Baylor College of Medicine, One Baylor Plaza, Houston, USA; E-mail: yamamasa1155@yahoo.co.jp

signaling systems. Regucalcin is the fisrt time finding as a suppressor protein in cell signaling. This chapter has been written to outline the recent advances that have been made concerning the role of regucalcin as a suppressor protein in cell signaling in liver, kidney, and other tissues.

Keywords: regucalcin, RGN, gene expression, calcium signaling, calmodulin, protein kinase C, nuclear function, apoptosis, cell proliferation

INTRODUCTION

Calcium ion (Ca^{2+}) plays a pivotal role in the regulation of many cell functions. Ca^{2+} can regulate muscle contraction, neurotransmission, hormone secretion, cell mitosis, and gene expression. A role as second messengers of Ca^{2+} in cells for hormonal stimulation comes into notice. Calcium signal is transmitted to intracellular responses, which are mediated through a family of calcium-binding protein and protein kinase C [1, 2]. Cell metabolism is regulated by an increase in Ca^{2+} in the cytoplasm of cells after hormonal stimulation [3-5]. The effect of Ca^{2+} is amplified through calmodulin and protein kinase C in the regulation of cell metabolism [1-5]. Liver has been shown to participate in the regulation of calcium metabolism through hepatic bile system in rats, and bile calcium excretion is increased by hormonal stimulation [6,7]. On the basis of this finding, it was found that a novel calcium-binding protein, which differs from calmodulin and other calcium-related proteins, was present in the hepatic cytoplasm of rats [8-10]. The name of regucalcin was proposed for this calcium-binding protein, which regulates various Ca^{2+}- or Ca^{2+}/calmodulin-dependent enzyme activations in liver cells [11- 15].

Regucalcin (RGN) and its gene (*rgn*) are identified in over 15 species consisting of regucalcin family [16-21]. Comparison of the nucleotide sequences of regucalcin from vertebrate species is highly conserved in their coding region with throughout evolution. The regucalcin gene is localized on the chromosome X, and the organization of the regucalcin gene consists of seven exons and six introns [21-23]. AP-1, NF1-A1, RGPR-p117, NF-κB, β-catenine, and other factors, which are a transcription factor, have been known to bind to the promoter region of the regucalcin gene and regulate transcription activity of regucalcin gene expression that is mediated through calcium and other signalings [21]. The expressions of regucalcin mRNA and protein are pronounced in the liver and kidney cortex of rats, although it is present only slightly in other tissues (including the duodenum, testis, spleen, lung, smooth muscle, heart, brain, and other tissues) [Reviewes in 17, 25-30].

The role of regucalcin in cell regulation is demonstrated in the liver and kidney cells in detail [reviewed in Refs. 31-37]. Regucalcin, which is localized into the nucleus, plays an important role as a regulatory protein for many signaling pathways from the cytoplasm to the nuclei in many cell types [38]. Regucalcin has been demonstrated to suppress activation of cell signaling-related protein kinases and protein phospatyases and DNA and RNA synthesis in the nucleus *in vitro* [25-30]. Overexpression of endogenous regucalcin has been shown to suppress apoptotic cell death and cell proliferation induced by various signaling factors *in vitro*. Moreover, regucalcin suppresses the over-proliferation in regenerating liver cells with partial hepatoectomy and the development of carcinogenesis in liver tissues *in vivo*, which are related to activation of cell signaling. Regucalcin is proposed to play a pivotal role in maintaining of cell homeostasis and function. The revelation of these effects is related to role of regucalcin as a suppressor in various cell-signaling systems. Thus suppressor protein was not known in cell signaling system, although many stimulating factors are known in siganal transduction in many cell types. Regucalcin was the first time finding as a novel suppressor in cell signal transduction system.

This chapter has been written to outline the recent advances that have been made concerning the role of regucalcin as a suppressor protein in cell signaling in liver, kidney, and other tissues: its role in regulation of intracellular Ca^{2+} homeostasis, inhibition of Ca^{2+}-dependent enzyme activations, regulation of nuclear calcium signaling, and inhibitory effects in cell apoptosis and cell proliferation induced by various signaling factors.

MOLECULAR ASPECT OF REGUCALCIN

Properties of Regucalcin

The molecular weight of rat regucalcin is estimated as 33,388 Da, composing of 299 amino acid residues from the cloning of rat regucalcin cDNA [16]. The regucalcin molecule does not contain the EF-hand motif as a calcium-binding domain [16]. The isoelectric point of regucalcin is 5.20 [16]. From the experimental data by Scatchard plot, the apparent association constant (Kf) for calcium ion (Ca^{2+}) of regucalcin is found to be 4.19×10^5 M^{-1} by equilibrium dialysis with a correlation coefficient of 0.99, and there are 6.52 high-affinity sites per molecule of protein [8-10]. Regucalcin appears to have six or seven high-affinity binding sites for Ca^{2+} per molecule of protein [9]. Extrapolation of the regression line to infinite Ca^{2+} concentration indicated a maximal Ca^{2+}-binding of 2.28×10^{-4} mol of Ca^{2+} per gram of protein. It is known that calmodulin exists as a monomer with a molecular weight of 17,000 and contains 4 Ca^{2+}-binding sites [1].

The conformational changes induced by binding of Ca^{2+} to regucalcin are investigated by means of the ultraviolet (UV) absorption spectrum [10]. UV of regucalcin showed a maximum at 278 nm in the range from 240 to 330 nm. In the presence of Ca^{2+} (0.1 and 1.0 mM), a decrease in absorption at 278 nm was observed [10]. This negative UV difference is similar to the Ca^{2+}-induced absorption changes in calmodulin. The spectrum can be attributed to charges in both tyrosine and tryptophan residues. Changes in the environment of both aromatic amino acids occur upon Ca^{2+}-binding [10].

Fluorescence spectroscopy is used to study the effect of Ca^{2+} on the conformation of regucalcin [10]. The spectral emission was quenched after the addition of 1.0 mM Ca^{2+}. Known fluorescence emission properties of isolated tyrosine and tryptophan residues suggest, as indicated above, that changes in the environment of these two aromatic amino acids occur upon Ca^{2+}-binding. These observations demonstrate that Ca^{2+}-binding induces conformational changes in regucalcin. These changes may result in increasing the hydrophobicity of regucalcin [10, 16].

The conformation of the polypeptide backbone of regucalcin is studied by circular dichroism (CD) spectroscopy [10]. The presence of 1.0 mM Ca^{2+} caused clear alterations in the CD spectrum. The apparent α-helical content of regucalcin in Ca^{2+}-free buffer was estimated to be 34%, and the presence of 1.0 mM Ca^{2+} decreased this by 4.5%. Thus, conformational changes are induced by Ca^{2+} binding to regucalcin. This binding also loosens the conformation of regucalcin. The apparent α-helical content of calmodulin is 30%, and it is increased after Ca^{2+} binding and its conformation is tightened. [10,16]. The increase in hydrophobicity after Ca^{2+} binding represents the mechanism that calmodulin activates its target proteins. However, regucalcin reverses the activation of many enzymes by Ca^{2+} /calmodulin. The role of regucalcin may be different from that of calmodulin in cells.

Regucalcin differs entirely from other calcium-binding proteins of the EF-hand type: calmodulin, calcineurin, parvalbumin, S-l00a, S-100b proteins, caligulin, calregulin, calbindin, calreticulin, and annexins. This is supported from the results of the molecular cloning and sequencing of the cDNA coding for a regucalcin from various mamarlian livers [16, 20]. The nucleotide and amino acid sequences of regucalcin does not have statistically significant homology, as compared with the registered sequences which are found in the EMBL and GenBank detabases (D14327 and D86217) [16].

The hydropathy profile of regucalcin shows that there is a hydrophobic sequence in both N-terminal and C-terminal regions of the regucalcin molecules [16]. Regucalcin shows a hydrophilic character as molecule [16, 20]. The most common EF-hand is composed of the helix-loop-helix-domain. The prototype loop consists of 12 amino acids, of which five have a carboxyl (or a hydroxyl group) in their side chain, precisely spaced so as to coordinate the Ca^{2+}. Analysis of the structure of the EF-hand from the regucalcin sequence did not give the expected pattern of amino acids conforming to the typical EF-hand structure of a calcium-binding site.

Amino acid analysis shows that regucalcin has a relatively high content of glycine and much lower amounts of glutamic acid, valine, aspartic acid, and lysine [16, 20]. Regucalcin contains about 20% (mol%) glutamic and aspartic acids and about 17% amide residues (lysine, histidine, and arginine) and thus a high proportion of charged residues: regucalcin molecule contains aspartic acid (24 residues) and glutamic acid (16 residues) [16]. Acidic amino acids comprise approximately one-fifth of the regucalcin molecule, a prevalence that appears to be characteristic of the composition of all Ca^{2+}-binding proteins. The significance of the high di-carboxylic acid content is that it permits binding of Ca^{2+} to protein. These amino acids may be related to Ca^{2+} binding.

The result of crystal structure with X-ray diffraction data shows that regucalcin contains the metal site bound with either a Ca^{2+} or a Zn^{2+} atom, suggesting that the Ca^{2+}-bound form may be physiologically relevant for stressed cells with an elevated Ca^{2+} level [39]. This supports our finding for regucalcin as a calcium-binding protein.

Regulation of Regucalcin Gene Expression

The rat regucalcin gene is localized on the proximal end of the rat chromosome Xq11.1-12 [22], and the gene are demonstrated in human, mouse, cow, monkey, dog, rabbit, and chicken but not yeast [17]. The amino acid sequence of mouse regucalcin had 94% homology as compared with that of rat regucalcin [20]. Regucalcin may be a protein that is highly differentiated. The organization of the rat regucalcin gene seems to be about 18 kb in size, and consisted of seven exons and six introns [24]. There are many regulatory elements (AP-1, NF1-A1, RGPR-p117, β-catenin, NF-κB, and other factors) in the 5'-flanking region [21, 40-48]. The promoter activity of the rat regucalcin gene is enhanced by treatment with Bay K

8644, dibutyryl cyclic AMP, phorbol esters, insulin, and dexamethasone [42]. Using gel mobility shift assays, it is found that nuclear proteins from rat liver cells and rat hepatoma H4-II-E cells specifically bind to the 5'-flanking region of the rat regucalcin gene [40-42]. Treatment with Bay K 8644, dibutyryl cyclic AMP, phorbol esters, and insulin stimulates the binding of nuclear factors to the 5'-flanking region of the rat regucalcin gene in H4-II-E cells. These factors-inducible nuclear proteins are related to enhance promoter activity of the regucalcin gene [42].

Regucalcin mRNA expression has been demonstrated to mediate through signaling pathway of Ca^{2+}/calmodulin-dependent protein kinase, protein kinase C, and tyrosine kinase in the cells [49-51]. AP-1 factor binds to the 5'-flanking region of the rat regucalcin gene that is mediated through the Ca^{2+} response [41]. AP-1 factor is complex of c-fos/c-jun that is phosphorylated by protein kinases [52, 53]. Calcium signaling system is an important pathway in the stimulation of regucalcin mRNA expression.

Regucalcin mRNA is mainly present in liver and renal cortex with a size of 1.8kb [25]. The expression of regucalcin mRNA in the liver and renal cortex is clearly stimulated through an increase in the cellular Ca^{2+} levels following an oral administration of calcium chloride in rats in vivo [54-56]. Hepatic regucalcin mRNA expression is increased with fetal development and its expression is stimulated after the intake of dietary calcium to maternal rats in vivo [57]. Liver regucalcin concentration is increased after an oral administration of calcium in rats [55].

Hepatic regucalcin mRNA expression is also stimulated after a single subcutaneous administration of calcitonin [58], insulin [59], and estrogen [60], suggesting that the expression of regucalcin mRNA is enhanced through various hormonal stimulation. Regucalcin mRNA expression is increased in regenerating rat liver, suggesting its role in the proliferation of liver cells [61]. Aging has been shown to decrease liver regucalcin mRNA expression [62].

ROLE OF REGUCALCIN IN INTRACELLULAR CA^{2+} HOMEOSTASIS

Regucalcin has been demonstrated to have a role as suppressor in cell Ca^{2+} signaling. Regucalcin plays a cell physiological role to depress elevation of intracellular Ca^{2+} levels. Intracellular Ca^{2+} homeostasis is regulated through plasma membrane (Ca^{2+}-Mg^{2+})-adenosine 5'-triphosphatease (ATPase), microsomal Ca^{2+}-ATPase, mitochondrial Ca^{2+} uptake, and nuclear Ca^{2+} transport in the cells. Regucalcin has been demonstrated to regulate Ca^{2+}-transporting systems in the liver, renal cortex cells, heart, and brain tissues, suggesting its role in the regulation of intracellular Ca^{2+} homeostasis.

Role of Regucalcin in Ca^{2+} Homeostasis in Liver Cells

The high-affinity (Ca^{2+}-Mg^{2+})-ATPase is located on the plasma membranes of liver cells [63, 64]. This enzyme acts as a Ca^{2+} pump to exclude the metal ion from the cytoplasm of liver cells. Addition of regucalcin into the reaction mixture in vitro caused an increase in

(Ca^{2+}-Mg^{2+})-ATPase activity in the plasma membranes isolated from rat liver, suggesting a role in the regulation of Ca^{2+} pump activity [65]. Regucalcin directly activates (Ca^{2+}-Mg^{2+})-ATPase independently of Ca^{2+}-stimulated phosphorylation of the enzyme [65-67], and it has been shown to stimulate ATP-dependent calcium transport across the plasma membrane vesicles of rat liver after addition of $^{45}Ca^{2+}$ into the reaction mixture in vitro [68]. Regucalcin-enhanced ATP-dependent $^{45}Ca^{2+}$ uptake in the plasma membrane vesicles is completely inhibited in the presence of N-ethylmaleimide or digitonin [67]. Regucalcin has been shown to bind the lipid components of liver plasma membrane, and it acts on the sulfhydryl (SH) groups that are an active site of (Ca^{2+}-Mg^{2+})-ATPase [67]. The mechanism of regucalcin in activating (Ca^{2+}-Mg^{2+})-ATPase may be not involved on GTP-binding protein that modulates the receptor-mediated hormonal effect (including calcitonin, epinephrine, phenylephrine, and insulin) in liver plasma membranes [69]. The effect of hormonal signaling factors (inositol-glycan, dibutyryl cyclic AMP, and inositol 1,4,5-trisphosphate) on the regucalcin-increased (Ca^{2+}-Mg^{2+})-ATPase activity in rat liver plasma membranes is examined [70]. Inositol-glycan, which is generated by insulin [70], can directly activate the plasma membrane (Ca^{2+}-Mg^{2+})-ATPase, and its effect is modulated by regucalcin. Cross talk with signaling factors may be seen in the regulation of Ca^{2+} pump activity in the plasma membranes of liver cells.

The physiological role of regucalcin in the regulation of (Ca^{2+}-Mg^{2+})-ATPase activity in liver plasma membranes is examined after an orall administration of calcium chloride solution in rats [71]. Calcium administration caused an increase in (Ca^{2+}-Mg^{2+})-ATPase activity in liver plasma membranes [71]. This increase was abolished in the presence of anti-regucalcin antibody, suggesting an involvement of endogenous regucalcin that is distributed in the cytoplasm. Regenerating rat liver with a proliferative cells significantly increased liver calcium content and plasma membrane (Ca^{2+}-Mg^{2+})-ATPase activity between 12 and 48 hours after partial hepatectomy [72]. This increase was completely abolished in the presence of anti-regucalcin antibody, indicating an involvement of endogenous regucalcin [72]. Activating effect of regucalcin on hepatic plasma membrane (Ca^{2+}-Mg^{2+})-ATPase was impaired in liver injury with carbon tetrachloride administration in rats [73]. Regucalcin plays a role as an activator protein for Ca^{2+} pump enzyme in the hepatic plasma membranes.

Regucalcin binds Ca^{2+} in the cytoplasm of liver cells, and the metal is subsequently transported into the organell dependent on ATP [74]. Regucalcin may not tightly bind cytosolic Ca^{2+} of lower levels, since its calcium-binding constant is $4.19x10^5$ M^{-1} [9]. Regucalcin may regulate cytoplasmic Ca^{2+} levels by activating Ca^{2+} pump enzyme in the plasma membranes of liver cells.

Regucalcin can stimulate the uptake of Ca^{2+} by rat liver mitochondria [75, 76]. The effect of regucalcin on mitochondrial Ca^{2+} uptake is inhibited in the presence of ruthenium red or lanthanum chloride [75], which is an inhibitor of mitochondrial Ca^{2+} transport. Regucalcin may have a role in the reduction of cytoplasmic Ca^{2+} levels due to activating mitochondrial Ca^{2+} uptake.

Regucalcin has also been demonstrated to activate Ca^{2+} pump enzymes (Ca^{2+}-ATPase) and to stimulate ATP-dependent $^{45}Ca^{2+}$ uptake by the liver microsomes [77, 78], suggesting a role in the regulation of cytoplasmic Ca^{2+} levels. The effect of regucalcin in increasing Ca^{2+}-ATPase activity in the microsomes is inhibited in the presence of thapsigargin, a specific inhibitor of microsomal Ca^{2+} pump enzyme. Regucalcin acts on the SH groups of microsomal Ca^{2+}-ATPase due to binding on the membranous lipids [78]. The components of Ca^{2+} uptake (Ca^{2+}-ATPase) and Ca^{2+} release (Ca^{2+}-channels) are located at separate sites on liver microsomes. Interestingly, regucalcin has been found to stimulate Ca^{2+} release from rat liver microsomes [79]. The mechanism is related to the inositol 1, 4, 5-triphosphate (IP3)-induced Ca^{2+} release [80]. Regucalcin may bind to IP3 receptors on the microsomes. This cell physiological significane of regucalcin is unknown. Presumably, regucalcin stimulates microsomal Ca^{2+} uptake when cytosolic Ca^{2+} concentration is raised. Also, regucalcin regulates Ca^{2+} storage in the endoplasmic reticulum of liver cells: it stimulates Ca^{2+} release from the microsomes to restore the microsomal calcium accumulation to regulate Ca^{2+}-related microsomal functions. Regucalcin has been shown to have the reversible effect on liver microsomal glucose-6-phosphatase activity increased by Ca^{2+} addition [11].

The existence of an ATP-stimulated Ca^{2+}-sequestration system is also found in the liver nuclei, and it generates a net increase in nuclear matrix free Ca^{2+} concentration [81]. This system may play an important role in the regulation of intranuclear Ca^{2+}-dependent processes [82]. ATPase, which is stimulated by Ca^{2+} in the presence of Mg^{2+}, exists in the nuclei of rat liver, and the Ca^{2+}-stimulated ATPase activity is involved in the nuclear Ca^{2+} uptake [83]. Regucalcin increases Ca^{2+}-ATPase activity in rat liver nuclei [84]. Regucalcin has also been shown to stimulate Ca^{2+} release from liver nuclei [85]. Presumably, regucalcin has a role in the regulation of liver nuclear function through the effect on Ca^{2+} transporting system in the nuclei. Regucalcin has been shown to lower intracellular Ca^{2+} levels by modulating plasma membrane Ca^{2+}-pumping activity in the cloned human hepatoma HepG2 cells that overexpress regucalcin [86].

Role of Regucalcin in Ca^{2+} Homeostasis in Kidney Cells

Regucalcin largely is present in the kidney cortex of rats [27]. Kidney plays a physiological role in the regulation of calcium homeostasis in blood through re-absorption of urinary calcium [87, 88]. Renal cortex cells play a role in the re-absorption of urinary calcium. Regucalcin mRNA is expressed in the kidney cortex but not the medulla of rats [56], and its expression is stimulated after calcium administration in vivo [56]. The binding of kidney nuclear proteins to the 5'flanking region of the rat gene for regucalcin has been shown to enhance through Ca^{2+}/calmodulin signaling [89, 90]. Regucalcin mRNA expression is stimulated after the administration of dexamethasone in rats [91], and its expression is suppressed in hypertensive state in rats [92-94]. The specific nuclear factor binds to the NF1-like sequence in the promoter region of regucalcin gene in the kidney cortex of rats [95], and the nuclear factor binding and regucalcin mRNA expression are suppressed after administration of cisplatin that induces kidney damage [95, 96].

Ca^{2+}-ATPase system has been shown to exceed the capacity of the Na^{+}/Ca^{2+} exchanger, and it plays a primary role in Ca^{2+} homeostasis of rat kidney cortex cells [97, 98]. Regucalcin has been demonstrated to play a role as an activator of the ATP-dependent Ca^{2+} pumps in the basolateral membranes isolated from rat kidney cortex [99]. The effect of regucalcin in increasing Ca^{2+} pump enzyme (Ca^{2+}-ATPase) activity in the basolateral membranes is completely inhibited in the presence of N-ethylmaleimide, indicating that regucalcin may act on the SH groups of Ca^{2+}-ATPase [99].

Regucalcin has also been shown to increase Ca^{2+}-ATPase activity and ATP-dependent calcium uptake in the microsomes of rat kidney cortex [100]. These increases are clearly decreased in the presence of N-ethylmaleimide, suggesting that regucalcin acts on the SH groups of Ca^{2+}-ATPase in the microsomes [98].

The finding, that regucalcin increases Ca^{2+}-ATPase activity in the basolateral membranes and microsomes of rat renal cortex, suggests a physiological role of regucalcin in the regulation of the Ca^{2+} homeostasis in renal cells. Regucalcin may be responsible for ATP-dependent transcellular Ca^{2+} transport, and it participates in the promotion of Ca^{2+} re-absorption in the nephron tubulo of kidney cortex. Regucalcin may play a physiological role in the regulation of calcium metabolism in blood through re-absorption of urinary calcium in kidney.

Role of Regucalcin in Ca^{2+} Homeostasis in Heart Cells

The Ca^{2+} current is one of the most important components in cardiac excitation-contraction coupling [102]. This coupling mechanism is based on the regulation of intracellular Ca^{2+} concentration by Ca^{2+} pump in the sarcoplasmic reticulum of heart muscle [102]. The role of regucalcin in the regulation of heart muscle function is shown. Regucalcin mRNA is expressed in rat heart [103]. The result with Western blot analysis indicates that regucalcin is present in the cytoplasm of heart muscle cells [103]. Regucalcin concentration in the heart muscle tissues has been shown to be about 3.86 $x10^{-8}$M [27]. Regucalcin mRNA expression in the hearts of rats is decreased with increasing age [104], and free radical stress has a suppressive effect on its gene expression [104]. Overexpression of regucalcin in transgenic rats has been found to accelerate free radical stress-induced death of rats [104].

Regucalcin has been found to increase Ca^{2+}-ATPase activity and ATP-dependent Ca^{2+} uptake, which regulates intracellular Ca^{2+} concentration related to cardiac excitation-contraction coupling, in rat heart microsomes, suggesting its role in the regulation of heart muscle function [103]. The effect of regucalcin in increasing heart microsomal Ca^{2+}-ATPase activity is inhibited in the presence of thapsigargin [103], a specific inhibitor of the sarcoplasmic reticulum Ca^{2+} pump enzyme (Ca^{2+}-ATPase) [105], indicating that regucalcin activates Ca^{2+} pump enzyme in the sarcoplasmic reticulum. It is suggested that regucalcin binds to the lipids at the close site of Ca^{2+}-ATPase in heart misrosomes, and that it acts on the SH group which may be an active site of the enzyme and stimulates Ca^{2+}-dependent phosphorylation of Ca^{2+}-ATPase [103].

Phospholamban has been known to inhibit Ca^{2+} pump enzyme (Ca^{2+}-ATPase) in the sarcoplasmic reticulum (misrosomes) of heart muscle [106]. Ca^{2+}-ATPase is activated through cAMP-dependent phosphorylation of phospholamban following hormonal stimulation [106]. The endogenous activatory protein of sarcoplasmic reticulum Ca^{2+}-ATPase is unknown. Regucalcin, which is present in the cytoplasm of heart muscle, may play an important role as an endogenous activator in the regulation of sacroplasmic reticulum Ca^{2+}-ATPase activity in rat heart muscle [102]. Regucalcin may play a physiological role in the regulation of cardiac excitation-contraction coupling.

The role of regucalcin in the regulation of Ca^{2+}-ATPase activity in the heart mitochondria of rats is examined, moreover [107]. Regucalcin is found to be present in the mitochondria of normal rat heart [107], and this localization is increased in the heart of regucalcin transgenic rats as compared with that of normal rats [107]. The addition of regucalcin (10^{-11} to 10^{-8} M) in the enzyme reaction mixture caused a significant increase in Ca^{2+}-ATPase activity in the heart mitochondria in the presence of 50 μM $CaCl_2$ [107]. Ca^{2+}-ATPase activity was also increased in the heart mitochondria of regucalcin transgenic rats [107]. Regucalcin has an activating effect on Ca^{2+}-ATPase in rat heart mitochondria, suggesting its role in the regulation of heart mitochondrial function.

Regucalcin may play a role in the regulation of cytoplasmic Ca^{2+} levels due to activating Ca^{2+} pump activity in the sarcoplasmic reticulum and mitochondria in heart cells.

Role of Regucalcin in Ca^{2+} Homeostasis in Brain Tissues

Intracellular Ca^{2+} concentration in the neuronal cells of brain is regulated by various buffering and transport systems such as the membrane Na^+-Ca^{2+} exchanges, the membranous Ca^{2+}-ATPase, Ca^{2+}-binding proteins and intracellular Ca^{2+} uptake systems [108-112].

The changes in the neuronal Ca^{2+} homeostasis with aging may be implicated in age-related disturbance in cognitive functions [112]. The alteration in the neuronal Ca^{2+} regulation is also implicated in the pathology of Alzheimer's disease [112].

The expression of regucalcin mRNA is demonstrated in the brain tissues [29, 113]. Regucalcin concentration in the brain tissues has been shown to be about $5x10^{-9}$ M as measured by using enzyme-linked immuoadsorbent assay, and this level is lowered with aging [27]. Regucalcin is localized in the neurons isolated from rat brain tissues [29], suggesting its role in brain function.

Brain calcium accumulation has been shown to increase after oral administration of calcium in rats, and fasting enhances its accumulation [114]. The supply of glucose may be required in the regulation of brain Ca^{2+} homeostasis in rats *in vivo* [114], suggesting a physiological significance of energy-dependent mechanism in brain calcium metabolism. Fasting caused a significant increase in brain calcium content and Ca^{2+}-ATPase activity in the microsomes and mitochondria of brain tissues in young and aged rats, and these increases were restored after the supply of glucoce, supporting a physiologic significance of energy-dependent mechanism in the regulation of brain calcium in rats with different ages [115].

Aging causes a decrease in Ca^{2+}-ATPase activity in the brain plasma membranes of rats [116], suggesting that aging enhances the entry of Ca^{2+} into brain neuronal cells across the plasma membranes. In addition, aging induces an attenuation of Ca^{2+}-seguestrating system in the brain microsomes, supporting the view that a disturbance of the neuronal Ca^{2+} regulation is brought with increasing age [117].

Protein kinase C activates brain microsomal Ca^{2+}-ATPase in aged rats [118]. It is speculated that aging-induced increase in Ca^{2+}-ATPase activity results from the translocation to the microsomes of protein kinase C in brain cytosol [118].

The development of brain disease with aging may be partly related to the toxicity of brain calcium raised by the increase in microsomal Ca^{2+}-ATPase activity with aging. The disturbance of brain Ca^{2+} homeostasis may play a pivotal role in the revelation of brain disease. Regucalcin has been found to have an inhibitory effect on Ca^{2+}-ATPase activity in rat brain microsomes [113], suggesting that regucalcin plays a role in the regulation of microsomal Ca^{2+}-ATPase activity in rat brain. The concentration of regucalcin in the cerebral cortex and hippocampus of brain tissues is decreased with aging [113].

The suppressive effect of regucalcin on brain microsomal Ca^{2+}-ATPase activity was weakened in aged rats [113]. Presumably, the aging-induced elevation of brain microsomal

Ca^{2+}-ATPase activity is partly resulted from an attenuation of regucalcin action on the enzyme activity with aging. There may be a possibility that the translocation of protein kinase C to the brain microsomes of aged rats [118] is a cause of the attenuation of regucalcin effect on the enzyme activity. Regucalcin may play a physiological and pahophysiological role in the regulation of intracellular Ca^{2+} concentration in the brain tissues.

Regucalcin plays an activating role in the regulation of Ca^{2+}-ATPase activity in the mitochondria of brain tissues of rats [119]. The addition of regucalcin (10^{-10} to 10^{8} M), which is a physiological concentration in rat brain tissues, into the enzyme reaction mixture containing 25 μM calcium chlrodie caused a significant increase in Ca^{2+}-ATPase activity, while it did not significantly change in Mg^{2+}-ATPase activity [119]. Regucalcin levels are increased in the brain tissues or the mitochondria obtained from regucalcin transgenic rats. The mitochondrial Ca^{2+}-ATPase activity has been found to increase in regucalcin transgenic rats as compared with that of wild-type rats [119]. Endogenous regucalcin plays a role in the regulation of Ca^{2+}-ATPase activity in the brain mitochondria of rats. Thus, regucalcin plays a cell physiological role as a regulatory protein that is involved in the regulation of Ca^{2+} homeostasis in various cell types. The low cytoplasmic Ca^{2+} concentration of living cells is maintained through energy-requiring pumps.

These pumps either remove Ca^{2+} to the extracellular space by transport across the plasma membrane or accumulate it inside of intracellular organelles such as the mitochondria and endoplasmic reticulum (microsomes). Regucalcin stimulates the activity of these pumps to lower the cytoplasmic Ca^{2+} levels. This may provide the cellular mechanism of the inhibitory effect of regucalcin in the regulation of cell functions related to Ca^{2+} signaling.

Regucalcin Suppresses Ca^{2+} Signaling-Dependent Enzyme Activation

Regucalcin Suppresses Ca^{2+}/Calmodulin-Dependent Enzyme Activation

Ca^{2+} and calmodulin systems generally activate various enzymes in many cells. Regucalcin has an inhibitory effect on enzyme activation by Ca^{2+}/ calmodulin [25-29]. This section outlines the findings on the inhibitory effect of regucalcin on action of Ca^{2+} in cell metabolism. Liver metabolism is regulated through an increase in Ca^{2+} level in the cytoplasm of liver cells due to hormonal stimulation [3-5]. Regucalcin has an inhibitory effect on Ca^{2+}/calmodulin-dependent enzyme activity in vitro. The hormonal effect on fructose 1,6-diphosphatase, which promotes the conversion from fructose 1,6-diphosphate to glucose-6-phosphate in the hepatic cytoplasm of rats, is mediated through Ca^{2+}. This enzyme activity is activated through Ca^{2+}/calmodulin [11].

The activation of fructose 1,6-diphosphatase by Ca^{2+} /calmodulin was completely reversed after the addition of regucalcin in the enzyme reaction mixture [11]. Phosphorylase *a* activity in the liver particulate glycogen is increased after addition of Ca^{2+} (10 μM) [13]. This increase was completely reversed after addition of regucalcin in the enzyme reaction mixture [13]. Regucalcin (1.0 μM) reversed activations of pyruvate kinase [12] and glucose-6-phosphatase [11] after addition of Ca^{2+} in the enzyme reaction mixture. These findings suggest that regucalcin regulates glycogenolysis and gluconeogenesis that is stimulated by

Ca^{2+} in liver cells. The reversible effect of regucalcin is also shown in Ca^{2+}-induced inhibition of 5'-nucleotidase activity in liver plasma membranes [15] and deoxyuridine 5'-triphosphatase activity in hepatic cytosol [120].

The controlled use of energy to maintain homeostasis and cellular function is a basic property of all cells. The free energy of ATP is belived to be the major cytosolic intermediate in this process. ATP is produced from the energy obtained by the oxidation of metabolic substrates in glycolysis and oxidative phosphorylation. ATPase produces the energy from ATP in the cell cytosol. The cytosolic factors regulating ATPase activity is important. Regucalcin has been shown to play an inhibitory role in the regulation of ATPase activity in the brain cytosol of young and aged rats [121], suggesting a role in the regulation of energy conversion in brain tissues.

Thus, regucalcin has a reversible effect on the activation and inhibition of many enzymes by Ca^{2+}. The mechanism of the reversible effect of regucalcin on various enzyme activities, which are regulated through Ca^{2+}, has not been well known. However, action of regucalcin may be partly based on Ca^{2+} binding, because the protein has 6-7 high-affinity binding sites per molecule (Kf = 4.19 x 10^5 M^{-1}) [9]. Regucalcin may directly bind to Ca^{2+} and/or calmodulin. In addition, it is possible that regucalcin may bind enzymes and that affects on enzyme activity.

Ca^{2+}/calmosulin-dependent enzymes are localized in many tissues and cells. Regucalcin has been found to have an inhibitory effect on various Ca^{2+}/calmodulin-dependent enzyme activations. Cyclic adenosine monophosphate (AMP) is a second messenger for hormonal stimulation in many cells. Cyclic AMP is degraded by cyclic AMP phosphodiestrase in the liver cytosol [122]. The enzyme activity is increased through Ca^{2+}/calmodulin [1]. Regucalcin has been found to inhibit the activation of cyclic AMP phosphodiesterase by Ca^{2+}/calmodulin in the cytosols of liver and renal cortex [123, 124], suggesting a role of regucalcin in the regulation of cyclic AMP level in the cells.

Nitric oxide (NO) may be important as a signaling factor in many cells [125]. NO, which an unpaired electron reacts with protein, targets primarily through their thiol or heme groups, and acts as a messenger or modulator molecule in many biological systems. NO is produced from L-arginine with L-citrulline as a coproduct in a reaction catalyzed by NO synthase that requires Ca^{2+}/calmodulin [125]. Regucalcin has been shown to have a suppressive effect in the enhancement of NO synthase activity in the cytosols of liver [126, 127] and kidney cortex [128] of rats. Overexpression of regucalcin did not cause a significant alteration of NO synthase activity in the kidney cortex cytosol of regucalcin transgenic rats as compared with that of wild-type rats [128]. However, the effect of calcium chloride (10 μM) in increasing NO synthase activity in the kidney cortex cytosol of wild-type rats was weakened in regucalcin transgenic rats [128]. The presence of anti-regucalcin monoclonal antibody (25 or 50 ng/ml) in the reaction mixture caused a significant increase in NO synthase activity, and this increase was completely ablolished after the addition of regucalcin (10^{-7} M). Endogenous regucalcin has a suppressive effect on NO synthatse activity in the cytosol of various tissues of rats.

Regucalcin has also been found to suppress Ca^{2+}/calmodulin-dependent NO synthase activity in the heart cytosol of rats [129]. Regucalcin had an inhibitory effect on NO synthase activity in the presence of antagonist for calmodulin [129], indicating a direct effect of regucalcin on the enzyme independent on Ca^{2+}/calmodulin. The physiological significance of regucalcin inhibition of NO synthase in heart muscle cytosol is unknown. However, regucalcin may participate in the regulation of NO production in heart muscle cells.

NO acts as a messenger or modulator molecule in heart muscle. NO production may be stimulated through Ca^{2+} signaling due to hormonal stimulation in heart muscle cells. Regucalcin may have a suppressive effect on over-production of NO due to inhibiting NO synthase in heart muscle cells.

NO acts as a messenger or modulator molecule in brain neurons. Regucalcin has also been shown to reveal a suppressive effect on NO synthase activity in the brain cytosol of young and aged rats, even though regucalcin levels are reduced with increasing age [130]

A remarkable expression of regucalcin protein was seen in the cytosol and nucleus of the brain tissues of regucalcin tranasgenic rats as compared with that of wild-type rats [131]. NO synthtase activity was decreased in the brain cytosol of transgenic rats, and the prsence of anti-regucalcin monoclonal antibody (50 ng/ml) in the enzyme reaction mixture caused a significant increase in cytosolic NO synthase activity in the cytosol of brain tissues of wild-type rats [131]. Endogenous regucalcin plays a suppressive role in the regulation of brain neuronal NO synthase activity in rats.

Thus, regucalcin has been demonstrated to have a suppressive role on Ca^{2+}/calmodulin-dependent NO synthase activity in various tissues. Regucalcin may play a role as a suppressor protein in NO production in many cell types, and it may regulate many cellular events that are involved in NO signaling.

Superoxide dismutase (SOD) plays a role in the prevention of cell death and apoptosis in the heart. The decrease in Mn-SOD activity is associated with increased mitochondrial oxidative damage as demonstrated by a decrease in the activities of iron sulfhydryl proteins sensitive to oxygene stress [132]. Cu/Zn-SOD has been shown to play the role of protector against dexorubicin-induced cardiotoxicity in mice [133]. Meanwhile, NO has a role in the suppression of myocardial O_2 consumption in rats [134].

Regucalcin has been found to increase SOD activity in the cytosol of rat liver [135] and heart [136]. Regucalcin has an inhibitory effect on NO synthase activity in the heart cytosol [136]. Production of superoxide radicals is widely accepted as the cause of the cardiodamage. Presumably, regucalcin participates in the control of production of superoxide radicals in rat heart muscle cells.

Regucalcin Suppresses Ca^{2+}-Dependent Protein Kinase Activation

The multifunctional Ca^{2+}/calmodulin-dependent protein kinases play an important role in the response of the cells to a calcium signal [1, 137]. Regucalcin has been shown to inhibit Ca^{2+}/calmodulin-dependent protein kinase activity in the cytosol of rat liver [138], kidney cortex [139], and brain tissues of rats [140, 141]. An appreciable effect of regucalcin is seen at 0.5 μM, which is a cell physiological concentration. As Ca^{2+}/calmodulin-dependent protein kinase in the cytoplasm is activated through calcium signal, regucalcin may regulate a signal transduction for Ca^{2+}. The mechanism of action of regucalcin may be partly based on its binding to Ca^{2+}/calmodulin and/or enzyme. Regucalcin has been demonstrated to bind on calmodulin in analysis with sodium dodecyl sulfate-polyacrylamide gel electrophoresis (SDS-PAGE) using calmodulin-agarose beads [142].

Nishizuka discovered a diacylglycerol-activated Ca^{2+} and phospholipid-dependent protein kinase (protein kinase C) [2]. Protein kinase C is distributed widespread in the body, with amounts in the liver being intermediate between the high levels found in brain and spleen [2]. Protein kinase C is capable of phosphorylating cytoplasmic proteins. It is found that regucalcin inhibits protein kinase C activity in the cytoplasm of rat liver [143], kidney cortex [144], and brain cytosol and brain neurons [140, 141], supporting the view that regucalcin plays a role in the regulation of Ca^{2+}-dependent cellular functions. The presence of anti-regucalcin monoclonal antibody in the enzyme reaction mixture caused a significant elevation of protein kinase activity, indicating that the endogenous regucalcin has an inhibitory effect on the enzyme activity [140, 141].

The regulatory effect of regucalcin in rat brain function may be attenuated with aging. Increasing age enhances protein kinase activity in rat brain cytosol [140]. This enhancement may be partly involved in aging-decreased regucalcin in rat brain tissues [113]. It is speculated that the endogenous regucalcin plays a suppressive role in the activation of Ca^{2+}-dependent protein kinase in the brain cytosol, and that aging may weaken the effect of regucalcin. Regucalcin may play a pivotal role in the regulation of phosphorylation of the cytosolic proteins in brain tissues. Interestingly, it has been reported that regucalcin gene is localized on human chromosome X that encompasses the map location for a growing number of diseases with a genetic basis; these include syndromic and non-syndromic forms of X-linked mental retardation and X-linked neurpmuscler diseases [23]. Regucalcin may have a pathophysiological role in brain disease with aging.

As mentioned above, regucalcin plays an inhibitory role in signaling pathway that is mediated through cyclic AMP, NO, and Ca^{2+}-dependent protein kinases in many tissues and cell types.

Regucalcin Suppresses Ca^{2+}-Dependent Protein Phosphatase Activation

Protein phosphorylation-dephosphorylation is a universal mechanism by which numerous cellular events are regulated [145]. It has become apparent that there may exist many phosphatases that, like the kinases, are just elaborately and rigorously controlled [145, 146]. Protein phosphatase plays an important role in intracellular signal transduction due to hormonal stimulation [145]. Calcineurin, a calmodulin-binding protein, has been shown to possess a Ca^{2+}-dependent and calmodulin-stimulated protein phosphatase activity [147]. Regucalcin has been demonstrated to inhibit calcineurin activity in the cytosol of liver and renal cortex after it's binding to calmodulin [148, 149]. Protein phosphatases, which endogenous regucalcin acts in liver cytoplasm, may be insensitive to okadaic acid [150]. Protein phosphatase activity toward phosphotyrosine, phosphoserine, and phosphothreonine in the cytosol of rat liver was elevated in the presence of anti-regucalcin monocional antibody in the enzyme reaction mixture in vitro, suggesting its role of endogenous regucalcin [148]. Regucalcin may be a unique protein, which has inhibitory effects on protein tyrosine phosphatase and protein serine/threonine phosphatase. Regucalcin, which was localized in rat liver nuclei, has been shown to inhibit nuclear protein phosphatase activity [151].

Endogenous regucalcin plays a role in the regulation of protein phosphatase activity in the cytosol and nuclei of rat renal cortex [152-154]. Regucalcin has been found to be present in the cytosol and nuclei of rat kidney cortex using Western blot analysis [154]. The addition of regucalcin (50-250 nM) in the enzyme reaction mixture obtained from the cytoplasm and nuclei from rat kidney cortex caused a decrease in protein phosphatase activity toward phosphotyrosin, phosphoserine, and phosphothreonine [152, 153]. The effect of calcium (25 μM) and calmodulin (2.5 μg/ml) in increasing protein phosphatase activity was decreased after the addition of regucalcin. Protein phosphatase activity in the cytosol and nuclei was increased in the presence of anti-regucalcin monoclonal antibody (10-50 ng/ml) in the enzyme reaction mixture [152, 153]. Regucalcin plays a suppressive role in the regulation of protein phosphatase activity in the cytoplasm and nucleus of rat kidney cortex.

Kidney cortex calcium content and the cytosolic and nuclear regucalcin levels were increased at 0.5 - 5 hours after a single intraperitoneal administration of calcium chloride solution (10 mg Ca/100 g body weight) in rats [155]. The cytosolic and nuclear protein phosphatase activity, which is raised in calcium-administered rats, was found to enhance when anti-regucalcin monoclonal antibody was added in the enzyme reaction mixture [155]. The effect of antibody was completely abolished after the addition of regucalcin in the enzyme reaction mixture. Thus, endogenous regucalcin has a suppressive effect on the enhancement of protein phosphatase activity in the cytosol and nucleus of kidney cortex in calcium-administered rats.

Cardiac hypertrophy is induced by calcineurin, which dephosphorylates the teanscription factor NF-A3, enabling it to translocate to the nucleus [155]. Transgenic mice, that express activated forms of calcineurin or NF-AT3 in the heart, develop cardiac hypertrophy and heart failure that mimic human heart diaease [155], suggesting a hypertrophic signaling pathway. If regucalcin has a suppressive effect on calcineurin activity in the heart cytosol of normal and transgenic rats [156], overexpression of regucalcin may have a pathophysiologic role in the prevention of development of cardiac hypertrophy and heart faliure.

Regucalcin has also been shown to have an inhibitory effect on Ca^{2+}/calmodulin-dependent protein phosphatase activity toward phosphotyrosine, phosphoserine, and phosphothreonine in rat brain cytosol [157] and neurons [158]. The presence of anti-regucalcin monoclonal antibody in the enzyme reaction mixture caused a significant elevation of protein phosphatase activity in the brain cytosol, indicating that the endogenous regucalcin has a suppressive effect on the cytosolic enzyme activity [157]. Regucalcin has been shown to have an inhibitory role in the regulation of protein phosphatase activity in rat brain cytosol.

Regucalcin is localized in the microsomes of rat brain, and aging causes a decrease in its protein levels [159]. Regucalcin has also been shown to have a suppressive effect on protein tyrosine phosphatase activity in rat brain microsomes [159].

Aging caused an increase in protein tyrosine phosphatase activity in rat brain microsomes and the suppressive effect of regucalcin on the enzyme activity was weakened in aged rats [159], suggesting that the decrease in microsomal regucalcin with aging is partly involved in the enhancement of microsomal protein tyrosine phosphatase activity with aging [159].

Regucalcin has been found to be present in the nucleus of rat brain and the endogenpous regucalcin has a suppressive effect on the nuclear protein tyrosine phosphatase activity [160]. Increasing age has also been found to induce a reduction in rat brain nucleus and may lead to attenuation of the suppressive effect of regucalcin on the nuclear protein tyrosine phosphatase activity [160]. Regucalcin may play a pivotal role as a regulatory protein in the regulation of brain function that relates to protein phosphorylation-dephosphorylation. Thus, regucalcin may play a physiological role in the intracellular control of the hormonal stimulation for phosphorylation and dephosphorylation of many proteins in various type cells.

As mentioned above, regucalcin has been demonstrated to reverse the activity of many Ca^{2+}-activated enzymes (phosphorylase *a,* glucose-6-phosphatase, fructose 1,6-bis-phosphatase, pyruvate kinase, protein kinase C, Ca^{2+}/calmodulin-dependent protein kinase, protein phosphatase, and Ca^{2+}/calmodulin-dependent cyclic AMP phosphodiesterase) and of Ca^{2+}-inhibited enzymes (5'-nucleotidase and dUTPase).

The first action is that regucalcin binds Ca^{2+} and that inhibits the metal's effect on many enzymes. The effect of regucalcin, that reverses Ca^{2+} action on many enzymes, may be based on its binding of Ca^{2+}, since the protein has 6 to 7 high-affinity binding sites per molecule, and a Ca^{2+}-binding constant of $4.19 x 10^5$ M^{-1} [9]. The intrinsic significance of regucalcin action may be the binding of Ca^{2+} by its protein.

The second is that Ca^{2+}-binding regucalcin directly inhibits the function of enzymes and somewhat stimulates enzyme function. The direct action of Ca^{2+}-binding regucalcin or the protein itself may be decided by the protein structure of enzymes. Spectroscopical studies have clearly demonstrates that Ca^{2+}-binding induces conformational changes in regucalcin, which may then result in increased hydrophobicity of the protein, and loosening of the conformation of regucalcin [10].

Regucalcin can directly inhibit the enzymes that are activated through Ca^{2+}-calmodulin [11]; this also results from regucalcin that affects the binding of Ca^{2+} to calmodulin. Which of the two proteins binds Ca^{2+} may be decided by their relative concentrations of Ca^{2+} in hepatic cytosol, since the Ca^{2+}-binding constant of regucalcin is greater than that of calmodulin [9]. Calmodulin exists as a monomer of molecular weight 17,000 and contains 4 Ca^{2+}-binding sites [1]. In the enzyme assay system of Ca^{2+}/calmodulin-dependent cyclic AMP phosphodiesterase activity, the inhibitory effect of regucalcin on the enzyme activation through Ca^{2+}-calmodulin is completely blocked after the addition with increasing concentrations of Ca^{2+} [120, 121]. This further supports the view that the mechanism by which regucalcin inhibits Ca^{2+} action is based on the binding of Ca^{2+}.

Moreover, in the enzyme assay system of protein kinases (protein kinase C and Ca^{2+}/calmodulin-dependent protein kinase), the effect of regucalcin, which inhibits the activation of enzymes by Ca^{2+}, is seen with increasing concentrations of Ca^{2+}. This suggests that regucalcin directly suppresses the enzyme activation in addition to Ca^{2+}-binding. Also, it is possible that Ca^{2+}-binding regucalcin or the protein itself can directly inhibit the enzyme activity.

REGUCALCIN SUPPRESSES PROTEIN OUTPUT

Regucalcin has been shown to have a regulatory effect on prptein synthesis and protein degradation, suggesting that regucalcin plays a role in the regulation of protein turnover in cells.

Regucalcin Suppresses Protein Synthesis

Protein synthesis is depressed in a variety of eukaryoticx cell types exposed to conditions depleting Ca^{2+} but not Mg^{2+} [161]. It has also been proposed that hormones (vasopressin and α-adrenergic agonist), which have been known to mobilize sequestered Ca^{2+} within liver cells, inhibit amino acid incoporation by influencing a Ca^{2+} requirement associated with protein synthesis [162]. Moreover, vasopressin inhibits the rate of protein synthesis in isolated hepatocytes partially depleted of Ca^{2+} [163]. These investigations propose the hypothesis that a sequestered pool of intracellular Ca^{2+} is required for the maintenance of high rates of protein synthesis in liver cells [162, 163]. On the other hand, vasopressin and α-adrenergic agonist causes an increase in the intracellular free Ca^{2+} concentration of hepatocytes that are not depleted of Ca^{2+} [164, 165]. Whether the increase in intracellular Ca^{2+} influences hepatic protein synthesis is as well undefined, however. It may be important to clarify the effect of Ca^{2+} addition on hepatic protein synthesis in vitro.

It has been demonstrated that Ca^{2+}, of various metals, can uniquely inhibit in vitro protein synthesis using the 5500 g supernatant fraction (the microsomes and cytosl) of rat liver homogenate [166]. Its inhibition was seen after addition of 1.0 μM Ca^{2+}. Ca^{2+} addition caused a remarkable decrease of the activity of aminoacyl (leucyl)-tRNA synthetase, which is a rate-limiting enzyme of protein synthesis at translational process, in hepatic cytosol [166]. Ca^{2+} may directly inhibit hepatic protein synthesis in subcellular fraction of liver cells. Ca^{2+} is required for protein synthesis in hepatocytes exposed to conditions depleting the cation [166]. It is not clarified whether protein synthesis in hepatocytes not depleted of Ca^{2+} requires exogenous Ca^{2+}. The mechanism by which Ca^{2+} is required in protein synthesis of hepatocytes depleted of Ca^{2+} may be complex.

Calmodulin, which can amplify Ca^{2+} effects on enzymes [1], did not have an appreciable effect on in vitro protein synthesis using the 5500 g supernatant fraction of liver homogenate in the presence of Ca^{2+} (10 μM) [166]. The activity of aminoacyl-tRNA synthetase in hepatic cytosol was not altered through calmodulin [167]. Presumably, the protein synthesis is inhibited by Ca^{2+}, which is not bound to calmodulin.

The role of regucalcin in the regulation of *in vitro* protein synthesis using the 5500 g supernatant fraction of rat liver homogenate is investigated [166]. Regucalcin caused a remarkable inhibition of hepatic protein synthesis in vitro [166]. Regucalcin could not reverse the Ca^{2+}-induced inhibition of protein synthesis [166], although it has been shown that regucalcin can reverse the Ca^{2+} effect on many enzymes in liver cells. Since regucalcin can bind to liver cytosolic proteins and the binding is slightly enhanced with the coexistence of 0.1 μM Ca^{2+} [163], Ca^{2+}-binding regucalcin and/or Ca^{2+} free regucalcin may be able to inhibit hepatic protein synthesis. In fact, the presence of regucalcin (1 and 2 μM) could fairly decrease hepatic protein synthesis that was reduced after the addition of 10 μM Ca^{2+}[166]. Regucalcin itself may play a role in the regulation of protein synthesis in liver cells.

Regucalcin has been shown to inhibit hepatic aminoacyl-tRNA synthase activity [166]. The inhibitory effect of regucalcin was seen in the presence of Ca^{2+} (10 μM). The inhibitory effect of regucalcin on hepatic protein synthesis may be partly based on a remarkable decrease of aminoacyl-tRNA synthetase activity caused by regucalcin. Regucalcin may bind aminoacyl (leucyl)-tRNA synthetase in hepatic cytosol, since iodinated regucalcin can bind the proteins in hepatic cytosol [65]. Regucalcin may be able to regulate liver cell function that is not affected by cellular Ca^{2+}.

The role of endogenous regucalcin on protein synthesis has been examined using anti-regucalcin monoclonal antibody, moreover [167]. The presence of anti-regucalcin monoclonal antibody in the reaction mixture caused a significant increase in protein synthesis and [^{3}H] leucyl-tRNA synthetase activity in normal rat liver. These increases were completely prevented after the addition of exogenous regucalcin (1.0 μM). Liver cytosol contained about 16 μg of regucalcin per 1 mg of the cytosolic protein; the reaction mixture contained about 0.17-0.19 μM of endogenous regucalcin, because the cytosolic protein in the range 360 – 390 μg were added into the mixture of 1.0 ml. Endogenous regucalcinin may have a suppressive effect on protein synthesis in liver cells.

Hepatic protein synthesis has been shown to enhance in regenerating rat liver, which induces a proliferation of liver cells after partial hepatectomy [167]. This enhancement was reamarkable at 24 and 48 h after partial hepatectomy. Hepatic protein synthesis in regenerating liver was further enhanced in the presence of anti-regucalcin monoclonal antibody in the reaction mixture. Endogenous regucalcin has a suppressive role on the enhancement of protein synthesis in regenerating liver.

Regucalcin Stimulates Protein Degradation

Evidence for the role of Ca^{2+}-activated protease (calpains) is implicated in signal transduction [168]. Two neutral Ca^{2+}-requiring proteinases, differing in molecular size, have been isolated from rabbit liver cytosol [169]. Both are recovered as inactive proenzymes that can be converted to the active forms by high (0.1-1.0 mM) concentrations of Ca^{2+} in the absence of substrate or, in the presence of a protein substrate, by low (1-5 μM) concentrations of Ca^{2+} [169]. The activated proteinases required only 1-5 μM Ca^{2+} for maximal activity [169].

The addition of regucalcin (0.25 – 2.0 μM) into the enzyme reaction mixture has been found to induce a remarkable increase of neutral proteinase activity in the presence of 5.0 μM Ca^{2+} [170]. The effect of regucalcin was seen at 0.25 μM. Regucalcin may activate Ca^{2+}-requiring proteinase in rat liver cytosol. The effect of regucalcin in increasing liver cytosolic proteinase activity is also seen in the absence of Ca^{2+} [170]. This increase was remarkable as compared with that of Ca^{2+} addition. Regucalcin may activate Ca^{2+}-not requiring neutral proteinase in rat liver cytosol. Regucalcin has a reversible effect on the activation of various enzymes by Ca^{2+} and/or calmodulin [123, 124, 150]. The finding, that regucalcin can activate neutral proteinase in rat liver cytosol in the presence or absence of Ca^{2+}, was novel.

The activating effect of regucalcin on liver cytosolic proteinase activity was not seen in the presence of anti-regucalcin antiserum in the enzyme reaction mixture [170], suggesting a role of endogenous regucalcin in the activation of cytosolic proteinase.

The activating effect of regucalcin on neutral proteases in the liver cytosol is characterized [172]. Leupeptin is a potent inhibitor of SH-proteinase. The regucalcin-increased protenase activity was inhibited in the presence of leupeptin in the enzyme reaction mixture [172]. Regucalcin may activate neutral cysteinyl-proteinase in the liver cytosol. The effect of regucalcin in increasing proteolytic activity in rat liver cytosol was not abolished in the presence of diisopropylfluorophosphate (DFP), an inhibitor of serine-protease, although DFP alone had an inhibitory effect on the proteolytic activity [172]. Regucalcin does not act on serine-proteases in liver cytosol. The effect of regucalcin was also seen in the presence of a chelator of metal ions, suggesting that regucalcin does not activate metal-related proteases in liver cytosol.

The proteolytic activity in liver cytosol was markedly increased after the addition of dithiothreitol (DTT), a protecting reagent for SH group, in the enzyme reaction mixture and this increase was completely inhibited in the presence of N-ethylmaleiimide (NEM), a SH group-modifying agent [172]. The effect of regucalcin in increasing proteolytic acitivity was seen in the presence of DTT in liver cytosol, although it was completely inhibited in the presence of NEM [172]. Regucalcin may act on the SH group of cysteinyl-proteases in liver cytosol.

Two forms of Ca^{2+}-activated neutral proteases (calpines) have been identified in hepatocytes and other cells, *m*-calpaine and μ-calpaine [169]. Isolated μ-calpain requires micromolar concentrations of Ca^{2+} for activation, while *m*-calpain requires millimolar concentrations [173]. The activation of proteases in rabbit liver cytosol is required only 1-5 μM Ca^{2+} for maximal activity [172]. Ca^{2+} (10 μM)-increased proteolytic activity in rat liver cytosol was inhibited in the presence of NEM, indicating that neutral proteases (including calpains), which are activated by Ca^{2+}, exist in liver cytoplasm [172].

The activity of calpaines, which are thiol protease [177], increases in hepatocytes following addition of ATP [174]. The proteolytic activity in liver cytosol was decreased after the addition of calpastatin, a specific inhibitory of calpains, indicating the existence of calpains in the cytosol [172]. Regucalcin-increased proteolytic activity was abolished in the presence of calpastatin. Regucalcin may be an activator for calpains. *m*-Calpain isolated from rabbit skeletal muscle was activated by regucalcin independent on Ca^{2+}. This activation was inhibited after the addition of NEM [172]. Regucalcin acts on the SH group in *m*-calpain. Presumably, regucalcin may be able to activate both *m*- and μ-calpains.

The role of regucalcin in the regulation of neutral proteolytic activity in rat kidney cortex cytosol is also examined [175, 176]. Regucalcin had an activatory effect on neutral proteolytic activity in the kidney cortex cytosol [176]. This increase was abolished in the presence of anti-regucalcin monoclonal antibody, supporting the view that endogenous regucalcin plays a role as activator of proteases in the renal cortex ctrtosol [176]. The effect of regucalcin on proteolytic activity was not altered in the presence of calcium chloride (0.01 and 1.0 mM) or EGTA (1.0 mM), indicating that the effect of regucalcin was independent on Ca^{2+}. Regucalcin may activate both calpaines and other proteases in the kidney cortex cytosol. Regucalcin has also been shown to activate SH proteases in rat renal cortex cytosol. Presumably, regucalcin acts on the SH-groups of protease in rat renal cortex cytosol.

The activating effect of regucalcin on proteases is seen in the concentrations of 0.01 – 0.25 μM [175]. The concentration of regucalcin in rat kidney tissue has been found to be present about 5.3 μM. Regucalcin plays a physiological role in the activation of thiol proteases in renal cortex cells.

Regucalcin uniquely activates thiol proteases independent on Ca^{2+} in the liver cytosol, whereas it has no effect on serine proteases and metalloproteases [170, 172]. Such an effect of regucalcin was also found in the renal cortex cytosol [176]. Regucalcin may be an activator on thiol proteases in liver and kidney cells. Regucalcin also plays a role as an activator in other tissues that express regucalcin.

Calpains are ubiquitous, non-lysosomal, calcium-dependent proteases that may play important roles in Ca^{2+}-mediated intracellular processes [168, 169, 174]. The ability of calpain to alter the limited proteolysis, the activity or function of numerous cytoskeletal proteins, protein kinases, receptors and transcription factors suggests the involvement of the protease in various Ca^{2+}-regulated cellular functions [178]. Calpains may also play an integral role in modulating the activity of protein kinase C, a key protein in many signal transduction processes [179], since they convert the native Ca^{2+}/phospholipids-dependent kinase to a soluble form that does not require Ca^{2+} or phospholipids for activity [2]. Regucalcin can increase the activity of thiol proteases including calpain in rat liver and renal cortex cytosols. Regucalcin may play a pivotal role in the regulation of cellular functions related to Ca^{2+} mediated through thiol proteases. Presumably, regucalcin plays an important role in the regulation of signal transduction that is involved in proteases.

As described above, regucalcin may play a role as suppressor in the enhancement of cytoplasmic protein levels due to inhibiting protein synthesis and activating protein degradation.

REGUCALCIN REGULATES NUCLEAR FUNCTION

Mounting evidence suggests that Ca^{2+} is active in liver nuclear function [81-83]. Calmodulin exists in rat liver nuclei [82]. The existence of an ATP-stimulated Ca^{2+}-sequestration system in rat liver nuclei that requires calmodulin and generates a net increase in nuclear matrix free Ca^{2+} concentration has been reported [81]. Calmodulin stimulates DNA synthesis in liver cells [180], and the effect of calmodulin is mediated through α-adrenergic stimulation [181, 182]. There is growing evidence that regucalcin plays an important role in the regulation of nuclear functions.

Regucalcin Binds to Nuclear Protein or DNA

Regucalcin has been shown to localize in the nuclei of rat liver [38, 151]. Regucalcin can bind on calmodulin-agarose beads [142]. Liver nuclear extract were incubated with calmodulin-agarose beads, and calmodulin-agarose beads were applied to SDS-PAGE. Band that coincides with regucalcin was found on SDS-PAGE [142], suggesting that regucalcin is present in the nucleus [151]. Exogenous regucalcin has been shown to transport into the nucleus isolated from normal rat liver [38]. Endogenous regucalcin is present in the nuclei of rat liver using Western blotting analysis [38]. When isolated liver nuclei were incubated in the presence of exogenous regucalcin (50 μg/ml; 1.5 μM), potent band for regucalcin was found in the nucleus [38], supporting that the protein is translocated into the nucleus. This translocation was seemed to be an early event, since potent band for regucalcin was seen with only 10 min of incubation. A part of regucalcin, which is localized in the cytoplasm of liver cells, is translocated to the nucleus [38].

Nuclear regucalcin translocation was not appreciably changed in the presence of ATP (2 mM), guanosine 5'-triphosphate (GTP, 2 mM), and calcium chloride (0.1 mM), suggesting that its translocation is not mediated through nuclear localization signal [38]. ATP and GTP are required for nuclear import of proteins that are localized in the nuclei. ATP or GTP does not regulate the translocation of regucalcin into liver nuclei. Regucalcin is translocated independently of Ca^{2+}. Nuclear protein transport is blocked in the presence of the lectin wheat germ agglutinin (WGA) [83]. Nuclear regucalcin translocation was not appreciably changed in the presence of WGA in the reaction mixture [38]. This finding suggests that the nuclear translocation of regucalcin is not related to nuclear localization signal that is responsible for selection for intranuclear active transport. Presumably, regucalcin is passively transported to the nucleus through nuclear pore in liver cells, since a molecular weight of regucalcin is about 33 kDa [16].

Regucalcin has also been shown to localize in the nuclei of the cloned normal rat kidney proximal tubular epithelial NRK52E cells with immunocytochemical analysis [183]. The nuclear localization of regucalcin is enhanced through hormonal signaling process that is involved in protein kinase C [183].

Regucalcin has been shown to bind proteins in isolated rat liver nuclei using Far-Western blot analysis [184]. The results of the Far-Western analysis showed the existence of protein components that bind to regucalcin in the nucleus isolated from rat liver [184]. Regucalcin has also been demonstrated to bind DNA using Western blot analysis for regucalcin with DNA cellulose-binding assay [184]. These findings show that regucalcin binds proteins and DNA in liver nucleus. Regucalcin may have a regulatory effect on signaling pathways that modulate transcriptional activity in liver cells.

Regucalcin Suppresses Ca^{2+}-Stimulated Nuclear DNA Fragmentation

Isolated rat liver nucleus contains a DNA endonuclease activity dependent upon Ca^{2+} and Ca^{2+} results in extensive DNA hydrolysis [185]. The Ca^{2+} dependence of this endogenous DNA fragmentation process is based on DNA endonuclease activity dependent upon sub-micromolar Ca^{2+} when the nucleus is reconstituted with NAD^+ and ATP [185]. This endogenous endonuclease activity may be responsible for the DNA fragmentation occurring during programmed cell death (apoptosis) and certain forms of chemically induced cell killing [185, 186]. Ca^{2+} has been shown to stimulate *in vitro* DNA fragmentation in isolated rat liver nuclei [187]. This increase was seen after the addition of 1.0 μM Ca^{2+}, in agreement with previous work [187]. The presence of regucalcin (0.5-2.0 μM) completely inhibited the activation of liver nuclear DNA fragmentation when 10 μM Ca^{2+} was added. This inhibition was not seen in the presence of Ca^{2+} at 25 or 50 μM. Thus, regucalcin had an inhibitory effect on DNA fragmentation when a comparatively lower concentration of Ca^{2+} (5.0 and 10 μM) was added [187]. The inhibitory effect of regucalcin on DNA fragmentation may be partly based on binding of Ca^{2+} [9].

DNA fragmentation in rat liver nucleus has been reported to be stimulated through Ca^{2+}-calmodulin [185], which exists in liver nuclei [82]. Addition of calmodulin (10 and 20 μg/ml) did not enhance Ca^{2+} (10 μM)-activated DNA fragmentation in liver nuclei [187]; however, nuclear endogenous calmodulin may be able to enhance Ca^{2+}-activated DNA fragmentation in the nucleus. Regucalcin has been found to inhibit Ca^{2+}-activated DNA fragmentation after Ca^{2+} addition [187]. Such an inhibition was also seen in the presence of exogenous calmodulin [187]. Presumably, regucalcin can inhibit Ca^{2+}/calmodulin-dependent DNA fragmentation in the liver nuclei, since radio-iodinated regucalcin is found in the nuclei isolated from rat liver in the absence or presence of 1.0 mM Ca^{2+} [65].

Several studies have shown that Ca^{2+} plays an important role in the regulation of nuclear functions [81, 82]. Also, it has been found that a sustained increase in cytosolic Ca^{2+} level precedes the activation of DNA fragmentation that is characteristic of programmed cell death (apoptosis) and in certain forms of chemically induced cell killing [185, 186]. The finding, that regucalcin inhibits the activation of DNA fragmentation by Ca^{2+}, was the first time for a role of regucalcin in liver nuclear functions.

Regucalcin Suppresses Nuclear Signaling-Related Enzyme Activation

Smal GTPase Ran (ras-related nuclear protein) is required for protein export from the nucleus and protein import into the nucleus [188]. Liver nuclear GTPase activity was increased after calcium addition with a comparatively higher concentration in the enzyme reaction mixture, and this increase was not seen in the presence of TFP, an antagonist of calmodulin [189]. The effect of calcium in increasing nuclear GTPase activity may be related to endogenous calmodulin. Calmodulin has been shown to be present in liver nucleus [82]. GTPase, which is activated by Ca^{2+}/calmodulin, may also be localized in liver nucleus. The presence of exogenous regucalcin (0.5 μM) used in the enzyme reaction mixture caused an inhibitory effect on GTPase activity in liver nucleus [189]. This effect was also seen in the presence of EGTA, a chelator of Ca^{2+}. Presumably, the inhibitory effect of regucalcin on liver nuclear GTPase activity is revealed independent of Ca^{2+}/calmodulin in the nucleus. Regucalcin has been shown to have an inhibitory effect on the activation of enzymes by Ca^{2+}/calmodulin due to binding Ca^{2+} and/or calmodulin [138-141]. Regucalcin may directly inhibit GTPase activity in liver nucleus.

The physiological significance of the inhibitory effect of regucalcin on GTPase activity in liver nucleus is unknown. However, the presence of anti-regucalcin monoclonal antibody in the enzyme reaction mixture caused a significant increase in this activity in liver nucleus [189]. This increase was completely blocked after regucalcin addition, suggesting that endogenous regucalcin has a suppressive effect on GTPase activity in liver nucleus. Endogenous regucalcin, which is localized in liver nucleus, may participate in the regulation of nuclear functions that are related to hydrolysis of GTP.

Regucalcin may be able to regulate a process of signal transduction from the cytoplasm to nucleus in liver cells. This process is mediated through various protein kinases. Nuclear Ca^{2+}-dependent protein kinase and protein tyrosine phosphatase activities were increased in the presence of anti-regucalcin monoclonal antibody in the enzyme reaction mixture, and these increases were completely abolished after the addition of regucalcin [38]. The translocation of regucalcin in the nucleus may play a suppressive role in the regulation of protein kinase and protein tyrosine phosphatase in liver nucleus [38]. Endogenous regucalcin has been demonstrated to have a suppressive effect on the enhancement of protein kinase activity with a proliferation of liver cells [190]. Protein kinase activity is enhanced in the cytosol and nucleus of regenerating rat liver [190]. Regucalcin had a suppressive effect on tyrosine kinase, protein kinase C, and Ca^{2+}/calmodulin-dependent protein kinase in the cytoplasm and nucleus of regenerating rat liver [123, 138, 139, 190].

Nuclear phosphotyrosine phosphatase activity was increased after Ca^{2+} addition in the enzyme reaction mixture, although the enzyme activity was not altered by TFP, an inhibitor of calmodulin, or cyclosporine A, an inhibitor of calcineurin [151]. Nuclear phosphatase activity toward phosphotyrosine may be independent of calmodulin. Meanwhile, nuclear phosphatase activity toward phosphoserine was elevated after the addition of Ca^{2+}, while the enzyme activity was appreciably decreased by TFP and cyclosporine A, suggesting that the enzyme activity is partly involved in Ca^{2+}/calmodulin-dependent protein phosphatase (calcineurin). Nuclear phosphatase activity toward phosphothreonine was not altered after the addition of Ca^{2+}, TFP, and cyclosporine A in the enzyme reaction mixture. Vanadate caused an inhibition of nuclear phosphatase activity toward phosphotyrosine and phosphoserine but not phosphothreonin. Thus, different protein phosphatases toward phosphotyrosine, phosphoserine, and phosphothreonine have been shown to be present in the liver nucleus. The addition of regucalcin in the enzyme reaction mixture caused a significant decrease in phosphatase activity toward phosphotyrosine, phosphoserine, and phosphothreonine in the liver nuclei [151]. The concentration of endogenous nuclear regucalcin was estimated about 82.4-98.8 nM. Further addition of exogenous regucalcin (0.25 μM) caused a significant decrease in nuclear phosphatase activity, although the effect was saturated with increasing concentrations of regucalcin (0.5 μM) [151].

Nuclear phosphatase activity was elevated in the presence of anti-regucalcin monoclonal antibody (25 and 50 ng/ml of reaction mixture) [151]. This elevation was completely abolished after addition of regucalcin. Endogenous regucalcin may regulate protein phosphatase activity toward phosphotyrosine, phosphoserine, and phosphothreonine in liver nucleus. Endogenous regucalcin may have an inhibitory effect on various protein phosphatases in the liver nucleus.

Regucalcin Suppresses Nuclear DNA and RNA Synthesis

Regucalcin has been shown to have an inhibitory effect on DNA synthesis activity in the nuclei of normal rat liver [191]. The inhibitory effect of regucalcin was seen in the presence of EGTA, a chelator of Ca^{2+}, in the reaction mixture. Ca^{2+} is present in liver nucleus [191]. Liver nuclear DNA synthesis activity was increased in the presence of EGTA in the reaction mixture, suggesting that Ca^{2+} suppresses DNA synthesis activity in the nucleus. The effect of regucalcin in inhibiting nuclear DNA synthesis may be not related to Ca^{2+} in liver nucleus.

Liver nuclear DNA synthesis has been shown to stimulate in regenerating rat liver [192]. Nuclear DNA synthesis was markedly increased at 1 day after hepatectomy, and this increase was also seen at 3 days [192]. Nuclear DNA synthesis was enhanced in the presence of EGTA (0.4 mM) in the incubation mixture. The presence of Ca^{2+} (1.0 – 25 μM) caused a significant decrease in the nuclear DNA synthesis of normal rat liver. Regucalcin (0.25 and 0.5 μM) caused an inhibition of nuclear DNA synthesis of normal rat liver [192]. This inhibition was also seen in the presence of Ca^{2+} (1.0 μM). The suppressive effect of regucalcin was remarkable in regenerating rat liver nuclei in comparison with that of normal rat liver. Regucalcin has been shown to have a suppressive effect on nuclear DNA synthesis in regenerating rat liver. Regucalcin may have a suppressive role in the enhancement of nuclear DNA synthesis in liver cell proliferation.

Regucalcin has also been shown to have a suppressive effect on DNA synthesis activity in the nuclei isolated from rat renal cortex [193]. The addition of regucalcin (0.1 – 0.5 μM) in the reaction mixture containing either EGTA (1 mM) or calcium chloride (50 μM) had an inhibitory effect on nuclear DNA synthesis activity [194]. The presence of anti-regucalcin monoclonal antibody (10-50 ng/ml) in the reaction mixture caused a significant increase in nuclear DNA synthesis activity [194]. This increase was completely abolished in the presence of regucalcin (0.5 μM). Endogenous regucalcin has been found to have a suppressive effect on DNA synthesis in the nuclei of rat renal cortex [194].

Regucalcin has been shown to have an inhibitory effect on RNA synthesis in the nuclei isolated from control rat liver [194] and regenerating rat liver [195]. RNA symthesis in rat liver nuclei was stimulated after Ca^{2+} addition with a comparatively lower concentration, although the sub-micromolar concentration of Ca^{2+} evoked an inhibition of nuclear RNA synthesis [194]. The addition of Ca^{2+} with higher concentrations has been reported to have an inhibitory effect of nuclear RNA synthesis in rat liver cells [196]. Since liver nuclei contain a DNA endonuclease activity dependent upon Ca^{2+} in the submicromolar range and Ca^{2+} causes extensive DNA hydrolysis [185, 187], nuclear RNA synthesis may be suppressed with higher Ca^{2+} concentrations. Inactivation of RNA polymerase III transcription has been shown to be calcium dependent; the changes in Ca^{2+} concentration, the activation of calpains, and the consequent proteolytic degradation of RNA of transcription factors has been suggested to be involved in the regulation of RNA polymerase III transcription in the presence of 1 mM Ca^{2+} [197].

The effect of regucalcin in decreasing nuclear RNA synthesis activity in normal rat liver was not seen in the presence of α-amanitin, an inhibitor of RNA polymerase II and III [194], suggesting that the suppressive effect of regucalcin on nuclear RNA synthesis activity is partly resulted from its inhibitory action on RNA polymerase II and III [194, 195]. Meanwhile, it has been reported that Ca^{2+} has a stimulatory effect on RNA synthesis in liver nucleus [196, 197]. This effect may be partly mediated through Ca^{2+}-dependent protein kinase [196, 197]. The stimulatory effect of Ca^{2+} on nuclear RNA synthesis activity was completely blocked in the presence of regucalcin [194, 195]. Regucalcin has been shown to inhibit Ca^{2+}-dependent protein kinases in rat liver nucleus [190]. Presumably, the effect of regucalcin in decreasing RNA synthesis activity in liver nucleus is partly involved in its inhibitory action on the activities of both RNA polymerase II and III and Ca^{2+}-dependent protein kinases. Regucalcin has been proposed to have a role as a transcriptional factor in liver nucleus.

Thus, regucalcin has been found to have a suppressive effect on liver nuclear DNA and RNA synthesis [191-195]. The mechanism by which regucalcin inhibits nuclear DNA and RNA synthesis remains to be elucidated. It is speculated that regucalcin has an inhibitory effect on DNA and RNA polymerase activity. However, the effect of regucalcin in inhibiting nuclear RNA synthesis activity is observed in the presence of α-amanitin, an inhibitor of RNA polymerase II. Regucalcin can directly bind DNA [184]. Which base pairs of DNA bind regucalcin remains to be determined. It is possible that regucalcin binds DNA and it has an inhibitory effect on nuclear DNA and RNA synthesis activity.

Regucalcin Suppresses Apoptotic Cell Death Mediated Through Cell Signalings

Regucalcin Suppresses NO Synthase Activation That Is Related to Apoptosis

NO may be important as a signaling factor in many cells [125], and it plays a role in apoptosis of hepatoma cells [198]. NO mediates apoptosis by D-galactosamine in a primary culture of rat hepatocytes [199]. Regucalcin has been shown to inhibit NO synthase that is related to cell apoptosis [126], suggesting that regucalcin has a suppressive role in apoptosis [200]. Regucalcin has a suppressive effect on Ca^{2+}/calmodulin-dependent NO synthase in the cloned rat hepatoma H4-II-E cells [126]. The effect of regucalcin in decreasing NO synthase activity was also seen in the presence of TFP or EGTA. Presumably, regucalcin has an inhbitiory effect on NO synthase activity due to binding to calmodulin and/or the enzyme independently of Ca^{2+} in proliferative cells.

Overexpressing of regucalcin has also been shown to have a suppressive effect on NO synthase activity in H4-II-E cells (transfectants) [126]. This decrease was completely abolished in the presence of anti-regucalcin monoclonal antibody in the reaction mixture. Moreover, the effect of Ca^{2+}/calmodulin addition in increasing NO synthase activity in H4-II-E cells cells (wild type) was completely prevented in the transfectants. Endogenous regucalcin had a suppressive effect on NO synthase activity the cloned rat hepatoma H4-II-E cells. NO synthase activity was enhanced in H4-II-E cells cultured with 10% FBS as compared with that of 1% FBS [126], suggesting that the enzyme is induced in proliferative cells. The enhancement of NO synthase activity in H4-II-E cells cultured with 10% FBS was abolished in the presence of anti-regucalcin monoclonal antibody [126]. Regucalcin levels were elevated in H4-II-E cells with 10% FBS-culture [126]. Endogenous regucalcin may have a suppressive effect on the enhancement of NO synthase activity with proliferation of H4-II-E cells.

A high concentration of NO, which is produced from inducible NO synthase, has been shown to inhibit cell proliferation [200] and to induce cell apoptosis [201]. It is reported that a low concentration of NO, which is produced from endothelial NO synthase, protects against the cytotoxic effects of reaction oxygene species in cells [202]. Whether endogenous regucalcin suppresses NO production in H4-II-E cells is unknown at present. It is speculated, however, that regucalcin may inhibit NO production in H4-II-E cells, since regucalcin can decrease NO synthase activity in the cells [126]. Endogenous regucalcin may have an inhibitory effect on inducible and endothelial NO synthetases in hepatoma cells. Alternatively, regucalcin may have a physiological role in the regulation of NO-related cell functions.

Regucalcin Suppresses Apoptotic Cell Death Mediated Through Various Signaling Factors in Liver Cells

Tumor necrosis factor α (TNF-α) and NO mediate apoptosis by D-galctosamine in a primary culture of rat hepatocytes [199, 200]. TNF-α induces apoptosis in mammary adenocarcinoma cells by an increase in intranuclear free Ca^{2+} concentration and DNA fragmentation [200]. TNF-α (0.1-10 ng/ml) caused a significant decrease in the number of H4-II-E cells (wild type), inducing cell death. Overexpressing of regucalcin in H4-II-E cells (transfectants) has been found to prevent the effect of TNF-α in decreasing cell number [203]. Overexpressing of regucalcin had a preventive effect on cell death induced with the higher concentration of TNF-α (10 ng/ml). This finding demonstrates that overexpression of regucalcin has a suppressive effect on cell death induced after stimulation of TNF-α [203]. Culture with NAME, an inhibitor of NO synthase, had a significant preventive effect on TNF-α-induced cell death. Regucalcin inhibits Ca^{2+}/calmodulin-dependent NO synthase activity in H4-II-E cells [126]. The suppressive effect of regucalcin on cell death may be partly resulted from the inhibition of NO production, which can induce apoptosis, stimulated after TNF-α stimulation in H4-II-E cells. The effect of caspase inhibitor on TNF-α-mediated cell death in H4-II-E cells is examined [203]. TNF-α-induced cell death was prevented in culture with caspase inhibitor in wild-type cells and transfectants, suggesting that TNF-α-induced cell death is partly involved activation of caspases in H4-II-E cells. Regucalcin may have an inhibitory effect on activation of caspases in the cells.

Lypopolysacharide (LPS) has been shown to induce cell apoptosis [204, 205]. LPS caused a decrease in the number of H4-II-E cells (wild type), inducing cell death and apoptosis [206]. This decrease was completely prevented in the regucalcin cDNA-transfected hepatoma cells overexpressing regucalcin with culture for 12-48 hours [206]. Overexpression of regucalcin has a suppressive effect on LPS-stimulated cell death and apoptosis. LPS acts to modulate the expression of a large number of genes that favor apoptosis of fibroblastic cells that are dependent upon activation of caspase-8 [205]. There is evidence that LPS-induced cell death is mediated through accumulation of reactive oxygene species and activation of p38 in rat cortex and hippocampus [205]. Culture with LPS caused a significant decrease in Ca^{2+}/calmodulin-dependent NO synthase activity in H4-II-E (wild type) cells [206]. LPS-induced decrease in NO synthase activity was found to prevent in the transfectants overexpressing regucalcin [206]. LPS-induced cell death may be not resulted from NO production in hepatoma cells, and the suppressive effect of regucalcin on LPS-induced cell death is not involved in NO in the cells. Moreover, LPS-induced cell death was prevented in culture with caspase 3 inhibitor [206]. The effect of regucalcin in suppressing LPS-induced cell death is partly related to the inhibitory effect on caspase-3 in hepatoma cells.

An induction of apoptosis is partly mediated through pathway of protein kinase. The death of H4-II-E cells (wild type) has been found to be induced after culture with PD 98059, a ERK inhibitor, dibucaine, an inhibitor of Ca^{2+}-dependent protein kinase, or staurosporine, a potent inhibitor of protein serine/threonin kinases (protein kinase C), suggesting that various inhibitors-induced cell deaths are partly involved in the inhibition of protein kinases [206]. Overexpression of regucalcin in H4-II-E cells rescued cell death with PD 98059 or dibucaine [206]. This effect was not observed with staurosporine. PD 98059 induces apoptosis, which is in part due to the inactivation of Bcl-2 by increasing phosphorylated Bcl-2 in human prostate cancer cells [207]. Dibucaine has been shown to activate various caspases, such as caspase-3, -6, -8, and -9 (-like) activities, but not caspase-1 (-like) activity, and to induce mitochondrial membrane depolarization and the release of cytochrome C from mitochondria into the cytosol in leukemia cells (HL-60) [208]. Staurosporine induces apoptosis in Chang liver cells by a mitochondria-caspase-dependent pathway, which is closely correlated with a decrease in Bcl-2 and Bcl-XL levels in cancer cells [209]. Regucalcin may partly inhibit the inactivation of Bcl-2 or the activation of caspases in signaling mechanism that PD 98059 or dibucaine induces apoptosis.

Calcium channel blockers, the endoplasmic reticulum Ca^{2+}-ATPase inghibitor thapsigargin and calcium ionophores are potent to lead several cell types to apoptosis [211, 212]. Thapasigargin is an inhibitor of Ca^{2+}-ATPase in the endoplasmic reticulum (Ca^{2+} store) in cells, and treatment with thapsigargin causes an elevation of sustained Ca^{2+} concentration in cells and induces apoptosis in the hepatoma cells [213]. Experiments on nucleus isolated from cells clearly demonstrate the induction of Ca^{2+}-dependent endonuclease activity during triggering apoptosis events [213]. Rises in intracellular Ca^{2+} concentration are believed to activate this nuclease and to mediate DNA cleavages into oligonucleosome fragments [214]. Regucalcin has been shown to have an inhibitory effect on Ca^{2+}-activated DNA fragmentation in isolated rat liver nucleus [187], suggesting that the protein has an inhibitory effect on apoptosis in liver cells. Thapsigargin induces cell death and apoptosis causing DNA fragmentation [207]. Thapsigargin-induced DNA fragmentation in the hepatoma cells is not altered after culture with caspase inhibitor, suggesting that thapsigargin-mediated apoptosis is independent of activation of caspases [207]. Overexpression of regucalcin in the hepatoma cells has been found to suppress DNA fragmentation induced by thapsigargin [207]. This effect was not further enhanced in culture with caspase inhibitor [207]. Presumably, regucalcin has a suppressive effect on thapsigargin-mediated cell death due to preventing the rise in intracellular Ca^{2+} concentration in the hepatoma cells, since the protein can keep intracellular Ca^{2+} homeostasis due to activating Ca^{2+} pum enzymes in the plasma membranes, mitochondria, and endoplasmic reticulum of rat liver cells [31-34].

Calcium entry into cells induces cell death [207, 215]. Culture with Bay K 8644, an antagonist of Ca^{2+} entry in cells, caused a significant increase in the death of hepatoma H4-II-E cells (wild type) [207]. Culture with Bay K 8644 did not induce cell death of the transfectants (H4-II-E cells) overexpressing regucalcin [207]. Overexpression of regucalcin in H4-II-E cells was found to suppress DNA fragmentation induced by Bay K 8644. Regucalcin may have a suppressive effect on Ca^{2+} entry-induced stimulation of apoptosis in the hepatoma cells [207]. Regucalcin has a suppressive effect on Ca^{2+} entry-mediated cell death due to preventing the rise in intracellular Ca^{2+} concentration in the hepatoma cells. In addition, regucalcin may suppresse the effect of Ca^{2+} on DNA fragmentation in the nucleus of H4-II-E cells.

The effect of insulin or IGF-I on cell death and apoptosis in H4-II-E cells has not been well known. H4-II-E cells were cultured in a medium containing, either vehicle, insulin, insulin-like growth factor-I (IGF-I), epinephrine, or transforming growth factor-β in the absence of FBS [216]. The number of wild-type cells was decreased in the presence of insulin or IGF-I [216]. Agarose gel electrophoresis showed the presence of low-molecular-weight DNA fragments of adherent wild-type cells cultured with insulin or IGF-I [216]. The effect of insulin or IGF-I in stimulating cell death and DNA fragmentation H4-II-E cells (wild type) was suppressed in transfectants overexpressing regucalcin [216]. The effect of insulin in decreasing the number of H4-II-E cells was prevented in the presence of caspase-3 inhibitor [216]. The effect of IGF-I on cell death, however, was also observed in the presence of caspase-3 inhibitor [216]. These observations suggest that the effect of insulin on cell death is involved in activation of caspase-3, and that the effect of IGF-I is not dependent on caspase-3 in H4-II-E cells. The effect of IGF-I in inducing cell death in the presence of caspase-3 inhibitor was completely blocked in transfectants overexpressing regucalcin [216], suggesting that regucalcin inhibits signaling pathway of IGF-I-induced cell death that is not mediated through caspase-3 in H4-II-E cells.

The effect of insulin or IGF-I in inducing cell death and apoptosis of H4-II-E cells was not observed in the presence of NAME, an inhibitor of NO synthase [216], suggesting that insulin- or IGF-induced cell death is partly involved in production of NO in H4-II-E cells. Overexpression of regucalcin has been shown to have a suppressive effect on activation of Ca^{2+}/calmodulin-dependent NO synthase in H4-II-E cells [126].

The effect of IGF-I in inducing cell death of H4-II-E cells was also observed in the presence of Bay K 8644 [216]. Such an effect was not seen in the case of insulin [216]. The mode of IGF-I action differs from that of insulin. It is assumed that insulin induces cell death that is partly mediated through intracellular calcium-dependent signaling pathway in H4-II-E cells, and that IGF-I may not be mediathed through calcium-dependent signaling pathway in H4-II-E cells. The effect of IGF-I in inducing cell death in the presence of Bay K 8644 was not observed in the transfectants overexpressing regucalcin [216].

Genistein has an inhibitory effect on protein tyrosine kinases and it can produce cell cycle arrest and apoptosis in leukemic cells [217]. Genistein was found to induce cell death of H4-II-E cells, and the effect was not seen in the transfectants overexpressing regucalcin [216]. Genistein-induced cell death is partly mediated through inhibition of protein tyrosine kinase in H4-II-E cells. Regucalcin has an inhibitory effect on protein tyrosine kinase activity in the cytoplasm and nucleus of rat liver [190]. The effect of insulin in inducing cell death of H4-II-E cells was not seen in the presence of genistein [216], although this effect was not seen in the case of IGF-I. The effect of IGF-I on cell death in the presence of genistein was prevented in the transfectants overexpressing regucalcin [216]. Regucalcin has a suppressive effect on apoptotic cell death that is mediated through signaling pathways with dependent or independent on protein tyrosine kinase.

Vanadate is an inhibitor of protein tyrosine phosphatase in cells [145]. Regucalcin has been shown to have an inhibitory effect on protein tyrosine phosphatase activity in the cytoplasm and nucleus of rat liver [150]. Vanadate was found to induce cell death of H4-II-E cells [216], suggesting that cell death is not caused by mechanism that is mediated through inhibition of protein tyrosine phosphatase activity. Vanadate induced cell death for transfectants overexpressing regucalcin [216], suggesting that the suppressive effect of regucalcin on cell death of H4-II-E cells is independent on protein phosphatase. IGF-I had a stimulatory effect on cell death of H4-II-E cells overexpressing regucalcin in the presence of vanadate [216], suggesting that the effect of IGF-I is not mediated through protein tyrosine phosphatase in the transfectants.

The effect of insulin in inducing cell death may be partly mediated through signaling pathway which is involved in caspase-3, calcium, NO, protein tyrosine kinase, or protein tyrosine phosphatase in H4-II-E cells. The effect of IGF-I on cell death of H4-II-E cells may be mediated through NO and other molecules. Overexpression of regucalcin may have a suppressive effect on signaling mechanism by which insulin or IGF-I induces cell death of H4-II-E cells.

Sulforaphane is an isothiocyanate that is present naturally in widely consumed vegetables and has a particularly high concentration in broccoli. This compound has been shown to block the formation of tumors initiated by chemicals in the rat [218]. Sulforaphane has been shown to induce a cell cycle arrest, followed by cell death in HT29 human colon cancer cells [218]. Sulforaphane increases expression of the pro-apoptotic protein Bax, the release of cytochrome C from the mitochondria to the cytosol, and the proteolytic cleavage of poly (ADP-ribose) polymerase in HT29 human colon cancer cells [218]. In human T-cell leukemia, sulforaphane induces apoptosis due to increased p53 and Bax protein expression, and slightly affected Bcl-2 expression [219]. In cultured PC-3 human prostate cancer cells, moreover, sulforaphane-induced apoptosis is associated with up-regulation of Bax, down-regulation of Bcl-2 and activation of caspase-3, -9, and -8 [220]. Sulforaphane has been found to induce cell death and apoptosis in H4-II-E cells [221]. Caspase-3 inhibitor prevented the effect of sulforaphane, while it was not inhibited by NAME, an inhibitor of NO synthase, in H4-II-E cells [221]. Sulforaphane-induced cell death and apoptosis partly result from activation of caspase-3 in the hepatoma cells. Overexpression of regucalcin had a suppressive effect on cell death and apoptosis induced by sulforaphane in H4-II-E cells [221]. The suppressive effect of regucalcin on sulforaphane-induced cell death and apoptosis in H4-II-E cells may be partly involved in the molecules of Bax, cytichrome C, caspase, and Bcl-2. In addition, regucalcin may have an inhibitory effect on NO synthase and Ca^{2+}-dependent endonuclease activities in

H4-II-E cells [126]. Regucalcin has a suppressive effect on many signaling pathways that mediate apoptotic cell death.

As mentioned above, regucalcin has been shown to play a role in the regulation of cell death and apoptosis in H4-II-E cells [203, 206, 216]. Overexpression of regucalcin has been demonstrated to have a suppressive effect on apoptotic cell death induced by TNF-α, LPS, thapsigargin, Bay K 8644, dibucaine, or PD98059, an inhibitor of protein thyrosine kinase, insulin, or IGF-I in H4-II-E cells [203, 206, 216]. The signaling mechanisms that TNF-α, LPS, or other factors mediate cell death and apoptosis may be different. The suppressive effect of regucalcin on apoptotic cell death is related to its inhibitory effect on the activities of various protein kinases, NO synthase, caspase-3, or Ca^{2+}-dependent endonuclease, and its activating effect on Bcl-2. Regucalcin has a suppressive effect on many signaling pathways that mediate cell death and apoptosis and that the protein suppresses cell death and apoptosis mediated through many different signaling pathways in H4-II-E cells.

Regucalcin Suppresses Apoptotic Cell Death in Kidney Cells

Regucalcin has been shown to express in the cloned normal rat kidney proximal tubular epithelial NRK52E cells and its expression is enhanced after hormonal stimulation [222]. Nuclear localizetion of regucalcin is enhanced after hormone stimulation in NRK52E cells [183]. The role of regucalcin in cell death and apoptosis is examined using NRK52E cells overexpressing regucalcin [223]. The number of wild-type cells was decreased with culture for 42-72 hours in the presence of TNF-α, LPS, Bay K 8644, or thapsigargin [223]. These effects were suppressed in the transfectants overexpressing regucalcin. DNA fragmentation induced after culture with LPS, Bay K 8644, or thapsigargin were prevented in the transfectants overexpressing regucalcin [223]. Thus, overexpression of regucalcin has a suppressive effect on apoptotic cell death induced by TNF-α, LPS, Bay K 8644, or thapsigargin in kidney NRK52E cells. The effect of regucalcin in suppressing apoptotic cell death may be mediated through its action on many intracellular signaling pathways in NRK52E cells.

Bcl-2 is a suppressor in apoptotic cell death [224]. Apaf-1 participates in activation of caspase-3 [225]. Akt-1 involves in survival signaling pathway for cell death [226]. Overexpression of regucalcin caused a remarkable elevation of Bcl-2 mRNA expression in NRK52E cells, and it slightly stimulated Akt-1 mRNA expression in the cells. Apaf-1, caspase-3, or G3PDH mRNA expressions were not significantly altered in the transfectants [223]. Presumably, the enhancement of Bcl-2 mRNA expression contributes to the suppression of apoptotic cell death in NRK52E cells overexpressing regucalcin. Regucalcin may play a role in the regulation of Bcl-2 gene expression in NRK52E cells.

TNF-α enhanced the expression of caspase-3 mRNA in NRK52E cells [223]. This enhancement was found to suppress in the transfectants [223], suggesting that the mechanism by which regucalcin suppresses TNF-α-induced cell death is partly related to the decrease in caspase-3 mRNA expression in transfectants.

The presence of LPS caused a significant decrease in Bcl-2 mRNA levels in NRK52E cells, suggesting that this decrease is partly related to LPS-induced cell death [223]. The enhancement of Bcl-2 mRNA expression induced after overexpression of regucalcin was also seen in the presence of LPS [223]. LPS-stimulated expression of Apaf-1 mRNA was suppressed after overexpression of regucalcin [223]. This may partly involve in the suppression of LPS-induced cell death in NRK52E cells overexpressing regucalcin.

Culture with Bay K 8644 or thapsigargin was found to cause a significant increase in caspase-3 mRNA levels in wild-type cells, indicating that the increased gene expression partly contributes to inducing apoptotic cell death [223]. This increase was completely prevented in the transfectants. Regucalcin may have a suppressive effect on caspase-3 mRNA expression enhanced by Bay K 8644 or thapsigargin in NRK52E cells. Thus, regucalcin was found to regulate the expression of Bcl-2, caspase-3, and Akt-1 mRNAs in the cloned normal rat kidney NRK52E cells. The change in protein levels, however, remains to be elucidated.

Toxic factors have been reported to induce renal failure due to stimulating apoptotic cell death [227]. Overexpression of regucalcin was found to have a suppressive effect on apoptotic cell death induced by various factors (including TNF-α, LPS, Bay K 8644, or thapsigargin) in NRK52E cells. Regucalcin may have a role as a suppressor in the development of apoptotic cell death in kidney proximal tubular epithelial cells. Presumably, regucalcin plays a physiological role in the maintenance of homeostasis of cellular response for cell stimulation.

Regucalcin Suppresses Cell Proliferation

The expression of regucalcin mRNA is stimulated in H4-II-E cells after culture with addition of serum (10% FBS) [50, 51], suugesting that regucalcin plays a role in the regulation of the proliferation of cells. Regucalcin has been demonstrated to have a role as suppressor in the enhancemence of proliferation of liver cells *in vitro*. This section is described the mechanism by which regucalcin has a suppressive effect on cell proliferation using H4-II-E cells and NRK52E cells.

Regucalcin Suppresses Protein Kinase Activity in Proliferating Cells

The role of endogenous regucalcin in the regulation of protein kinase activity in the proliferation of H4-II-E cells is shown [228]. H4-II-E cells were cultured for 6-72 hours in the presence of FBS (1 or 10%). The number of cells and protein kinase activity in the 5500*g* supernatant of cell homogenate was increased at 24 and 48 hours after the culture with FBS (1 or 10%); culture with 10% FBS had potent effect as compared with that of 1% FBS [228]. Culture with FBS produced an increase in protein kinase activity and a corresponding elevation of cell number in H4-II-E cells [228]. The increase in protein kinase activity of the 5500*g* supernatant of cell homogenate preceded a significant elevation of cell number, suggesting that serum factors (including growth factors, cytokines, and hormones) stimulate cell proliferation that is partly mediated through cascades of protein kinases.

Serum-stimulated protein kinase activity in H4-II-E cells was further enhanced in the presence of calmodulin or dioctanoylglycerol in the presence of calcium chloride, and this increase was inhibited in the presence of trifluoperazine, staurosporine, or genistein in the enzyme reaction mixture [228], indicating that Ca^{2+}/calmodulin-dependent protein kinase, protein kinase C, and protein tyrosine kinase are present in H4-II-E cells [228]. Various protein kinases may be involved in the enhancement of cell proliferation with serum stimulation.

The presence of anti-regucalcin monoclonal antibody in the enzme reaction mixture containing the 5500*g* supernatant of cell homogenate of H4-II-E cells with FBS caused a significant increase in protein kinase activity [228]. This effect was completely abolished after the addition of exogenous regucalcin, which has an inhibitory effect on the enzyme activity [228]. This finding indicates that endogenous regucalcin plays a suppressive role in the enhancement of protein kinase activity in the cytoplasm in H4-II-E cells with cell proliferation. The anti-regucalcin monoclonal antibody-increased protein kinase activity in H4-II-E cells was inhibited in the presence of trifluoperzine, staurosporine, or genistein, suggesting that endogenous regucalcin inhibits Ca^{2+}/calmodulin-dependent protein kinase, protein kinase C, or protein tyrosine kinase activities [228].

Serum stimulation may lead to an increase in cell proliferation that is partly mediated through cascade for various protein kinases in in H4-II-E cells [228]. Regucalcin may have a suppressive effect for overexpression of cell proliferation due to inhibiting various protein kinases in the cytoplasm and nucleus of H4-II-E cells. Presumably, regucalcin plays a role as suppressor protein in cell proliferation of normal liver cells and hepatoma cells.

Regucalcin Suppresses Protein Phosphatase Activity in Proliferating Cells

Regucalcin has been shown to have an inhibitory effect on protein phosphatase activity in the cytoplasm of rat liver [150, 151]. H4-II-E cells were cultured for 3 days in a medium containing serum (10% FBS). The subconfluent cells were used for the assay of protein phosphatase activity toward phosphotyrosine. Ca^{2+}/calmodulin-dependent protein tyrosine phosphatases were present in H4-II-E cells [229]. Regucalcin had an inhibitory effect on Ca^{2+}/calmodulin-dependent protein tyrosine phosphatase activity in the cells [229]. Culture with Bay K 8644, an agonist of Ca^{2+} channels, caused an elevation of protein tyrosine phosphatase in H4-II-E cells, whereas dibutyryl cyclic AMP had no effect [229]. Bay K 8644-induced increase in protein phosphatase activity was inhibited in the presence of TFP, indicating that this increase is mediated through Ca^{2+}/calmodulin in hepatoma cells [229]. The effect of antibody was enhanced in the presence of TFP [229]. This enhancement may result from an increase in endogenous regucalcin in H4-II-E cells, since the expression of regucalcin mRNA in H4-II-E cells is stimulated after culture with Bay K 8644 [229]. This finding may support the view that endogenous regucalcin, which is enhanced through Ca^{2+} signaling, has a suppressive effect on Ca^{2+}/calmodulin-activated protein phosphatase activity in proliferative cells. The presence of anti-regucalcin monoclonal antibody in the enzyme reaction mixture caused a remarkanle elevation of protein tyrosine phosphatase activity in the cell homogenate (5,500*g* supernatant) of H4-II-E cells cultured with serum addition (1 or 10% FBS) [230]. This elevation is completely depressed after the addition of regucalcin [230]. Endogenous regucalcin may have a suppressive effect on the enhancement of protein tyrosine phosphatase activity in the proliferative cells. There may be many protein phosphatases in hepatoma H4-II-E cells. The antibody-increased protein tyrosine phosphatase activity was also inhibited by okadaic acid or vanadate, which is an inhibitor of protein phosphatases [145], although cyclosporine A, an inhitor of calcineurin (protein phosphatase) [232], had no effect [230]. Endogenous regucalcin may also act on okadaic acid or vanadate-sensitive protein tyrosine phosphatases in H4-II-E cells.

Culture with serum addition (10% FBS) caused an increase in proliferation of H4-II-E cells and a corresponding elevation of protein tyrosine phosphatase activity in the cells [230]. This finding suggests that the augmentation of protein tyrosine phosphatase activity is partly involved in the proliferation of H4-II-E cells, although protein phosphatase activity toward phosphoserine and phosphothreonine was not raised in the proliferating cells cultured with serum addition [230]. Processes that are reversibly controlled by protein phosphorylation require not only a protein kinase but also a protein phosphatase [145, 231]. Target proteins are phosphorylated at specific sites by one or more protein kinases and these phosphoproteins are removed by specific protein phosphatase [232]. As mentioned above, regucalcin may plays an important role as a suppressor for the enhancement of cell proliferation due to inhibiting various protein kinases and protein phosphatases activities that are raised in the proliferation of H4-II-E cells.

Regucalcin Suppresses DNA Synthesis in Proliferating Cells

Endogenous regucalcin has a suppressive effect on the enhancemence of DNA synthesis in the nuclei of H4-II-E with cell proliferation [233]. Cells were cultured for 6-96 hours in medium containing FBS (1 or 10%). Cell number was increased between 24 and 96 hours; cell proliferation was markedly stimulated after culture with 10% FBS as compared with that of 1% FBS [233]. Nuclear DNA synthesis activity *in vitro* was elevated 6 hours after culture with 10% FBS and its elevation was remarkable at 12 and 24 hours after the culture [233]. The increase in nuclear DNA synthesis activity preceded an elevation of the number of the cloned rat hepatoma cells H4-II-E cells cultured with FBS (1 or 10%) [233].

The increase in nuclear DNA synthesis activity at 12 and 24 hours after culture with FBS was inhibited in the presence of PD98059, an inhibitor of MAP kinase, staurosporine, an inhibitor of protein kinase C, and TFP, an antagonist of Ca^{2+}/calmodulin-dependent protein kinase, in the reaction mixture [233]. The increase in nuclear DNA synthesis activity after serum stimulation may be partly mediated through action of various protein kinases in the nuclei of H4-II-E cells. However, serum-stimulated increase in nuclear DNA synthesis activity was not related to various protein phosphatases [233].

Regucalcin has been shown to transport in the nucleus of rat liver [38, 151] and it can inhibit nuclear DNA synthesis of normal rat liver [191, 192]. The presence of regucalcin in the reaction mixture caused a decrease in DNA synthesis activity in the nuclei of H4-II-E cells cultured with FBS [233]. This effect was not altered in the presence of various protein kinase inhibitions. Regucalcin has been demonstrated to inhibit the activity of various protein kinases in the nuclei of liver cells [190, 228]. The effect of regucalcin in decreasing nuclear DNA synthesis activity may be partly mediated through the pathway of various protein kinases in H4-II-E cells.

Nuclear DNA synthesis activity was increased in the presence of anti-regucalcin monoclonal antibody in the reaction mixture containing the nucleus of H4-II-E cells cultured for 24 hours with 10% FBS [233]. This elevation was inhibited after addition of various protein kinase inhibitors in the reaction mixture [233]. These findings support the view that endogenous regucalcin suppresses DNA synthesis activity through mechanism by which inhibits nuclear protein kinases, and that it has a suppressive effect on DNA synthesis activity in the nuclei of H4-II-E cells with proliferation.

The transcriptional activity for the regucalcin gene has been shown to enhance in H4-II-E cells cultured with Bay K 8644, an agonist of Ca^{2+} entry in cells, in the presence of 10% FBS [44, 51]. Culture with Bay K 8644 caused an increase in regucalcin levels in H4-II-E cells that were cultured in the presence of FBS (1 and 10%) [233]. In this case, nuclear DNA synthesis activity was not changed after culture with Bay K 8644 [233]. The presence of anti-regucalcin monoclonal antibody in the reaction mixture containing the nuclei of H4-II-E cells cultured with Bay K 8644 resulted in an elevation of nuclear DNA synthesis activity [233], suggesting that nuclear DNA synthesis activity is suppressed through endogenous regucalcin that is increased in the nuclei of H4-II-E cells cultured with Bay K 8644.

Thus, regucalcin may have a suppressive effect on the enhancemence of nuclear DNA synthesis activity in proliferative cells, and it may play a suppressive role for overexpression of cell proliferation. This was further supported in H4-II-E cells overexpressing endogenous regucalcin stably [234].

The regucalcin content of regucalcin/pCXN2-transfected cells used in this study was 19.7-fold as compared with that of the parental wild-type H4-II-E cells and pCXN2 vector-transfected cells (mock type) [234]. Regucalcin/pCXN2 vector-transfected cells (transfectants) were cultured for 72 hours in the presence of FBS (10%) [234]. Cell numbers and DNA synthesis activity in the transfectants were found to suppress as compared with those of wild and mock type, suggesting that overexpression of endogenous regucalcin has a suppressive effect on cell proliferation [234].

The presence of anti-regucalcin monoclonal antibody in the reaction mixture caused an increase in DNA synthesis activity in the nuclei obtained from wild-type H4-II-E cells, mock-type cells, and transfectants with overexpression of regucalcin [234]. However, the augmentation of nuclear DNA synthesis activity was remarkable in the transfectants [234]. This may support the view that endogenous regucalcin has a great suppressive effect on nuclear DNA synthesis activity.

The expression of regucalcin mRNA has been shown to stimulate through a Ca^{2+}-signaling mechanism in H4-II-E cells [41, 51]. Regucalcin is translocated to the nucleus of liver cells [38,152]. Regucalcin inhibits nuclear protein kinase and protein phosphatase activities [228, 229], which are involved in siganal transduction to the nucleus, and it causes an inhibition in nuclear DNA synthesis in proliferating liver cells [233]. Regucalcin may play a suppressive role for the overexpression of proliferation of liver cells.

Regucalcin Suppresses Cell Cycle-Realted Gene Expression in Proliferating Cells

Regucalcin has a suppressive effect on liver cell proliferation [228-230]. Whether regucalcin suppress cell cycle-realted genes is examined in proliferating cells [235]. H4-II-E cells (wild type) and stable regucalcin/pCXN2 tranafectants were cultured for 72 hours in a medium containing 10% FBS to obtain subconfluent monolayers. The proliferation of cells was suppressed in the transfectants cultured for 24 -72 hours. The proliferation of wild-type cells was inhibited when the cells were cultured for 72 hours in a medium containing an inhibitor of transcriptional activity or protein synthesis. Such an effect was not seen in the transfectants [235]. Regucalcin has a suppressive effect on cytosolic protein synthesis [166, 167] and nuclear RNA synthesis [194, 195] in rat liver. If overexpression of regucalcin inhibits protein and RNA synthesis in the transfectants, the inhibitors of transcriptional activity or protein synthesis may not have additional effect. This suggests that the effect of regucalcin in suppressing cell prpliferation is partly mediated through its suppressive effect on protein and RNA synthesis in the cells.

Regucalcin has an inhibitory effect on various protein kinases in rat liver cytosol and nucleus [138,143, 190, 228]. The proliferation of H4-II-E cells (wild type and transfectants) was inhibited in the presence of PD98059, dibucaine, staurosporine, or genistein, which is an inhibvitor of various protein kinases [235]. The effect of regucalcin in suppressing cell proliferation may be partly related to its inhibitory effect on MAP kinase, Ca^{2+}/calmodulin-dependent kinase, and protein tyrosine kinase in H4-II-E cells.

Wortmannin is known to have an inhibitory effect on PI3-kinase. The proliferation of H4-II-E cells (wild type) was inhibited in the presence of wortmannin, an inhibitor of PI3-kinase, or vanadate, an inhibitor of protein tyrosine phosphatase [235]. These effects were not observed in the transfectants [236]. Regucalcin may inhibit PI3-kinase and protein tyrosine phosphatase activities, and those inhibitory effects may partly contribute to the suppression of proliferation in H4-II-E cells.

Bay K 8644 is an agonist of calcium entry into cells. The proliferation of H4-II-E (wild type) was inhibited in the presence of Bay K 8644 [235]. This effect of Bay K 8644 was not seen in the transfectants [235], because regucalcin has a role in the maintenance of intracellular calcium homeostasis in many cell types [Review in Ref. 236].

Overexpression of regucalcin has been found to suppress the inhibitory effect of various factors, which induce cell-cycle arrest, on the proliferatuon of H4-II-E cells (wild type) [235]. The effect of roscovitine, a potent and selective inhibitor of the cyclin-dependent kinase cdc2, cdk2m, and cdk5 [237], or sulforaphane, which can induce G2/M phase cell cycle arrest [238], in inhibiting the proliferation of wild-type cells was not observed in the transfectants [235]. Sulforaphane with a higher concentration caused a decrease in cell number of the transfectant, suggesting that the chemical induces cell death and apoptosis. Butyrate induced an inhibiton of the proliferation of wild-type cells and transfectants. Roscovitine can arrest in G1 and accumulate in G2 of cell cycle [237]. Butyrate induces an inhibition of G1 progression [241].

The inhibitory effect of roscovitine or sulforaphane on cell proliferation may not be seen in transfectants, if regucalcin has a suppressive effect on the same pathway which roscovitine or sulforaphane has an inhibitory effect on cell proliferation. Presumably, regucalcin induces G1 and G2/M phase cell cycle arrest in H4-II-E cells.

The expression of p21 mRNA has been found to enhance in the transfectant overexpressing regucalcin, although cdc2a and chk2 (checkpoint-kinase 2) mRNA levels are not changed in the transfectants [236]. P21 is an inhibitor of cyclin-dependent kinases (cdk) [239]. Regucalcin may enhance p21 expression and it inhibits G1 progression in H4-II-E cells. It cannot exclude the possibility; however, that regucalcin directly inhibits cdk activity in the cells.

Overexpression of regucalcin suppressed the expression of IGF-I mRNA in H4-II-E cells [235]. IGF-I is a growth factor in cell proliferation. Regucalcin may have a suppressive effect on IGF-I expression in H4-II-E cells, and theis suppression of IGF-I expression leads to retardation of cell proliferation.

Regucalcin Modulates Tumor-Related Gene Expression in Proliferating Cells

It is known that c-*myc,* c-*fos*, c-*jun,* and Ha-*ras* are tumor stimulator genes [240]. *p53* and *Rb* are tumor suppressor genes and *c-src* is oncogene [241]. The expression of c-*myc*, Ha-*ras,* or c-*src* mRNAs was found to suppresse in the transfectant overexpressing regucalcin [242, 243]. The expression of *p53* and *Rb* mRNAs was markedly enhanced in the transfectants overexpressing regucalcin [243]. Presumably, the suppression of c-*myc*, Ha-*ras* and *c-src* mRNAs expressions and the enhancement of *p53* and *Rb* mRNAs expression in the transfectants overexpressing regucalcin is partly involved in the retardation of proliferation of hepatoma H4-II-E cells.

The mechanism by which regucalcin regulates the expression of genes related to tumor is unknown. Regucalcin has been shown to translocate into the nucleus of rat liver [38, 151], and the protein has an inhibitory effect on RNA synthesis in isolated rat liver nucleus [194, 195]. Regucalcin can bind DNA and modulates nuclear transcriptional activity [184]. Regucalcin may bind to promoter region of tumor-related genes and it may suppress the expression of tumor stimulator gene or stimulate the expression of tumor suppressor gene in H4-II-E cells overexpressing regucalcin. As the result, the proliferation of hepatoma cells overexpressing regucalcin may be suppressed.

The expression of regucalcin is reduced in the cloned rat hepatica H4-II-E cells as compared with that of normal rat liver interestingly [228], suggesting an involvement of regucalcin in the suppression of carcinogenesis. The down-regulation of regucalcin expression in liver cells may lead to stimulation of cell proliferation with alteration of various tumor-related gene expressions. Regucalcin may play an important role as suppressor in cell proliferation and carcinogenesis of liver cells.

Regucalcin Suppresses Cell Proliferation in Normal Kidney Cells

Regucalcin has also been shown to have a role in the regulation of the proliferation of rat kidney proximal tubular epithelial NRK52E cells [244]. The regucalcin content of regucalcin/pCXN2-transfected cells was about 21-fold as comoared with that of the parental wild-type cells. The enhancement in cell proliferation was suppressed in the trensfectants overexpressing regucalcin [244]. The decrease in cell number of NRK52E (wild type) cells after culture with butyrate, rescovitine, and sulforaphane, which is an inhibitor of cell cycle, was not observed in the transfectants, suggesting that regucalcin induces G1 and G2/M phase cell cycle arrest in NRK52E cells [244].

The inhibition of the proliferation of NRK52E cells induced by PD98059, staurosporine, or dibucaine that is an inhibitor of various protein kinase inhibitors was not seen in the transfectants [245-247]. The suppressive effect of regucalcin on cell proliferation may result from the inhibitory effect of regucalcin on various protein kinases that are involved in stimulation of cell proliferation. The inhibition in proliferation of NRK52E cells by wortmannin, an inhibitor of PI3-kinase, was not observed in the transfectants [248], suggesting that regucalcin inhibits PI3-kinase and that it partly contributes suppression of cell proliferation in NRK52E cells.

Bay K 8644 is an agonist of calcium entry into cells. The proliferation of NRK52E cells (wild type) was inhibited in the presence of Bay K 8644. Overexpression of regucalcin had a preventive effect on Bay K 8644-induced inhibition of cell proliferation [244]. Regucalcin has a role in the maintenance of intracellular calcium homeostasis in many cell types.

The effect of regucalcin on the gene expression of proteins that are related to cell proliferation and cell cycle has been examined. The expression of c-jun and chk2 (checkpoint-kinase 2) mRNAs was found to suppress in the transfectants [244]. The expression of p53 mRNA was enhanced in the transfectants, while the exptression of c-myc, c-fos, cdc2, and p21mRNA was not changed in the transfectants [244]. The decrease in c-jun and chk2 mRNA expressions may partly contribute to suppress cell proliferation induced in NRK52E cells overexpressing regucalcin. The expression of the tumor suppressor gene p53 mRNA, which was enhanced with overexpression of regucalcin, may have a partial role in the retardation of proliferation of NRK52E cells. Regucalcin has been shown to localize in the nucleus of NRK52E cells [249]. Regucalcin may have a suppressive effect on cell proliferation due to regulating many gene expressions that are related to cell proliferation in normal kidney NRK52E cells.

Regucalcin, moreover, has been shown to have a regulatory effect on gene expression of proteins that regulate calcium transport system in kidney cells. Overexpression of regucalcin has been found to have suppressive effects on the gene expression of L-type Ca^{2+} channel and calcium-sensing receptor (CaR), which regulate intracellular Ca^{2+} signaling in NRK52E cells [249]. Overexpression of regucalcin caused an increase in rat outer medullary K^{+} channel (ROMK) mRNA expression in NRK52E cells, while it did not have an effect on Na, K-ATPase and epithelial sodium channel (ENaC) mRNA expressions [249]. The expression of Type II Na-Pi cotransporter (NaPi-IIa) and angiotensinogen mRNAs was not changed in NRK52E cells overexpressing regucalcin [249], suggesting that regucalcin does not have effects on NaPi-IIa and angiotensinogen mRNA expressions in kidney NRK52E cells.

The blockade of calcium influx through L-type calcium channels has been shown to attenuate mitochondrial injury and apoptosis in hypoxia renal tubular cells [250]. The entry of calcium through L-type Ca^{2+} channels induces mitochondrial disruption and cell death [250]. CaR participates in the regulation of renal Ca^{2+} transport [251]. It is speculated that RC regulates intracellular Ca^{2+}-signaling pathway through its suppressive effect on L-type Ca^{2+} channel or CaR mRNA expression in the kidney proximal tubular epithelial cells.

The expression of regucalcin mRNA in NRK52E cells has been shown to enhance after the treatment of PTH [222], suggesting that regucalcin partly mediates cellular response for PTH in kidney cells. Overexpression of regucalcin did not attenuate the expression of L-type Ca^{2+} channel or CaR mRNAs, which is decreased after PTH treatment, in NRK52E cells. Regucalcin decreased L-type Ca^{2+} channel or CaR mRNA expressions in NRK52E cells. Regucalcin may partly contribute as a mediator in cellular response for stimulation of PTH in NRK52E cells.

Whether handling of calcium in NRK52E cells is changed in the transfectants overexpressing regucalcin is unknown. Overexpression of regucalcin has been shown to have the suppressive effects on apoptosis with culture of Bay K 8644 in NRK52E cells [223]. Presumably, the effects of regucalcin on gene expression are not mediated through change in calcium handling in the transfectants. Regucalcin has been shown to play a role in the regulation of intracellular Ca^{2+} transport; the protein activates Ca^{2+}-pumping enzymes (Ca^{2+}-ATPase) in the basolateral membranes, mitochondria, and microsomes in rat kidney cortex

[99-101]. Regucalcin may regulate intracellular Ca^{2+} homeostasis in kidney proximal tubular epithelial cells so that Ca^{2+} is passed through transcellular transport. Moreover, regucalcin was found to suppress the expression of L-type Ca^{2+} channel or CaR mRNAs in NRK52E cells, supporting the view that regucalcin plays a physiological role in the regulation of intracellular Ca^{2+} homeostasis in kidney proximal tubular epithelial cells.

Overexpression of regucalcin has a suppressive effect on cell responses that are mediated through signaling process following stimulation with TNF-α or transforming growth factor-β1 (TGF-β1) in NRK52E cells [252]. Overexpression of regucalcin had a suppressive effect on apoptotic cell death induced by TNF-α or TGF-β1 that is mediated through caspase-3 in NRK52E cells [252]. Culture with TNF-α or TGF-β1 caused a remarkable increase in α-smooth muscle actin level in NRK52E cells. Such an increase was not seen in the transfectants. In addition, the expression of α-smooth muscle actin was markedly suppressed in the transfectants cultured without TNF-α or TGF-β1. These findings demonstrate that overexpression of regucalcin has a suppressive effect on the expression of α-smooth muscle actin in NRK52E cells cultured with TNF-α or TGF-β1, and that regucalcin regulates signaling pathway that is mediated through TNF-α or TGF-β1 to stimulate the expression of α-smooth muscle actin in NRK52E cells. TGF-β1 is a key mediator that regulates transdifferentiation of NRK52E cells into myofibroblasts expressing α-smooth muscle actin [253]. This may contribute to renal fibrosis associated with overexpression of TGF-β1 within the diseased kidney [253]. Regucalcin may regulate transdifferentiation to renal fibrosis in NRK52E cells with TGF-β1 or TNF-α.

Overexpression of regucalcin caused a remarkable increase in the expression of mRNA of Smad 2, which is involved in signal transduction of TGF-β1 [258], or NF-κB, which is related to signaling of TNF-α [259], in NRK52E cells. Such an increase was not seen in Smad 3 mRNA expression in the transfectants.

This finding suggests that regucalcin stimulates the gene expression of Smad 2 or NF-κB, which is related to signaling mechanism of TNF-α or TGF-β1. Regucalcin may have a suppressive effect on signaling pathway by which TNF-α or TGF-β1 stimulates gene expression of NF-κB or Smad 2 in NRK52E cells. These cytokines may not have enhancing effects on the gene expression of NF-κB or Smad 2 in the transfectants. Presumably, the suppressive effects of regucalcin on apoptotic cell death and α-smooth muscle actin expression may be not involved in the expressions of NF-κB or Smad 2 that is stimulated by TNF-α or TGF-β1 in NRK52E cells.

As mentioned above, regucalcin has been demonstrated to have a suppressive effect on hepatoma H4-II-E cella and normal kidney NRK52E cells. The suppressive effect of regucalcin on cell proliferation is related to its inhibitory effect on the activities of various protein kinases and protein phosphatases, calcium-dependent signaling factors, protein synthesis, nuclear DNA and RNA synthesis, and IGF-I expression and its activating effect on p21, an inhibitor of cell cycle-related protein kinases. Moreover, regucalcin has been shown to suppress the expressions of c-*myc*, Ha-*ras* and *c-src* mRNAs and to enhance the expressions of of *p53* and *Rb* mRNAs, which are related tumorigenesis of liver cells. *p53* is also known to stimulate p21 mRNA expression to induce cell-cycle arrest. Regucalcin has a suppressive effect on many intracellular signaling pathways that are related to cell proliferation after hormonal stimulation as summarized in Figure 1. The suppressive effect of regucalcin on apoptotic cell death and cell proliferation is based on the mechanism by which the protein inhibits cellular events that are mediated through many intracellular signaling factors.

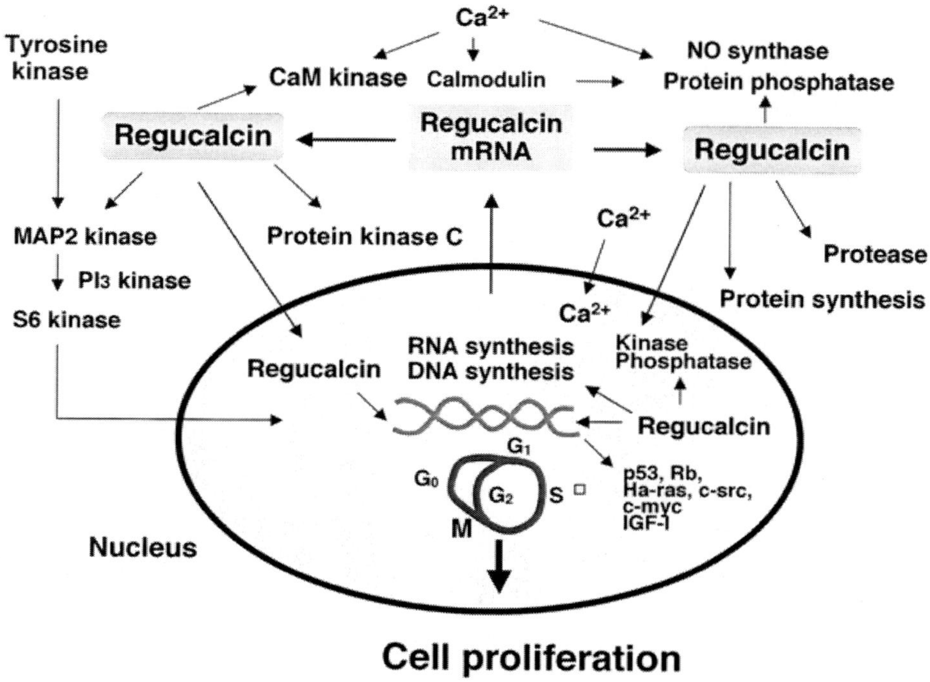

Figure 1. Regucalcin has a suppressive effect on the enhancement of cell proliferation. Regucalcin mRNA expression is stimulated through the pathway of signaling mechanism concerning Ca^{2+}/calmodulin (CaM)-dependent protein kinase (CaM kinase), protein kinase C (C kinase), protein kinase A (A kinase), and tyrosine kinase due to hormal stimulation. Regucalcin inhibits the activities of various protein kinases and protein phosphatases in the cytoplasm and nucleus of cells, and it also can inhibit Ca^{2+}/calmodulin-dependent enzyme activity (including cyclic AMP phosphodiesterase, NO synthase, superoxide dismutase (SOD), and others. Cytoplasmic regucalcin translocates into nucleus. Regucalcin inhibits nuclear DNA and RNA synthesis. Regucalcin has an inhibitory effect on the expression of *c-myc*, Ha-*ras*, and c-*src* mRNAs, which are tumor-stimulator genes. Regucalcin also stimulates the expression of *p53* and *Rb* mRNAs that are tumor-suppressor genes. Moreover, regucalcin can inhibit protein synthesis and it can stimulate protein degradation. Regucalcin induces G1 and G2/M phase cell cycle arrest in cells. The suppressive effect of regucalcin on cell proliferation is mediated through regulating many signaling systems.

Regucalcin may play an important role as suppressor protein in the maintenance of homeostasis of cellular response for cell stimulation. Regucalcin has been demonstrated to have a suppressive effect on cell death and apoptosis and cell proliferation induced by stimulation of various factors in hepatoma H4-II-E cells and normal kidney NRK52E cells. Presumably, regucalcin plays a physiological role in maintaining cell homeostasis of cellular response for various stimulating factors. Regucalcin may be a key molecule as a suppressor protein in cell regulation.

REGUCALCIN SUPPRESSES CELL PROLIFERATION IN REGENERATING LIVER *IN VIVO*

Adult rat hepatocytes are normally quiescent *in vivo*. However, 20-30 hours after partial hepatectomy (about 70%) they undergo a synchronous wave of DNA synthesis and cell division and continue to divide until the original mass of the liver is regenerated 3-7 days late [256]. The liver weight at 1 day after hepatoectomy is increased about 50% of that of sham-operated rats, and it reached to the same levels as sham operated at 3 days after hepatectomy, indicating that the increase in cell proliferation is induced at 1 day after partial hepatectomy [257]. Regenerating liver is a good model of liver cell proliferation *in vivo*.

Hepatocyte growth factor, which can promote liver regenerating after partial hepatectomy [258], has been shown to increase calcium concentration in rat hepatocyes [259]. Calmodulin synthesis in regenerating rat liver is increased at 8 hours after partial hepatoectomy [260].

Regenerating rat liver has been shown to enhance Ca^{2+}-ATPase activity in the nuclei between 1 and 5 days after partial hepatectomy [257]. Liver nuclear Ca^{2+}-ATPase is related to Ca^{2+} uptake by the nuclei and the metal accumulates nuclear matrix [81, 84, 85]. The increase in the nuclear Ca^{2+}-ATPase activity in regenerating liver caused a corresponding augmentation of the nuclear calcium content [257]. Liver calcium content was increased between 1 and 5 days after partial hepatectomy, although the nuclear calcium content was a slight increase at 1 day after partial hepatectomy [257]. Calcium, which increased in the cytoplasm of regenerating liver, may be transported into the nuclei with the enhancement of nuclear Ca^{2+}-ATPase activity.

Protein kinase C and Ca^{2+}/calmodulin-dependent protein kinase exists in liver nuclei [191]. Nuclear Ca^{2+}-ATPase activity of regenerating rat liver was decreased after the addition of protein kinase inhibitor into the enzyme reaction mixture [257]. The increase in nuclear Ca^{2+}-ATPase activity in regenerating liver may be partly activated through Ca^{2+}-dependent kinases. DNA content in the nuclei of regenerating rat liver was increased from 1 day after partial hepatectomy, while the nuclear DNA fragmentation activity was decreased in regenerating liver [257]. The nuclear DNA fragmentation activity in regenerating liver was not altered in the presence of Ca^{2+} chelator (EGTA) [257]. Presumably, the increased nuclear calcium in regenerating liver is involved in the regulation of nuclear DNA synthesis rather than the activation of nuclear DNA fragmentation.

Liver regucalcin mRNA expression has been shown to enhance 1-5 days after hepatectomy as compared with that of sham-operated rats, which are at the time point during activation of liver proliferation with a longer time (5 days) [62]. The expression of regucalcin mRNA in regenerating liver has been suggested to mediate through Ca^{2+}/calmodulin in rat liver, since hepatocyte growth factor, which promotes liver regenerating after partial hepatectomy, has been shown to increase calcium concentration in rat hepatocytes [260]. The increase in regucalcin mRNA expression in regenerating liver was independent on the decomposition of mRNA [260]. The process that transcripts to regucalcin mRNA may be stimulated in regenerating liver. The increase in regucalcin may be related to the suppression of cell proliferation in regenerating rat liver.

The role of regucalcin in the regulation of the nuclear function of regenerating rat livers is examined. Protein kinase activity was elevated in the liver nuclei obtained at 6 to 48 hours after partial hepatectomy [190], suggesting that protein kinases, which are involved in Ca^{2+}

signaling, participate in the enhancement of nuclear phosphorylation in regenerating liver with proliferative liver cells. Regucalcin was found to have a suppressive effect on the enhancement of Ca^{2+}/calmodulin-dependent protein kinase and protein kinase C activities in the nuclei of regenerating rat livers [190].

The phosphorylation in the nuclei of regenerating rat livers was increased in the presence of anti-regucalcin antibody in the reaction mixture [190]. This stimulation was completely abolished in the presence of exogenous regucalcin [190], suggesting that endogenous regucalcin is involved in the suppression of protein kinase activity in the nuclei. The net increase in protein kinase activity after anti-regucalcin antibody addition was about two-fold in the nuclei of regenerating rat liver as compared with that of normal rat liver [190]. Endogenous regucalcin, which is enhanced in regenerating liver, may be translocated to the nucleus *in vivo*.

Protein phosphatase activity toward phosphotyrosine is also found in the liver nuclei, whereas the nuclear enzyme activity toward phosphoserine and phosphothreonine is only slightly [150, 151]. Protein phosphotyrosine phosphatase activity in the cytoplasm and nuclei of liver cells was increased in regenerating rat liver [150]. This increase was enhanced in the presence of anti-regucalcin monoclonal antibody in the enzyme reaction mixture from regenerating liver [150].

Regucalcin has an inhibitory effect on Ca^{2+}/calmodulin-dependent protein kinases, protein kinase C, and protein phosphatases [142, 143]. Presumably, regucalcin participates in the regulation of cellular function due to inhibiting processes of phosphorylation and dephosphorylation of various proteins in proliferating liver cells. This role may be important intensively in the suppression of over-proliferation of liver cells *in vivo*.

The suppressive effect of regucalcin on protein synthesis in regenerating rat liver is also shown [167]. Regenerating liver induced an increase in protein synthesis. This increase was enhanced in the presence of anti-regucalcin monoclonal antibody [167]. This elevation was completely prevented after the addition of regucalcin. Endogenous regucalcin has a suppressive effect on the enhancement of protein synthesis in regenerating rat liver with proliferating cells [167]. Regucalcin may have a role as an inhibitor in protein synthesis in proliferating liver cells *in vivo*.

Liver nuclear DNA synthesis has been shown to stimulate in regenerating rat liver [191, 192]. Liver nuclear DNA synthesis was markedly enhanced at 1 and 3 days after partial hepatectomy [191]. Regucalcin was found to have an inhibitory effect on nuclear DNA synthesis in regenerating rat liver [192]. This effect was not resulted from nuclear DNA hydrolysis, because regucalcin can inhibit Ca^{2+} activated DNA fragmentation in rat liver nuclei [192]. DNA synthesis activity in the nucleus of regenerating rat liver was enhanced in the presence of anti-regucalcin monoclonal antibody in the reaction mixture [192]. This increase was completely abolished after the addition of regucalcin [192]. The endogenous regucalcin in liver nucleus may play a suppressive role in the enhancement of nuclear DNA synthesis activity in regenerating liver.

Nuclear functions in regenerating liver may be stimulated through many signaling factors. The increase in nuclear DNA synthesis activity in regenerating rat liver was prevented in the presence of strurosporine, TFP, and PD98059, which are various protein kinase inhibitors, in the reaction mixture [192]. This suggests that the nuclear DNA synthesis activity in regenerating liver with proliferating cells is partly stimulated through signaling pathway, which is related to protein kinase C, Ca^{2+}/calmodulin-dependent protein kinase, and

MAPK kinase in the nucleus. The effect of anti-regucalcin monoclonal antibody in enhancing nuclear DNA synthesis in regenerating liver was blocked in the presence of strurosporine, TFP, and PD98059 [192]. This finding suggests that the suppressive effect of regucalcin on nuclear DNA synthesis activity in regenerating liver may be partly mediated through the inhibitory action of regucalcin on various protein kinases. In addition, regucalcin may directly inhibit nuclear DNA synthesis activity in liver cells.

Regucalcin has also been shown to inhibit RNA synthesis in the nuclei isolated from control rat liver [194] and from regenerating rat liver [195]. The presence of anti-regucalcin monoclonal antibody in the reaction mixture caused a significant increase in RNA synthesis activity in the nuclei of normal rat liver [195]. This increase was completely abolished after the addition of regucalcin. The effect of anti-regucalcin monoclonal antibody in increasing nuclear RNA synthesis activity in regenerating rat liver was enhanced in the presence of the antibody in the reaction mixture [195].

The enhancement of nuclear RNA synthesis activity in regenerating rat liver is partly dependent on the activation of protein kinases and protein phosphatases, which are involved intracellular signaling pathway of hormone and growth factors [195]. The effect of anti-regucalcin monoclonal antibody in enhancing RNA synthesis activity in the nuclei of regenerating rat liver was inhibited in the presence of α-amanitin, PD98059, staurosporine, or vanadate in the reaction mixture [195]. The enhancement of nuclear RNA synthesis activity in regenerating liver is partly suppressed through signaling pathways that endogenous regucalcin inhibits RNA polymerase II and III, MAPK kinase, protein kinase C, and protein phosphatases in the liver nucleus. Regucalcin could inhibit RNA synthesis activity in the nuclei of normal and regenerating rat livers. Regucalcin may have a role as the transcription-related factor in liver nucleus. Whether regucalcin can directly bind to the promoter region in the gene is unknown, however. As mentioned above, regucalcin mRNA expression in liver cells is stimulated through the pathway of signaling mechanism concerning protein kinases and protein phosphatases that are increased in regenerating liver with proliferative cells *in vivo*, and endogenous regucalcin has a suppressive effect on protein, DNA and RNA synthesis, which are mediated through calcium siganaling, in the cells. Presumably, regucalcin plays a suppressive role in the overexpression of cell proliferation in vivo. Regucalcin may have an important role as a suppressor protein in the differentiation and proliferation of liver cells. Interestingly, new markers for the liver pre-neoplastic foci are investigated in rats treated with diethylnitrosamine and then 2-acetylaminofluorene combined with partial hepatectomy [261]. Transaldolase, rat aflatoxin B1 aldehyde reductase, and gamma-glutamylcysteine synthetase are found as up-regulated gene, and regucalcin is found as a down-regulated gene, in line with findings for hepatocellular carcinomas [261]. The suppression of regucalcin expression may play a role in the development of carcinogenesis in liver cells. If the chemical feeding-induced suppression of regucalcin expression influences on Ca^{2+}-signaling regulation in liver cells, it may generate a possible circumstance for carcinogenesis. In this aspect, the pathophysiological role of regucalcin in carcinogenesis with chemical feeding may be significant *in vivo*. These observations support the view that regucalcin plays a pivotal role as a suppressor in signaling pathways which are related to cell proliferation with carcinogenesis *in vivo*.

Conclusion

Regucalcin, which its gene is localized on the X chromosome in human, rat, and mouse, is thought as a protein that is highly differentiated because of a great conservation of the regucalcin genes throughout evolution in vertebrates specifies. Regucalcin plays a multifunctional role in cellular regulation: a role in maintaining intracellular Ca^{2+} homeostasis, an inhibitory role on activation of various Ca^{2+}-dependent enzymes, protein kinases and protein phosphatases, suppression of protein synthesis, and nuclear DNA and RNA synthesis, which are mediated through various signaling systems in many type cells *in vitro*. Moreover, regucalcin suppresses an enhancement of cell proliferation and apoptotic cell death that are induced by various signaling factors. Regucalcin has been proposed to play a pivotal role as a suppressor protein of intracellular signaling system in maintaining cell homeostasis.

The regucalcin gene expression is upregulated to depress over-proliferation of liver cells in regenerating liver with partial hepatectomy of rat liver *in vivo*. Also, the expression of the regucalcin gene is downregulated in the development of carcinogenesis in liver cells *in vivo*. The suppressed regucalcin expression may lead to promotion of tumor cells, supporting the view that regucalcin may play a role as a suppressor protein in various signaling pathways which are related to carcinogenesis *in vivo*. Thus, regucalcin plays a pivotal role as a suppressor protein in cell signal transduction. This was finding with the first time.

References

[1] Cheung WY (1980) Calmodulin plays a pivotal role in cellular regulation. *Science* 202:19-27.

[2] Nishizuka Y (1986) Studies and perspectives of protein kinase C. *Science* 233: 305-31.

[3] Williamson JR, Cooper RK, Hoek JB (1981) Role of calcium in the hormonal regulation of liver metabolism. *Biochim. Biophys. Acta* 639: 243-295.

[4] Reinhart PH, Taylor WM, Bygrave FL (1984) The role of calcium ions in the mechanisms of action of α-adrenergic agonists in rat liver. *Biochem. J.* 223: 1-13.

[5] Kraus-Friedman N, Feng L (1996) The role of intracellular Ca^{2+} in the regulation of gluconeogenesis. *Metabolism* 48: 389-403.

[6] Yamaguchi M, Takei Y, Yamamoto Y (1975) Effect of thyrocalcitonin on calcium concentration in liver of intact and thyroparathyroidectomized rats. *Endocrinology* 96: 1004-1008.

[7] Yamaguchi M, Yoshida H (1985) Participation of Ca^{2+} channel in liver calcium regulation by calcitoninn in rats. *Acta Endocrinol.* 110: 239-243.

[8] Yamaguchi M, Yamamoto T (1978) Purification of calcium binding substance from soluble fraction of normal rat liver. *Chem. Pharm. Bull.* 26: 1915-1918.

[9] Yamaguchi M, Sugii K (1981) Properties of calcium-binding protein isolated from the soluble fraction of normal rat liver. *Chem. Pharm. Bull.* 29: 567-570.

[10] Yamaguchi M (1988). Physicochemical properties of calcium-binding protein isolated from rat liver cytosol: Ca^{2+}-induced conformational changes. *Chem. Pharm. Bull.* 36: 286-290, 1988.

[11] Yamaguchi M, Yoshida H (1985) Regulatory effect of calcium-binding protein isolated from rat liver cytosol on activation of fructose 1,6-diphosphatase by Ca^{2+}-calmodulin. *Chem. Pharm. Bull.* 33: 4489-4493.

[12] Yamaguchi M, Shibano H (1987) Calcium-binding protein isolated from rat liver cytosol reverses activation of pyruvate kinase by Ca^{2+}. *Chem. Pharm. Bull.* 35: 2025-2029.

[13] Yamaguchi M, Shibano H (1987) Effect of calcium-binding protein on the activation of phosphorylase *a* in rat hepatic particulate glycogen by Ca^{2+}. *Chem. Pharm. Bull.* 35: 2581-2584.

[14] Yamaguchi M, Shibano H (1987) Reversible effect of calcium-binding protein on the Ca^{2+}-induced activation of succinate dehydrogenase in rat liver mitochondria. *Chem. Pharm. Bull.* 35: 3766-3770.

[15] Yamaguchi M, Mori S (1988) Effect of Ca^{2+} and Zn^{2+} on 5'-nucleotidase activity in rat liver plasma membranes: Hepatic calcium-binding protein (regucalcin) reverses the Ca^{2+} effect. *Chem. Pharm. Bull.* 36: 321-325.

[16] Shimokawa N, Yamaguchi M (1993) Molecular cloning and sequencing of the cDNA coding for a calcium-binding protein regucalcin from rat liver. *FEBS Lett.* 327: 251-255.

[17] Shimokawa N, Isogai M, Yamaguchi M (1995) Specific species and tissue differences for the gene expression of calcium-binding protein regucalcin. *Mol. Cell. Biochem.* 143: 67-71.

[18] Misawa H, Yamaguchi M (2000) Transcript heterogeneity of the human gene for Ca^{2+}-binding protein regucalcin. *Int. J. Mol. Med.* 5: 283-287.

[19] Murata T, Yamaguchi M (1997) Molecular cloning of the cDNA coding for regucalcin and its mRNA expression in mouse liver: The expression is stimulated by calcium administration. *Mol. Cell. Biochem.* 173: 127-133.

[20] Misawa H, Yamaguchi M (2000) The gene of Ca^{2+}-binding protein regucalcin is highly conserved in vertebrate species. *Int. J. Mol. Med.* 6: 191-196.

[21] Yamaguchi M (2011) The transcriptional regulation of regucalcin gene expression. *Mol. Cell. Biochem.* 346:147-171.

[22] Shimokawa N, Matsuda Y, Yamaguchi M (1995) Genomic cloning and chromosomal assignment of rat regucalcin gene. *Mol. Cell. Biochem.* 151:157-163.

[23] Thiselton DL, McDowall J, Brandau O, Ramser J, d'Esposito F, Bhattacharga SS, Ross MT, Hardcastle AJ, Meindl A (2002) An integrated, functionally annotated gene map of the DXS8026-ELK1 internal on human Xp11.3-Xp11.23: Potential hotspot for neurogenetic disorders. *Genomics* 79: 560-572.

[24] Yamaguchi M, Makino R, Shimokawa N (1996) The 5'end seguences and exon organization in rat regucalcin gene. *Mol. Cell. Biochem.* 165: 145-150.

[25] Shimokawa N, Yamaguchi M (1992) Calcium administration stimulates the expression of calcium-binding protein regucalcin mRNA in rat liver. *FEBS Lett.* 305: 151-154.

[26] Yamaguchi M, Isogai M, Kato S, Mori S (1991) Immunohistochemical demonstration of calcium-binding protein regucalcin in the tissues of rats: The protein localizes in liver and brain. *Chem. Pharm. Bell.* 36: 1601-1603.

[27] Yamaguchi M, Isogai M (1993) Tissue concentration of calcium-binding protein regucalcin in rats by enzyme-linked immunoadsorbent assay. *Mol. Cell. Biochem.* 122: 65-68.

[28] Yamaguchi M, Nakajima R (2002) Role of regucalcin as an activator of sarcoplasmic reticulum Ca^{2+}-ATPase activity in rat heart muscle. *J. Cell. Biochem.* 86: 184-193.
[29] Yamaguchi M, Hamano T, Misawa H (2000) Expression of Ca^{2+}-binding protein regucalcin in rat brain neurons: Inhibitory effect on protein phosphatase activity. *Brain Res. Bull.* 52: 343-348.
[30] Yamaguchi M, Misawa Y, Uchiyama S, Morooka Y Tsurusaki Y (2002) Role of endogenous regucalcin in bone metabolism: Bone loss is induced in regucalcin transgenic rats. *Int. J. Mol. Med.* 10: 377-383.
[31] Yamaguchi M (1992) A novel Ca^{2+}-binding protein regucalcin and calcium inhibition. Regulatory role in liver cell function. In: Calcium Inhibition. K Kohama (ed.) *Japan Sci Soc Press*, Tokyo and CRC Press, Boca Raton pp. 19-41.
[32] Yamaguchi M (1998) Role of calcium-binding protein regucalcin in regenerating rat liver. *J. Gastroen. Hepatol.* 13 (Suppl.): S106-S112.
[33] Yamaguchi M (2000) Role of regucalcin in calcium signaling. *Life Sci.* 66: 1769-1780.
[34] Yamaguchi M (2000). The role of regucalcin in nuclear regulation of regenerating liver. *Biochem. Biophys. Res. Commun.* 276: 1-6.
[35] Yamaguchi M (2002) Impact of aging on calcium channels and pumps. In: Calcium Homeostasis and Signaling in Aging. MP Mattson (ed.) Elsevier, Amsterdam pp 47-65.
[36] Yamaguchi M (2012) Yamaguchi M: Role of regucalcin in brain calcium signaling: Involevement with aging. *Integr. Biol. (Camb)*, 4: 825-837.
[37] Yamaguchi M (2012) Regucalcin: Genomics, Cell Regulation and Disease. Nova Science Publishers, Inc., New York, USA.
[38] Tsurusaki Y, Misawa H, Yamaguchi M (2000) Translocation of regucalcin to rat liver nucleus: Involvement of nuclear protein kinase and protein phosphatase regulation. *Int. J. Mol. Med.* 6: 655-660.
[39] Chakraborti S, Bahnson BJ (2010) Crystal structure of human senescence marker protein 30: insights linking structural, enzymatic, and physiological functions. *Biochemistry 49*: 3436-3444.
[40] Murata T, Yamaguchi M (1998) Tissue-specific binding of nuclear factors to the 5'-flanking region of the rat gene for calcium-binding protein regucalcin. *Mol. Cell. Biochem.* 178: 305-310.
[41] Murata T, Yamaguchi M (1998) Ca^{2+} administration stimulates the binding of AP-1 factor to the 5'-flanking region of the rat gene for the Ca^{2+}-binding protein regucalcin. *Biochem. J.* 329: 157-163.
[42] Murata T, Yamaguchi M (1999) Promoter characterization of the rat gene for Ca^{2+}-binding protein regucalcin. Transcriptional regulation by signaling factors. *J. Biol. Chem.* 274: 1277-1285.
[43] Misawa H, Yamaguchi M (2000) Involvement of hepatic nuclear factor I binding motif in transcriptional regulation of Ca^{2+}-binding protein regucalcin gene. *Biochem. Biophys. Res. Commun.* 269: 270-278.
[44] Misawa H, Yamaguchi M (2000) Intracellular signaling factors-enhanced hepatic nuclear protein binding to TTGGC sequence in the rat regucalcin gene promoter: Involvement of protein phosphorylation. *Biochem. Biophys. Res. Commun.* 279: 275-281.

[45] Misawa H, Yamaguchi M (2002) Indentification of transcription factor in the promoter region of rat regucalcin gene: Binding of nuclear factor I-A1 to TTGGC motif. *J. Cell Biochem.* 84: 795-802.

[46] Misawa H, Yamaguchi M (2001) Molecular cloning and sequencing of the cDNA coding for a novel regucalcin gene promoter region-related protein in rat, mouse and human liver. *Int. J. Mol. Med.* 8: 513-520.

[47] Misawa H, Yamaguchi M (2002) Gene expression for a novel protein RGPR-p117 in various species: The stimulation by intracellular signaling factors. *J. Cell. Biochem.* 87: 188-193.

[48] Yamaguchi M, Misawa H, Ma ZJ (2003) Novel protein RGPR-p117: The gene expression in physiologic state and the binding activity to regucalcin gene promoter region in rat liver. *J. Cell. Biochem.* 88: 1092-1100.

[49] Murata T, Shinya N, Yamaguchi M (1997) Expression of calcium-binding protein regucalcin mRNA in the cloned human hepatoma cells (Hep G2): Stimulation by insulin. *Mol. Cell. Biochem.* 175: 163-168.

[50] Nakajima M, Murata T, Yamaguchi M (1999) Expression of calcium-binding protein regucalcin mRNA in the cloned rat hepatoma cells (H4-II-E) is stimulated through Ca^{2+} signaling factors: Involvement of protein kinase C. *Mol. Cell. Biochem.* 198: 101-107.

[51] Yamaguchi M, Nakajima M (1999) Involvement of intracellular signaling factors in the serum-enhanced Ca^{2+}binding protein regucalcin mRNA expression in the cloned rat hepatoma cells (H4-II-E). *J. Cell. Bichem.* 74: 81-89.

[52] Vinson CR, Sigler PB, McKnight SL (1989) Scissors-grip model for DNA recognition by a family of leucine zipper proeins. *Science* 246: 911-916.

[53] O'Shea EK, Rutkowski R, Stafford WF III, Kim PS (1989). Preferential heterodimer formation by isolated leucine zippers from fos and jun. *Science* 245: 646-648.

[54] Shimokawa N, Yamaguchi M (1993) Expresion of hepatic calcium-binding protein regucalcin mRNA is mediated through Ca^{2+}/calmoclulin in rat liver. *FEBS Lett.* 316: 79-84.

[55] Isogai M, Yamaguchi M (1995) Calcium administration increases calcium-binding protein regucalcin concentration in the liver of rats. *Mol. Cell. Biochem.* 143: 53-58.

[56] Yamaguchi M, Kurota H (1995) Expression of calcium-binding protein regucalcin mRNA in the kidney cortex of rats: The stimulation by calcium administration. *Mol. Cell. Biochem.* 146: 71-77.

[57] Yamaguchi M, Ueoka S (1998) Expression of calcium-binding protein regucalcin mRNA in fetal rat liver is stimulated by calcium administration. *Mol. Cell. Biochem.* 178: 283-287.

[58] Yamaguchi M, Kanayama Y, Shimokawa N (1994) Expression of calcium-binding protein regucalcin mRNA in rat liver is stimulated by calcitonin: the hormonal effect is mediated through calcium. *Mol. Cell. Biochem.* 136: 43-48.

[59] Yamaguchi M, Oishi K, Isogai M (1995) Expression of hepatic calcium-binding protein regucalcin mRNA is elevated by refeeding of fasted rats: Involvement of glucose, insulin and calcium as stimulating factors. *Mol. Cell. Biochem.* 142: 35-41.

[60] Yamaguchi M, Oishi K (1995) 17β-Estradiol stimulates the expression of hepatic calcium-binding protein regucalcin mRMA in rats. *Mol. Cell. Biochem.* 143: 137-141.

[61] Yamaguchi M, Kanayama Y (1995) Enhanced expression of calcium-binding protein regucalcin mRNA in regenerating rat liver. *J. Cell. Biochem.* 57: 185-190.

[62] Ueoka S, Yamaguchi M (1998) Sexual difference of hepatic calcium-binding protein regucalcin mRNA expression in rats with different ages: Effect of ovarian hormone. *Biol. Pharm. Bull.* 21: 405-407.

[63] Lotersztajn S, Hanoune J, Pecker F (1981) A high affinity calcium-stimulated magnesium-dependent ATPase in rat liver plasma membranes. Dependence on an endogenous protein activator distinct from calmodulin. *J. Biol. Chem.* 256:11209-11215.

[64] Chen K-M, Junger KD (1983) Calcium transport and phosphorylated intermediate of (Ca^{2+}-Mg^{2+})-ATPase in plasma membranes of rat liver. *J. Biol. Chem.* 258: 4404-4410.

[65] Yamaguchi M, Mori S, Kato S (1988) Calcium-binding protein regucalcin is an activator of (Ca^{2+}-Mg^{2+})-adenosine triphosphatase in the plasma membranes of rat liver. *Chem. Pharm. Bull.* 36: 3532-3539.

[66] Takahashi H, Yamaguchi M (1993) Regulatory effect of regucalcin on (Ca^{2+}-Mg^{2+})-ATPase in rat liver plasma membranes: comparison with the activation by Mn^{2+} and Co^{2+}. *Mol. Cell. Biochem.* 124: 169-174.

[67] Takahashi H, Yamaguchi M (1994) Activatory effect of regucalcin on (Ca^{2+}-Mg^{2+})-ATPase in rat liver plasma membranes: relation to sulfhydryl group. *Mol. Cell. Biochem.* 136:71-76.

[68] Takahashi H, Yamaguchi M (1997) Stimulatory effect of regucalcin on ATP-dependent calcium transport in rat liver plasma membranes. *Mol. Cell. Biochem.* 168: 149-153.

[69] Takahashi H, Yamaguchi M (1993) Regucalcin modulates hormonal effect on (Ca^{2+}-Mg^{2+})-ATPase activity in rat liver plasma membranes. *Mol. Cell. Biochem.* 125: 171-177.

[70] Takahashi H, Suzuki S, Yanaguchi M (1995) Stimulatory effect of hormonal signaling factors on (Ca^{2+}-Mg^{2+})-ATPase activity in rat liver plasma membranes: Cross talk with regucalcin. *Mol. Cell. Biochem.* 151: 1-7.

[71] Takahashi H, Yamaguchi M (1995) Increase of (Ca^{2+}-Mg^{2+})-ATPase activity in hepatic plasma membranes of rats administered orally calcium: The endogenous role of regucalcin. *Mol. Cell Biochem.* 144:1-6.

[72] Takahashi H, Yamaguchi M (1996) Enhancement of plasma membrane (Ca^{2+}-Mg^{2+})-ATPase activity in regenerating rat liver: Involvement of endogenous activating protein regucalcin. *Mol. Cell Biochem.* 162:133-138.

[73] Takahashi H, Yamaguchi M (1996) Activatory effect of regucalcin on hepatic plasma membrane (Ca^{2+}-Mg^{2+})-ATPase is impaired by liver injury with carbon tetrachloride administration in rats. *Mol. Cell. Biochem.* 158: 9-16.

[74] Yamaguchi M (1985) Mitocondrial uptake of $^{45}Ca^{2+}$ bound to calcium-binding protein isolated from rat liver cytosol. *Chem. Pharm. Bull.* 33: 3390-3394.

[75] Mori S, Yamaguchi M (1991) Calcium-binding protein regucalcin stimulates the uptake of Ca^{2+} by rat liver mitochondria. *Chem. Pharm. Bull.* 39: 224-226.

[76] Takahashi H, Yamaguchi M (2000) Stimulatory effect of regucalcin on ATP-dependent Ca^{2+} uptake activity in rat liver mitochondria. *J. Cell. Biochem.* 78: 121-130.

[77] Yamaguchi M, Mori S (1989) Activation of hepatic microsomal Ca^{2+}-adenosine triphophatasde by calcium-binding protein regucalcin. *Chem. Pharm. Bull.* 37: 1031-1034.

[78] Takahashi H, Yamaguchi M (1999) Role of regucalcin as an activator of Ca^{2+}-ATPase activity in rat liver microsomes. *J. Cell. Biochem.* 74: 663-669.

[79] Yamaguchi M, Mori S (1989) Effect of the calcium-binding protein regucalcin on the Ca^{2+} transport system in rat liver microsomes: The protein stimulates Ca^{2+} release. *Chem. Pharm. Bull.* 37: 3037-3041.

[80] Yamaguchi M, Mori S (1990) Regucalcin-induced Ca^{2+} release from rat liver microsomes: The effect is inhibited by heparin. *Chem. Pharm. Bull.* 38: 2305-2307.

[81] Nicotera P, McConkey DJ, Jones DP, Orrenius S (1989) ATP stimulates Ca^{2+} uptake and increases the free Ca^{2+} concentration in isolated rat liver nuclei. *Proc. Natl. Acad. Sci. USA* 86: 453-457.

[82] Bachs O, Carafolli E (1995) Calmodulin and calmodulin-binding proteins in liver cell nuclei. *J. Biol. Chem.* 262, 10786-10790.

[83] Csermely P, Schnaider T, Szanto I (1995) Signalling and transport through the nuclear membrane. *Biochim. Biophys. Acta* 1241: 425-452.

[84] Tsurusaki Y, Yamaguchi M (2000) Role of endogenous regucalcin in the regulation of Ca^{2+}-ATPase activity in rat liver nuclei. *J. Cell. Biochem.* 78: 541-549.

[85] Yamaguchi M (1992) Effect of calcium-binding protein regucalcin on Ca^{2+} transport system in rat liver nuclei: stimulation of Ca^{2+} release. *Mol. Cell. Biochem.* 113: 63-70.

[86] Fujita T, Inoue H, Kitamura T, Sato N, Shimosawa T, Maruyama N (1998) Senescence marker protein-30 (SMP30) rescues cell death by enhancing plasma membrane Ca^{2+}-pumping activity in Hep G2 cells. *Biochem. Biophys. Res. Commun.* 250:374-380.

[87] Van Os CH (1985) Transcellular calcium transport in intestinal and renal epithelial cells. *Biochim. Biophys. Acta* 906: 195-222.

[88] Taylor CW (1985) Calcium regulation in vertebrates: An overview. *Comp. Biochem. Physiol.* 82A: 249-255.

[89] Murata T, Yamaguchi M (1999) Binding of kidney nuclear proteins to the 5'flanking region of the rat gene for Ca^{2+}-binding protein regucalcin: Involvement of Ca^{2+}/calmodulin signaling. *Mol. Cell. Biochem.* 199: 35-40.

[90] Misawa H, Yamaguchi M (2001) Involvement of nuclear factor-1 (NF1) binding motif in the regucalcin gene expression of rat kidney cortex: The expression is suppressed by cisplatin administration. *Mol. Cell Biochem.* 219: 29-37.

[91] Kurota H, Yamaguchi M (1996) Steroid hormonal regulation of calcium-binding protein regucalcin mRNA expression in the kidney cortex of rats. *Mol. Cell. Biochem.* 155: 105-111.

[92] Shinya N, Kurota H, Yamaguchi M (1996) Calcium-binding protein regucalcin mRNA expression in the kidney cortex is suppressed by saline ingestion in rats. *Mol. Cell. Biochem.* 162: 139-144.

[93] Shinya N, Yamaguchi M (1997) Alteration in Ca^{2+}-ATPase activity and calcium-binding protein regucalcin mRNA expression in the kidney cortex of rats with saline ingestion. *Mol. Cell. Biochem.* 170: 17-22.

[94] Shinya N, Yamaguchi M (1998) Stimulatory effect of calcium administration on regucalcin mRNA expression is attenuated in the kidney cortex of rats ingested with saline. *Mol. Cell. Biochem.* 178: 275-281.

[95] Misawa H, Yamaguchi M (2001) Involvement of nuclear factor-I (NFI) binding motif in the regucalcin gene expression of rat kidney cortex: The expression is suppressed by cisplatin administration. *Mol. Cell. Biochem.* 219:29-37.

[96] Kurota H, Yamaguchi M (1995) Suppressed expression of calcium-binding protein regucalcin mRNA in the renal cortex of rats with chemically induced kidney damage. *Mol. Cell. Biochem.* 151: 55-60.

[97] Agus ZS, Chiu PJS, Goldberg M (1997) Regulation of urinary calcium-excretion in the rat. *Am. J. Physiol.* 232: F545-F549.

[98] Kennedy MB, Bennett MK, Erondu NE, Miller SG (1987) Calcium/calmodulin-dependent protein kinases. In: Calcium and Cell Function. Cheung WY (ed). Acadamic Press Inc, New York, pp 61-107

[99] Kurota H, Yamaguchi M (1997) Activatory effect of calcium-binding protein regucalcin on ATP-dependent calcium transport in the basolateral membranes of rat kidney cortex. *Mol. Cell. Biochem.* 169: 149-156.

[100] Kurota H, Yamaguchi M (1997) Regucalcin increases Ca^{2+}-ATPase activity and ATP-dependent calcium uptake in the microsomes of rat kidney cortex. *Mol. Cell. Biochem.* 177: 201-207.

[101] Xue JH, Takahashi H, Yamaguchi M (2000) Stimulatory effect of regucalcin on mitochondrial ATP-depemdemt calcium uptake activity in rat kidney cortex. *J. Cell. Biochem.* 80: 285-292.

[102] Langer GA (1992) Calcium and the heart: Exchange at the tissue, cell, and organelle levels. *FASEB J.* 6: 893-902.

[103] Yamaguchi M, Nakajima R (2002) Role of regucalcin as an activator of sarcoplasmic reticulum Ca^{2+}-ATPase activity in rat heart muscle. *J. Cell. Biochem.* 86: 184-193.

[104] Akhter T, Nakagawa T, Kobayashi A, Yamaguchi M (2007) Suppression of regucalcin mRNA expression in the hearts of rats administered with free radical compound: The administration-induced death is accelerated in regucalcin transgenic rats. *Int. J. Mol. Med.* 19:653-658.

[105] Thastrup O, Culler PJ, Drbbak BK, Hanley MR, Dawson AP (1990) Thapsigargin, a tumor promoter, discharges intracellular Ca^{2+} stores by specific inhibition of the endoplasmic reticulum Ca^{2+}-ATPase. *Proc. Natl. Acad. Sci. USA* 87: 2466-2470.

[106] Tada M, Kadoma M (1989) Regulation of the Ca^{2+} pump ATPase by cAMP-dependent phosphorylation of phospholamban. *Bioessays* 10: 157-163

[107] Akther T, Sawada N, Yamaguchi M (2006) Regucalcin increases Ca^{2+}-ATPase activity in the heart mitochondria of normal and regucalcin transgenic rats. *Int. J. Mol. Med.* 18:171-176.

[108] Dahan D, Spanier R, Rahaaminoff H (1991) The modulation of rat brain Na^{+}-Ca^{2+} exchange by K^{+}. *J. Biol. Chem.* 266: 2067-2075.

[109] MacDermott AB, Dale N (1986) Receptors, ion channels and synaptic potential underlying the intergrative actions of excitatory ami acids. *Trends Neurosic* 10: 280-284.

[110] Cambray-Deakin MA, Burgoyne RD (1992) Intracellular Ca^{2+} and N-methyl-D-aspartate-stimulated neuritogenesis in rat cerebellar granule cell cultures. *Dev. Brain. Res.* 66: 25-32.

[111] Treves S, De Mattei M, Lanfred M, Villa A, Green NM, MacLennan DH, Meldolesi J, Pozzan T (1994) Calreticulin is a candidate for a calsequestrin-like function in Ca^{2+}-storage compartment (calcitosomes) of liver and brain. *Biochem. J.* 271: 473-480, 1990.

[112] Hartman H, Eckert A, Muller WE: Disturbances of the neuronal calcium-homeostasis in the aging nervous system. *Life Sci.* 55: 2011-2018.

[113] Yamaguchi M, Hanahisa Y, Murata T (1999) Expression of calcium-binding protein regucalcin and microsomal Ca^{2+}-ATPase regulation in rat brain: Attenuation with increasing age. *Mol. Cell. Biochem.* 200: 43-49.
[114] Hanahisa Y, Yamaguchi M (1996) Characterrization of calcium accumulation in the brain of rats administered orally calcium: The significance of energy-dependent mechanism. *Mol. Cell. Biochem.* 158: 1-7.
[115] Hanahisa Y, Yamaguchi M (1997) Increase in calcium content and Ca^{2+}-ATPase activity in the brain of fasted rats: Comparison with different ages. *Mol. Cell Biochem.* 171: 127-132.
[116] Hanahisa Y, Yamaguchi M (2001) Decrease in Ca^{2+}-ATPase activity in the brain plasma membrane of rats with increasing age: Involvement of brain calcium accumulation. *Int. J. Mol. Med.* 7: 407-411.
[117] Hanahisa Y, Yamaguchi M (1999) Brain microsomal calcium accumulation in rats with increasing age: Involvement of thapsigargin sensitive Ca^{2+}-ATPase. *Int. J. Mol. Med.* 4: 627-631.
[118] Hanahisa Y, Yamaguchi M (1998) Increase of Ca^{2+}-ATPase activity in the brain microsomes of rats with increasing ages: Involvement of protein kinase C. *Brain Res. Bull.* 46: 329-332.
[119] Yamaguchi M, Takakura Y, Nakagawa T (2008) Regucalcin increases Ca^{2+}-ATPase activity in the mitochondria of brain tissues of normal and transgenic rats. *J. Cell. Biochem.* 104:795-804.
[120] Yamaguchi M, Sakurai T (1992) Reversible effect of calcium-binding protein regucalcin on the Ca^{2+}-induced inhibition of deoxyuridine 5'-triphosphatase activity in rat liver cytosol. *Mol. Cell. Biochem.* 110: 25-29.
[121] Hanahisa Y, Yamaguchi M (1999) Effect of calcium-binding protein on adenosine 5'-triphosphatase activity in the brain cytosol of rats of different ages: The inhibitory role of regucalcin. *Biol. Pharm. Bull.* 22: 313-316.
[122] Rasmussen J (1970) Cell communication, calcium ion, and cyclic adenosine monophosphate. *Science* 170: 404-412.
[123] Yamaguchi M, Tai H (1991) Inhibitory effect of calcium-binding protein regucalcin on Ca^{2+}/calmodulin-dependent cyclic nucleotide phosphodiesterase activity in rat liver cytosol. *Mol. Cell. Biochem.* 106: 25-30.
[124] Yamaguchi M, Kurota H (1997) Inhibitory effect of regucalcin on Ca^{2+}/calmodulin-dependent cyclic AMP phosphodiesterase activity in rat kidney cytosol. *Mol. Cell. Biochem.* 177: 209-214.
[125] Lowenstein CJ, Dinerman JL and Snyder SH: Nitric oxide: A physiologic messenger. *Ann. Intern. Med.* 120: 227-237, 1994.
[126] Izumi T, Tsurusaki Y, Yamaguchi M (2003) Suppressive effect of endogenous regucalcin on nitric oxide synthase activity in cloned rat hepatoma H4-II-E cells overexpressing regucalcin. *J. Cell. Biochem.* 89: 800-807.
[127] Yamaguchi M, Takahashi H, Tsurusaki Y: (2003) Suppressive role of endogenous regucalcin in the enhancement of nitric oxide synthase activity in liver cytosol of normal and regucalcin transgenic rats. *J. Cell. Biochem.* 88: 1226-1234.
[128] Ma ZJ, Yamaguchi M (2003) Regulatory effect of regucalcin on nitric oxide synthase activity in rat kidney cortex cytosol: Role of endogenous regucalcin in transgenic rats. *Int. J. Mol. Med.* 12:201-206.

[129] Ma ZJ, Yamaguchi M (2002) Suppressive role of endogenous regucalcin in the regulation of nitric oxide synthase activity in heart muscle cytosol of normal and regucalcin transgenic rats. *Int. J. Mol. Med.* 10:761-766.

[130] Tobisawa M, Yamaguchi M (2003) Inhibitory role of regucalcin in the regulation of nitric oxide synthase activity in rat brain cytosol: involvement of aging. *J. Neurol. Sci.* 209: 47-54.

[131] Tobisawa M, Yamaguchi M (2003) Role of endogenous regucalcin in brain function: Suppression of cytosolic nitric oxide synthase and nuclear protein tyrosine phosphatase activities in brain tissue of transgenic rats. *Int. J. Mol. Med.* 12:581-585.

[132] Van Remmen H, Williams MD, Guo Z, Estlack L, Yang H, Carlson EJ, Epstein CJ, Huang TT, Richaardson A (2001) Knockout mice heterozygous for Sod2 show alterations in cardiac mitocondrial function and apoptosis. *Am. J. Physol. Heart. Cric. Physiol.* 281: H1422-1432.

[133] Den Hartog GJ, Haenen GR, Boven E, van der Vijgh WJ and Bast A (2004) Lecithinized copper, zinc-superoxide dismutase as a protector against dexorubicin-induced cardiotoxicity in mice. *Toxicol. Appl. Pharmacol.* 194: 180-188.

[134] Adler A, Messina E, Sherman B, Wang Z, Huang H, Linke A, Hintze TH (2003) NAD(P)H oxidase-generated superoxide anion accounts for reduced control of myocardial O_2 consumption by NO in old Fisher 344 rats. *Am. J. physiol. Heart Circ. Physiol.* 285: H1015-1022.

[135] Fukaya Y, Yamaguchi M (2004) Regucalcin increases superoxide dismutase activity in rat liver cytosol. *Biol. Pharm. Bull.* 27: 1444-1446.

[136] Ichikawa E, Yamaguchi M (2004) Regucalcin increases superoxide dismutase activity in the heart cytosol of normal and regucalcin transgenic rats. *Int. J. Mol. Med.* 14: 691-695.

[137] Connelly PA, Sisk RB, Schulman H, Garrison JC (1987) Evidence for the activation of the multifunction Ca^{2+}/calmodulin-dependent protein kinase in response to hormones that increase intracellular Ca^{2+}. *J. Biol. Chem.* 262: 10154-10163.

[138] Mori S, Yamaguchi M (1990) Hepatic calcium-binding protein regucalcin decreases Ca^{2+}/calmodulin-dependent protein kinase activity in rat liver cytosol. *Chem. Pharm. Bull.* 38: 2216-2218.

[139] Kurota H, Yamaguchi M (1997) Inhibitory effect of regucalcin on Ca^{2+}/calmodulin-dependent protein kinase activity in rat renal cortexcytosol. *Mol. Cell. Biochem.* 177: 239-243.

[140] Hamano T, Hanahisa Y, Yamaguchi M (1999) Inhibitory effect of regucalcin on Ca^{2+}-dependent protein kinase activity in rat brain cytosol: Involvement of endogenous regucalcin. *Brain Res. Brain* 50: 187-192.

[141] Hamano T, Yamaguchi M (2001) Inhibitory role of regucalcin in the regulation of Ca^{2+}-dependent protein kinase activity in rat brain tissues. *J. Neurol. Sci.* 183: 33-38.

[142] Omura M, Yamaguchi M (1998) Inhibition of Ca^{2+}/ calmodulin-dependent phosphatase activity by regucalcin in rat liver cytosol: Involvement of calmoduling binding. *J. Cell. Biochem.* 71: 140-148.

[143] Yamaguchi M, Mori S (1990) Inhibitory effect of calcium-binding protein regucalcin on protein kinase C activity in rat liver cytosol. *Biochem. Med. Metab. Biol.* 43: 140-146.

[144] Kurota H, Yamaguchi M (1998) Inhibitory effect of calcium-binding protein regucalcin on protein kinase C activity in rat renal cortex cytosol. *Biol. Pharm. Bull.* 21: 315-318.

[145] Hunter T (1995) Protein kinases and phosphatases: the Yin and Yang of protein phosphorylation and signaling. *Cell* 80: 225-236.

[146] Wang Y, Santini F, Qin K, Huang CY (1995) A Mg^{2+}-dependent, Ca^{2+}-inhibitable seruin/throsine protein phosphatase from bovine brain. *J. Biol. Chem.* 270: 25607-25612.

[147] Pallen CJ and Wang JH: Calmodulin-stimulated dephosphorylation of *p*-nitrophenylphosphate and free phosphotyrosine by calcineurin. *J. Biol. Chem.* 258: 850-855, 1983.

[148] Omura M, Yamaguchi M (1999) Effect of anti-regucalcin antibody on neutral phosphatase activity in rat liver cytosol: Involvement of endogenous regucalcin. *Mol. Cell. Biochem.* 197: 25-29.

[149] Omura M, Kurota H, Yamaguchi M (1998) Inhibitory effect of regucalcin on Ca^{2+}/calmodulin-dependent phosphatase activity in rat renal cortex cytosol. *Biol. Pharm. Bull.* 21: 440-443, 1998.

[150] Omura M, Yamaguchi M (1999) Enhancement of neutral phosphatase activity in the cytosol and nuclei of regenerating rat liver: Role of endogenous regucalcin. *J. Cell. Biochem.* 73: 332-341.

[151] Omura M and Yamaguchi M: Regulation of protein phosphatase activity by regucalcin localization in rat liver nuclei. *J. Cell. Biochem.* 75: 437-445, 1999.

[152] Morooka Y, Yamaguchi M (2001) Suppressive role of endogenous regucalcin in the regulation of protein phosphatase activity in rat renal cortex cytosol. *J. Cell. Biochem.* 81: 639-646.

[153] Morooka Y, Yamaguchi M: (2001) Inhibitory effect of regucalcin on protein phosphatase activity in the nuclei of rat kidney cortex. *J. Cell. Biochem.* 83: 111-120.

[154] Morooka Y, Yamaguchi M (2002) Endogenous regucalcin suppresses the ehmancement of protein phosphatase activity in the cytosol and nucleus of kidney cortex in calcium-administered rats. *J. Cell. Biochem.* 85: 553-560.

[155] Molkentin JD, Lu JR, Antos CL, Markham B, Richardson J, Robbins, Grant SR, Olson EN (1998) A calcineurin-dependent transcriptional pathway for cardiac hypertrophy. *Cell* 17: 215-228.

[156] Ichikawa E, Tsurusaki Y, Yamaguchi M (2004) Suppressive effect of regucalcin on protein phosphatase activity in the heart cytosol of normal and regucalcin transgenic rats. *Int. J. Mol. Med.* 13: 289-293.

[157] Hamano T, Yamaguchi M (1999) Inhibitory effect of regucalcin on Ca^{2+}/calmodulin-dependent protein phosphatase activity in rat brain cytosol. *Int. J. Mol. Med.* 3: 615-619.

[158] Yamaguchi M, Hamano T, Misawa H (2000) Expression of Ca^{2+}- binding protein regucalcin in rat brain neurons: Inhibitory effect on protein phosphatase activity. *Brain Res. Bull.* 52: 343-348.

[159] Tobisawa M, Tsurusaki Y, Yamaguchi M (2003) Decrease in regucalcin level and enhamcement of protein tyrosine phosphatase activity in rat brain microsones with increasing age. *Int. J. Mol. Med.* 12: 577-580.

[160] Tobisawa M, Yamaguchi M (2003) Suppressive effect of endogenous regucalcin on protein tyrosine phosphatase activity in the nucleus of rat brain: Attenuation with increasing age. *Int. J. Mol. Med.* 11: 205-210.

[161] Brostrom CO, Bocckino SB, Brostrom MA, Galuska EM (1983) Identification of a Ca^{2+} requirement for protein synthesis in eukaryotic cells. *J. Biol. Chem.* 258:14390-14399.

[162] Brostrom CO, Bocckino SB, Brostrom MA, Galuska EM (1986) Regulation of protein synthesis in isolated hepatocytes by calcium-mobilizing hormones. *Mol. Pharmacol.* 29:104-111.

[163] Menaya J, Parvilla R, Ayuso MS (1988) Effect of vasopressin on the regulation of protein synthesis initiation in liver cells. *Biochem.* 254:773-779.

[164] Thomas AP, Alexander J, Williamson JR (1984) Relationship between inositol polyphosphate production and the increase of cytosolic free Ca^{2+} induced by vasopressin in isolated hepatocytes. *J. Biol. Chem.* 259:5574-5584.

[165] Berthon B, Binet A, Mauger J-P, Claret M (1984) Cytosolic free Ca^{2+}in isolated rat hepatocytes as measured by quin 2. Effects of noradrenalin and vasopressin. *FEBS Lett.* 167:19-24.

[166] Yamaguchi M, Mori S (1990) Effect of calcium-binding protein regucalcin on hepatic protein synthesis: Inhibition of aminoacyl-tRNA synthetase activity. *Mol. Cell. Biochem.* 99:25-32.

[167] Tsurusaki Y, Yamaguchi M (2000) Suppressive effect of endogenous regucalcin on the enhancement of protein synthesis and aminoacyl-tRNA synthetase activity in regenerating rat liver. *Int. J. Mol. Med.*, 6:295-299.

[168] Wang KK, Villalobo A, Ronfogalis BD (1989) Calmodulin-binding protein as calpain substrates. *Biochem. J.* 262:693-706.

[169] Pontremoli S, Melloni E, Salamino F, Sparator B, Michetti M, Horecker BL (1984) Cytosolic Ca^{2+}-dependent neutral proteinases from rabbit liver: Activation of the proenzymes by Ca^{2+} and substrate. *Proc. Natl. Acad. Sci. USA* 81:53-56.

[170] Yamaguchi M, Tai H (1992) Calcium-binding protein regucalcin increases calcium-independent proteolytic activity in rat liver cytosol. *Mol. Cell Biochem.*, 12:89-95.

[171] Valle B, Williams RJP (1968) Metalloenzymes: the entatic nature of their active sites. *Proc. Nsatl. Acad. Sci. USA* 59:498-505.

[172] Yamaguchi M, Nishina N (1995) Characterization of regucalcin effect on proteolytic activity in rat liver cytosol: relation to cysteinyl-proteases. *Mol. Cell. Biochem.*, 148:67-72.

[173] Mellgren RI (1987) Calcium-dependent proteases: an enzyme system active at cellular membranes? *FASEB J.* 1:110-115.

[174] Rosser BG, Powers SP, Gores GJ (1993) Calpain activity increases in hepatocytes following addition of ATP. Demonstration by a novel fluorescent approach. *J. Biol. Chem.* 268:23593-23600.

[175] Baba T, Yamaguchi M (1999) Stimulatory effect of regucalcin on proteolytic activity in rat renal cortex cytosol: Invovement of thiol proteases. *Mol. Cell. Biochem.* 195: 87-92.

[176] Baba T, Yamaguchi M (2000) Stimulatory effect of regucalcin on proteolytic activity is impaired in the kidney cortex cytosol of rats with saline ingestion. *Mol. Cell. Biochem.* 206: 1-6.

[177] Bode W, Huber R (1992) Natural protein proteinase inhibitors and their interaction with proteinase. *Eur. J. Biochem.* 204:433-451.

[178] Croall DE, demartino GN (1991) Calcium-activated neutral protease (calpain) system: Structure, function, and regulation. *Physiol. Rev.* 71:813-847.
[179] Melloni E, Pontremoli S, Michetti M, Sacco O, Sparatore B, Horecker BL (1986) The involvement of calpain in the activation of protein kinase C in neutrophils stimulated by phorbol myristic acid. *J. Biol. Chem.* 261:4101-4105.
[180] Boynton AL, Whitfield JF, MacManus JP (1980) Calmodulin stimulates DNA synthesis by rat liver cells. *Biochem. Biohys. Res. Commun.* 95:745-749.
[181] Cruise J, Houck KA, Michalopoulos GK (1985) Induction of DNA synthesis in cultured rat hepatocytes through stimulation of alpha 1 adrenoreceptor by norepinephrine. *Science* 227:749-751.
[182] Pujol MJ, Soriano M, Alique R, Carafoli E, Bachs O (1989) Effect of alpha-adrenergic blockers on calmodulin associate with the nuclear matrix of rat liver cells during proliferative activation. *J. Biol. Chem.* 264:18863-18865.
[183] Nakagawa T, Yamaguchi M (2008) Nuclear localization of regucalcin is enhanced in culture with protein kinase C activation in cloned normal rat kidney proximal tubular epithelial NRK52E cells. *Int. J. Mol. Med.* 21:605-610.
[184] Tsurusaki Y, Yamaguchi M (2004) Role of regucalcin in liver nuclear function: Binding of regucalcin to nuclear protein or DNA and modulation of tumor-related gene expression. *Int. J. Mol. Med.* 14:277-281.
[185] Jones DP, McConkey DJ, Nicotera P, Orrenius S (1989) Calcium-activated DNA fragmentation in rat liver nuclei. *J. Biol. Chem.* 264: 6398-6403.
[186] Farber JL (1981) The role of calcium in cell death. Life Sci. 29: 1289-1295.
[187] Yamaguchi M, Sakurai T (1991) Inhibitory effect of calcium-binding protein regucalcin on Ca^{2+}-activated DNA fragmentation in rat liver nuclei. *FEBS Lett.*, 279:281-284.
[188] Moroianu J, Blobel G (1995) Protein export from the nucleus requires the GTPase Ran and GTP hydrolysis. *Proc. Natl. Acad. Sci. USA* 92:4318-4322.
[189] Tsurusaki Y, Yamaguchi M (2001) Suppressive effect of endogenous regucalcin guanosine triphosphatase activity in rat liver nucleus. *Biol. Pharm. Bull.,* 24:958-961.
[190] Katsumata T, Yamaguchi M (1998) Inhibitory effect of calcium-binding protein regucalcin on protein kinase activity in the nuclei of regenerating rat liver. *J. Cell. Biochem.* 71:569-576.
[191] Yamaguchi M, Kanayama Y (1996) Calcium-binding protein regucalcin inhibits deoxyribonucleic acid synthesis in the nuclei of regenerating rat *liver. Mol. Cell. Biochem.* 162:121-126.
[192] Tsurusaki Y, Yamaguchi M (2002) Suppressive role of endogenous regucalcin in the enhancement of deoxyribonucleic acid synthesis activity in the nucleus of regenerating rat liver. *J. Cell. Biochem.* 85:516-52.
[193] Morooka Y, Yamaguchi M (2002) Suppressive effect of endogenous regucalcin on deoxyribonucleic acid synthesis in the nuclei of rat renal cortex. *Mol. Cell. Biochem.* 229: 157-162.
[194] Yamaguchi M, Ueoka S (1997) Inhibitory effect of calcium-binding protein regucalcin on ribonucleic acid synthesis in isolated rat liver nuclei. *Mol. Cell Biochem.* 173: 169-175.
[195] Tsurusaki Y, Yamaguchi M (2002) Role of endogenous regucalcin in nuclear regulation of regenerating rat liver: Suppression of the enhanced ribonucleic acid synthesis activity. *J. Cell. Biochem.* 87: 450-457.

[196] Pardo JP, Fernandez F (1982) Effect of calcium and calmodulin on RNA synthesis in isolated nuclei from rat liver cells. *FEBS Lett.* 143:157-160.

[197] Sturges MR, Peck LJ (1994) Calcium-dependent inactivation of RNA polymerase III transcription. *J. Biol. Chem.* 269:5712-5719.

[198] Liu S, Shia D, Liu G, Chen H, Liu S, Hu Y (2000) Roles of Se and NO in apoptosis of hepatoma cells. *Life Sci.* 68:603-610.

[199] Abou-Elella AM, Siendones E, Padillo J, Montero JL, De la Meta M, Relat JM (2002) Tumour necrosis factor-alpha and nitric oxide mediate apoptosis by D-galactosamine in a primary culture of rat hepatocytes: Exacerbation of cell death by cocultured Kupper cells. *Can. J. Gastroenterol.* 16:791-799.

[200] Belloma G, Perotti M, Taddei F, Mirabelli F, Finardi G, Nicotera P, Orrenius S (1992) Tumor necrosis factor α induces apoptosis in mammary adenocarcinoma cells by an increase in intranuclear free Ca^{2+} concentration and DNA fragmentation. *Cancer Res.* 52: 1342-1346.

[201] Hukkanen M, Hughes FJ, Buttery LD, Gross SS, Evans TJ, Seddon S, Riveros-Moreno V, MacIntyre I, Polak JM (1995) Cytokine stimulated expression of inducible nitric oxide synthase by mouse, rat, and human osteoblast-like cells and its functional role in osteoblast metabolic activity. *Endocrinology* 136:5445-5453.

[202] Wink DA, Hanbauer I, Laval F, Cook JA, Krishna MC, Mifchell JB (1994) Nitric oxide protects against the cytotoxic effects of reactive oxygene species. *Ann. N. Y. Acad. Sci.* 738:265-278.

[203] Izumi T, Yamaguchi M (2004) Overexpression of regucalcin suppresses cell death in cloned rat hepatoma H4-II-E cells induced by tumor necrosis factor-α or thapsigargin. *J. Cell. Biochem.* 92: 296-306.

[204] Alikhani M, Alikhani Z, He H, Liu R, Popek BI, Graves DT (2003) Lipopolysaccharides indirectly stimulate apoptosis and global induction of apoptic genes in fibroblasts. *J. Biol. Chem.* 278:52901-52908.

[205] Nalan Y, Vereker E, Lynch AM, Lynch MA (2003) Evidence that lipopolysaccaride-induced cell death is mediated by accumulation of reactive oxygene species and activation of p38 in rat cortex and hippocampus. *Exp. Neurol.* 184:794-804.

[206] Izumi T, Yamaguchi M (2004) Overexpression of regucalcin suppresses cell death and apoptosis in cloned rat hepatoma H4-II-E cells induced by lipopolysaccharide, PD98059, dibucaine, or Bay K 8644. *J. Cell. Biochem.* 93: 598-608.

[207] Zelivianshi S, Spellman M, Kellerman M, Kakitelashvilli V, Zhou XW, Lugo E, Lee MS, Taylor R, Daris TL, Hauke R, Lin MF (2003) ERK inhibitor PD 98059 enhances docetaxel-induced apoptosis of androgen-independent human prostate cancer cells. *Int. J. Cancer* 107: 478-485.

[208] Arita K, Utsumi T, Kato A, Kanno T, Kobuchi H, Inoue B, Akiyama J, Utsumi M (2000) Mechanism of dibucaine-induced apoptosis inpromyelocytic leukemia cells (HL-60). *Biochem. Pharmacol.* 60: 905-915.

[209] Giuliano M, Bellacvia G, Lauricella M, D'Anneo A, Vassallo B, Vento R, Tesoriere G (2004) Staurosporine-induced apoptosis in Chang liver cells is associated with down-regulation of Bcl-2 and Bcl-XL. *Int. J. Mol. Med.* 13: 565-571.

[210] Balakuraman A, Campbell GA, Molsen MT (1996) Calcium channel blockers induces thymic apoptosis *in vivo* in rats. *Toxicol. Appl. Pharmacol.* 139: 122-127.

[211] Christensen SB, Andersen A, Kromann H, Treiman M, Tombal B, Denmeads S, Isaacs JT (1999) Thapsigargin analogues for targeting programmed death of androgen-independent prostate cancer cells. *Bioorg. Med. Chem.* 7: 1273-1280.

[212] Tombal B, Weeraratna AT, Denmeade SR, Isaacs JT (2000). Thapsigargin induces a calmodulin/calcineurin-dependent apoptotic cascade responsible for the death of prostatic cancer cells. *Prostate* 43: 303-317.

[213] Cohen JJ, Duke RC (1984) Glucocorticoid activation of a calcium-dependent endonuclease in thymocytes nuclei leads to cell death. *J. Immunol.* 132:38-42.

[214] Pereira M, Millot J-M, Sebille S, Manfait M (2002) Inhibitory effect of extracellular Mg^{2+} on intracellular Ca^{2+} dynamic changes and thapsigargin-induced apoptosis in human cancer MCF7 cells. *Mol. Cell. Biochem.* 229:163-171.

[215] Cano-Abad MF, Villarroya M, Garcia AG, Gabilan NH, Lopez MG (2001) Calcium entry through L-type calcium channels causes mitocondrial disruption and chromaffin cell death. *J. Biol. Chem.* 276: 39695-39704.

[216] Fukaya Y, Yamaguchi M (2005) Overexpression of regucalcin suppresses cell death and apoptosis in cloned rat hepatoma H4-II-E cells induced by insulin or insulin-like growth factor-I. *J. Cell. Biochem.* 96:145-154.

[217] Spinozzi F, Pagliacci MC, Migliorati G, Moraca R, Grignami F, Riccardi C, Nicoletti I (1994) The natural tyrosine kinase inhibitor genistein produces cell cycle arrest and apoptosis in Jurkat T-leukemia cells. *Leuk. Res.* 18:431-439.

[218] Gamet-Payrastre L, Li P, Lumeau S, Cassar G, Duport NA, Chevolleau S, Gasc N, Tulliez J, Terce F (2000) Sulforaphane, a naturally occurring isothiocyanate, induces cell cycle arrest and apoptosis in HT29 human colon cancer cells. *Cancer Res.* 60:1426-1433.

[219] Fimognari C, Nusse M, Cesari R, Iori R, Cantelli-Forti G, Hrelia P (2002) Growth inhibition, cell-cycle arrest and apoptosis in human T-cell leukemia by the isothiocyanate sulforaphane. *Carcinogenesis* 23:581-586.

[220] Singh AV, Xiao D, Lew KL, Dhir R, Singh SV (2004) Sulforaphane induces caspase-mediated apoptosis in cultured PC-3 human prostate cancer cells and retards growth of PC-3 xenografts in vivo. *Carcinogenesis 25*:83-90.

[221] Fukaya Y, Yamaguchi M (2005) Overexpression of regucalcin suppresses apoptotic cell death in the cloned rat hepatoma H4-II-E cells induced by a naturally occurring isothiocyanate sulforaphane. *Int. J. Mol. Med.* 15:853-857.

[222] Nakagawa T, Yamaguchi M (2005) Hormonal regulation on regucalcin mRNA expression in cloned normal rat kidney proximal tubular epithelial NRK52E cells. *J. Cell. Biochem.* 95:589-597.

[223] Nakagawa T, Yamaguchi M (2005) Overexpression of regucalcin suppresses apoptotic cell death in cloned normal rat kidney proximal tubular epithelial NRK52E cells: Change in apoptosis-related gene expression. *J. Cell. Biochem.* 96:1274-1285.

[224] Vogelstein B, Lane D, Levine AJ (2000) Surfing the p53 network. *Nature* 408:307 – 310.

[225] Zou H, Hanzel WJ, Liu X, Lutschg A, Wang X (1997) Apaf-l, a human protein homologous to C. elegans CED-4, participates in cytochrome c-dependent activation of caspase-3. *Cell* 90:405 – 413.

[226] Widmann C, Gibson S, Johnson GL. 1988. Caspase-dependent cleavage of signaling proteins during apoptosis. A turn-off mechanism for anti-apoptotic signals. *J. Biol. Chem.* 273:7141 – 7147.

[227] Dieguez-Acuna FJ, Polk WW, Ellis ME, Simmonds PL, Kushleika JV, Woods JS. (2004) Nuclear factor kappaB activity determines the sensitivity of kidney epithelial cells to apoptosis: Implications for mercury-induced renal failure. *Toxicol. Sci.* 82:1114-1123.

[228] Inagaki S, Yamaguchi M (2001) Suppressive role of endogenous regucalcin in the enhancement of protein kinase activity with proliferation of cloned rat hepatoma cells (H4-II-E). *J. Cell. Biochem.* (Supplement) 36: 12-18.

[229] Inagaki S, Misawa H, Yamaguchi M (2000) Role of endogenous regucalcin in protein tyrosine phosphatase regulation in the cloned rat hepatoma cells (H4-II-E). *Mol. Cell. Biochem.* 213: 43-50.

[230] Inagaki S, Yamaguchi M (2000) Enhancement of protein tyrosine phosphatase activity in the proliferation of cloned rat hepatoma H4-II-E cells: Suppressive role of endogenous regucalcin. *Int. J. Mol. Med.* 6:323-328.

[231] Cohen P, Cohen PTW (1989) Protein phosphatase come of age. *J. Biol. Chem.* 264:21435-21438.

[232] Mackintosh C, Malintosh RW (1997) Inhibitors of protein kinases and phosphatases. *Trends Biochem. Sci.* 19:444-448.

[233] Inagaki S, Yamaguchi M (2001) Regulatory role of endogenous regucalcin in the enhancement of nuclear deoxyribonucleic acid synthesis with proliferation of cloned rat hepatoma cells (H4-II-E). *J. Cell. Biochem.* 82: 704-711.

[234] Misawa H, Inagaki S, Yamaguchi M (2002) Suppression of cell proliferation and deoxyribonucleic acid synthesis in cloned rat hepatoma H4-II-E cells overexpressing regucalcin. *J. Cell. Biochem.* 84: 143-149.

[235] Yamaguchi M, Daimon Y (2005) Overexpression of regucalcin suppresses cell proliferation in cloned rat hepatoma H4-II-E cells: Involvement of intracellular signaling factors and cell cycle-related genes. *J. Cell. Biochem.* 95:1169-1177.

[236] Yamaguchi M (2005) Role of regucalcin in maintaining cell homeostasis and function (Review) *Int. J. Mol. Med.* 15:371-389.

[237] Meijer L, Borgne A, Mulner O, Chhong JP, Blow JJ, Inagaki N, Inagaki M, Delcros JG, Moulinoux JP (1997) Biochemical and cellular effects of roscovitine, a potent and selective inhibitor of the cyclin-dependent kinases cdc2, cdk2 and cdk5. *Eur. J. Biochem.* 243:527-536.

[238] Singh SV, Herman-Antosiewice A, Singh AV, Lew KL, Srivastava SK, Kamath R, Brown KD, Zhang L, Baskaran R (2004) Sulforaphane-induced G2/M phase cell cycle arrest involves checkpoint kinase 2-mediated phosphorylation of cell division cycle 25C. *J. Biol. Chem.* 279:25813-25822.

[239] Charollais RH, Buquet C, Mester J (1990) Butyrate blocks the accumulation of CDc2 mRNA in late G1 phase but inhibits both early and late G1 progression in chemically transformed mouse fibroblasts BP-A31. *J. Cell. Physiol.* 145:46-52.

[240] Curran T (1991) Fos and Jun: Intermediary transcription factors. In: Cohen P, Foulkes JG, editors. The hormonal control of gene transcription. New York: Elservier Science Publisher, pp. 295-308.

[241] Hulla JE, Schneider RP (1993) Structure of the rat *p53* tumor suppressor gene. *Nucleic. Acids Res.* 21:713-717.

[242] Tsurusaki Y, Yamaguchi M (2003) Overexpression of regucalcin modulates tumor-related gene expression in cloned rat hepatoma H4-II-E cells. *J. Cell. Biochem.* 90: 619-626.

[243] Tsurusaki Y, Yamaguchi M (2004) Role of regucalcin in liver nuclear function: Binding of regucalcin to nuclear protein or DNA and modulation of tumor-related gene expression. *Int. J. Mol. Med.* 14:277-281.

[244] Nakagawa T, Sawada N, Yamaguchi M (2005) Overexpression of regucalcin suppresses cell proliferation of cloned normal rat kidney proximal tubular epithelial NRK52E cells. *Int. J. Mol. Med.* 16:637-643.

[245] Dang ZC, Lowik CW (2004) Differential effects of PD98059 and UO126 on osteogenesis and adipogenesis. *J. Cell. Biochem.* 92:525-533.

[246] Tamaoki T, Nomoto H, Takahashi I, Kato Y, Morimoto M, Tomita E (1986) Staurosporine, a potent inhibitor of phospholipids/Ca^{2+}-dependent protein kinase. *Biochem. Biophys. Res. Commun.* 135:397-402.

[247] Vincenzi FF (1982) Pharmacology of calmodulin antagosism. In: Calcium Modulators. Godfranid T, Albertini A, Paoletti R (eds), Elservier Biomedical Press, Amsterdam, pp. 67-80.

[248] Park YC, Lee CH, Kang HS, Chung HT, Kim HD (1997) Wortmannin, a specific inhibitor of phosphatidylinositol-3-kinase, enhances LPS-induced NO production from murine peritoneal macrophages. *Biochem. Biophys. Res. Commun.* 240:692-696.

[249] Nakagawa T, Yamaguchi M (2006) Overexpression of regucalcin enhances its nuclear localization and suppresses L-type Ca^{2+} channel and calcium-sensing receptor mRNA expressions in cloned normal rat kidney proximal tubular epithelial NRK52E cells. *J. Cell Biochem.* 99:1064-1077.

[250] Tanaka T, Nangaku M, Miyata T, Inagi R, Ohse T, Ingelfinger JR, Fujita T (2004) Blockade of calcium influx through L-type calcium channels attenuates mitochondrial injury and apoptosis in hypoxic renal tubular cells. *J. Am. Soc. Nephrol.* 15:2320 – 2333.

[251] Ba J, Friedman PA. 2004. Calcium-sensing receptor regulation of renal mineral ion transport. *Cell Calcium* 35:229 – 237.

[252] Nakagawa T, Yamaguchi M (2007) Overexpression of regucalcin suppresses cell response for tumor necrosis factor-α or transforming growth factor-β1 in cloned normal rat kidney proximal tubular epithelial NRK52E cells. *J. Cell. Biochem.* 100:1178-1190.

[253] Fan JM, Ng Y-Y, Hill PA, Nikolic-Paterson DJ, Mu W, Atkins RC, Lan HY (199) Transforming growth factor-β regulates tubular epithelial-myofibroblast transdifferentiation in vitro. *Kidney Int.* 56:1455 – 1467.

[254] Zhang YQ, Kanzaki M, Furukawa M, Shibata H, Ozeki M, Kojima I (1999) Involvement of Smad proteins in the differentiation of pancreatic AR42J cells induced by activin A. *Diabetologia* 42:719 – 727.

[255] Hammar EB, Irminger J-G, Richenback K, Parnaud G, Ribaux P, Bosco D, Rouiller DG, Halban PA (2005) Activation of NF-kB by extracellular matrix is involved in apreading and glucose-stimulated insulin secretion of pancreatic beta cells. *J. Biol. Chem.* 280: 30630 – 30637.

[256] Higgins GM, Anderson RM (1931) Experimental pathology of the liver. Restoration of the liver of the white rat following partial surgical removal. *Arch. Pathol.* 12:186-202.
[257] Kanayama Y, Yamaguchi M (1995) Enhancement of nuclear Ca^{2+}-ATPase activity in regenerating rat liver: involvement of nuclear DNA increase. *Mol. Cell. Biochem.* 146:179-186.
[258] Higuchi O, Nakamura T (1991) Identification and change in the receptor for hepatocyte growth factor in rat liver after partial hepatectomy or induced hepatitis. *Biochem. Biophys. Res. Commun.* 176:599-607.
[259] Baffy G, Yang L, Michalopoulos GK, Williamson JR (1992) Hepatocyte growth factor induces calcium mobilization and inositol phosphate production in rat hepatocytes. *J. Cell. Physiol.* 153:332-339.
[260] Pinol MR, Berchtold MW, Backs O, Heizmann CW (1988) Increased calmodulin synthesis in the pre-replicative phase of rat liver regeneration. *FEBS Lett.* 231:445-450.
[261] Suzuki S, Asamoto M, Tsujimura K, Shirai T (2004) Specific differences in gene expression profile revealed by cDNA microarray analysis of glutathione S-transferase placental form (GST-P) immunohistochemically positive rat liver foci and surrounding tissue. *Carcinogenesis* 25: 439-443.

In: New Developments on Signal Transduction Research
Editor: Masayoshi Yamaguchi
ISBN: 978-1-62618-230-1

Chapter 7

ISOLATION AND PARTIAL CHARACTERIZATION OF REACTIVE OXYGEN SPECIES (ROS) RESISTANT MUTANTS WITH HIGH YIELDING IN *ORYZA SATIVA* CV KOSHIHIKARI

Kohji Hasunuma,[1,*] Yusuke Yoshida,[1] Haruna Matsuya,[1] Ken Nomura,[2] Osamu Miyoshi[2] and Md. Emdadul Haque[1]

[1]Kihara Institute for Biological Research, Yokohama City University, Maioka-cho, Totsuka-ku, Yokohama, Kanagawa, Japan
[2]Kanagawa Agricultural Technology Center, Kamiyoshizawa, Hiratsuka-shi, Kanagawa, Japan

ABSTRACT

By use of Paraquat (PQ) treatment during germination of pea seeds we obtained a high yielding mutants in the preceding works [Haque et al. (2008) Plant Biotechnol. Rep. 2: 21-31, Haque et al. (2010) Planta, 232:367-382]. The germinating seeds of *Oryza sativa* cv Koshihikari were treated in the presence of 4 μM of PQ, and we isolated PQ resistant mutants, R4-2-4 and R4-5-8 with high yielding. The two mutant lines, R4-2-4 and R4-5-8 showed significantly increased tiller numbers, ear numbers and ear weights and the leaves were deeper in green color than those of original wild type line. The mutant R4-2-4 and R4-5-8 lines showed resistant phenotype to PQ at the concentrations of 1-4 μM exhibiting good growth ability with green color of cotyledon. However, at the same concentration of PQ the seeds of original wild type showed germination with white cotyledon with no further growth ability. The leaves from both mutants were isolated and

* Corresponding author: Kohji Hasunuma. Professor Emeritus; Kihara Institute for Biological Research, Yokohama City university, 641-12 Maioka-cho, Totsuka-ku, Yokohama 244-0813, Japan. Chairman; Institute of High Yielding Crop, Co., Ltd. 438-7 Tsuruma, Machida-shi, Tokyo 194-0004, Japan. Tel: +81-42-795-1430; Fax: +81-42-799-6045; E-mail: waxy19orezt2@u08.itscom.nct; Mobilc: 080-2007-0162.

the crude extracts were prepared and these fractions were phosphorylated followed by the SDS-PAGE fractionation.

The increase in the phosphorylation bands in low molecular weights corresponding to the precursor of nucleoside diphospate kinase (NDPK) were observed in both mutant lines, and also increase in the phosphorylation bands in the higher molecular weights corresponding to high molecular weight histidine kinases could be detected in both of the mutant lines.

However, in the original wild type increased amounts of putative degradation product of phosphorylated precursor of NDPK could be detected. By the further fractionation of the soluble, chloroplast and mitochondrial fractions, the increase in the phosphorylation of protein bands corresponding to putative precursor of NDPK and higher molecular weight histidine kinases in the fractions of chloroplast could be observed. The nucleotide sequence analyses of the precursor of NDPK1, 2 and 3 showed no difference in the nucleotide sequence with those of original wild type line and mutant lines, R4-2-4 and R4-5-8. However, we detected that 78 amino acid sequence spanning in the transit peptide of Nipponbare NDPK2 was deleted in the Koshihikari transit peptide of NDPK2. The yield tests of original wild type and a mutant R4-2-4 with Kinuhikari as a reference line performed in Kanagawa Agricultural Technology Center demonstrated higher yielding of 1.11-, 1.12- and 1.35-fold than those of original line.

We estimate that higher molecular weight histidine kinase and the putative precursor of NDPKs in cytosol and on chloroplast may have capacity to function in the process of detoxification of reactive oxygen species including singlet oxygen ($^1O_2$).

Keywords: Histidine kinase Light signal transduction Oryza sativa cv Koshihikari Nucleoside diphosphate kinase Catalases Reactive oxygen species Singlet oxygen

ABBREVIATIONS

FAD	Flavin Adenine Dinucleotide
FMN	Flavin Mononucleotide
NDPK	Nucleoside diphosphate kinase
OsR424	R4-2-4 mutant line
OsR458	R4-5-8 mutant line
$^1O_2$	Singlet oxygen
PQ	Paraquat
ROS	Reactive oxygen species
SDS-PAGE	Sodium Dodesyl Sulphate-Polyacryl Amide Gel Electrophoresis

INTRODUCTION

In the processes to establish the molecular basis of light signal transduction by use of *Neurospora crassa* and *Pisum sativum* cv Alaska, we isolated PQ resistant mutants in *Neurospora crassa* [1] and also in *Pisum sativum* cv Alaska [2, 3]. Most of the PQ resistant mutants of *Neurospora* showed inability to form female sexual organ of protoperithecia. However, these mutants have capacity to function as males showing fertility to wild type protoperithecia. The results of the analysis of the progenies of this crossing of mutant conidia into wild type protoperithecia resulted in the abnormal phenomena that only mutant lines were recovered in the progenies. There is no progeny showing wild type phenotype, which indicates that in the process isolating PQ resistant mutant of *Neurospora crassa*, we obtained several of mutants showing altered mitochondrial inheritance from mother inheritance to father inheritance [1]. Fundamentally these PQ resistant mutants showed lack of formation of sexual organ, protoperithecia, which will support to produce ascospore, and therefore this process is an important process for the formation of ascospores. We also remind that the process of spore formation in *Bacillus subtillis*, high molecular weight histidine kinases and phosphate transferring processes are surely included in the process of spore formation [4]. These parallel relationship between histidine kinase and spore formation suggest that high molecular weight histidine kinases may also contribute to the process of ROS detoxification. We also isolated mutants showing resistant phenotype to ROS including singlet oxygen in Alaska pea [3]. The mutant, *R3-1* showed a phenotype to resist to ROS including singlet oxygen. The mutant included two mutations in the precursors of NDPK2 and NDPK3, which are directed to localize in chloroplast and in mitochondria as NDPK2 and NDPK3, respectively. The precursor of NDPK2 located in the cytosol and localized on the surface of chloroplast will have different function to control singlet oxygen emitted from chloroplast under strong sunlight. After excitation of chlorophyll in the chloroplast by strong sunlight the energy will be transferred to triplet oxygen forming singlet oxygen, which will have lifetime of 0.5 ~ 4 μsec indicating to make flight to outside of chloroplast. In the cytosol there are large amounts of catalases having capacity to capture singlet oxygen. We have demonstrated that C-terminal half of NDPK will interact with catalases by use of two hybrid methods and also native PAGE and gel filtration [3, 5, 6]. Histagged-NDPK-1 in *Neurospora crassa* had the capacity to bind NADH. However, Histagged-NDPK-1^{P72H} showed no capacity to bind NADH [7], which has very low activity to be autophosphorylated. These results indicated that strong sunlight will excite chlorophyll to exited states and transfer the energy to ambient triplet oxygen generated by photolysis of H_2O. The transfer of the energy from exited chlorophyll to triplet oxygen will make flight of singlet oxygen to outside of chloroplast separating singlet oxygen from proton. This process will be the first step in the process of photosynthesis. The singlet oxygen making flight of around 500 μm will be captured by catalases in the cytosol, which will make interaction with NDPK binding NADH [1, 3].

In ROS resistant high yielding mutant of *Pisum sativum* cv Alaska we detected two mutations in the transit peptide and C terminal part of NDPK2, Ile12Leu and Glu205Lys, which showed increased ability to be autophosphorylated and higher activity of nucleoside diphosphate kinase [3].

In the present work we have tried to isolate mutants by the mutagenesis of germinating seeds of *Oryza sativa* cv Koshihikari in the presence of PQ, and the isolated 2 lines of rice mutants resistant to PQ, which showed high yielding under strong sunlight. In Koshihikari NDPK2 localizing to chloroplast the transit peptide of it had large deletion in Koshihikari compared with Nipponbare, which provided whole DNA sequences in *Oryza sativa*. In the PQ resistant mutants the degradation of NDPK2 precursor seemed to be blocked.

Materials and Methods

Origin of Seeds

Seeds of *Oryza sativa* cv Koshihikari was obtained from Yokohama Nursery Co., Ltd. in Yokohama, Japan. The seeds were used throughout this work. *Oryza sativa* cv Kinuhikari was provided from Kanagawa Agricultural Technology Center.

Methods to Isolate ROS Resistant Mutants

Four sets of 50 seeds per Petri dish were prepared for each treatment sets of 0, 4, 8, 40 and 80 μM Paraquat in the sterilized MS medium at Aug. 2, 2005. Seeds of original wild type were sterilized by 25% NaClO for 10 min, and then the seeds were washed for 20 min by sterilized deionized water for 5-times. The sterilized seeds were placed in a refrigerator overnight, and then sterilized MS media (25 ml) with 0, 4, 8, 40 and 80 μM of Paraquat were added to each 4 plates including seeds, respectively. The Petri dishes were placed in a incubator at 8 °C with 160 μmol/m^2/sec of illumination under 8 h light and 16 h dark regime for 7 days. After the incubation period the germinating seeds were sown in a 2.5 litter of vermiculite bed soften by adding 2 litter of 0.2 % Hyponex in a plastick basin (24 cm wide, 29 cm length and 7 cm high). The 40 beds including 25 seeds were placed in a glass green house at 23 °C under sunlight. At Aug. 18, 2005 the growing seedlings were transplanted in a round bath (45 cm in diameter, with 17 cm high) with soil filled with water, which is shown in Figure 1. Four plants were transplanted per each bath and the bathes were placed in the field. In this experiment we obtained R4-2-4 mutant line as shown in Figure 1. In the second experiment started in Jan. 2006, we obtained mutant line, R4-5-8.

a b

Figure 1. Growth pattern of original wild type and the first step putative mutant R4-2-4 line just before emersion of ear. a) Original wild type with dispersed stems. b) Mutant R4-2-4 line with multiple condensed stems (right side). Under strong sunlight R4-2-4 leaves showed deeper green color than those of original wild type.

Characterization of R4-2-4 and R4-5-8 Mutant Lines

In 2006, we obtained putative homozygous mutant, R4-2-4 line, and at the early 2007, we obtained putative homozygous R4-5-8 line. At Mar. 2007 we sowed seeds in a plastic basin and at Apr. 16, 2007 we transplanted the rice seedlings at a density of 4 seedlings per a plastic bath with soil and water. At Oct. 30, 2007 we checked final characteristics of original wild type and mutant R4-2-4 and R4-5-8 lines. In the sample obtained only bathes with 4 plants were characterized to consider the mutual competition between plants. Four rice plants in one bath were counted for their characteristics and original wild type (9 bathes), R4-2-4 mutant line (23 bathes) and R4-5-8 (6 bathes). The data obtained on tiller numbers, ear numbers and total ear weights were treated by test of significance.

Assay for the Resistance to Paraquat

After sterilization of seeds from original line and mutant lines, 10 seeds were placed in a Petri dish and sterilized MS medium with various concentrations of Paraquat was added. The plates were placed in growth chamber at 7 °C for 3 days at a light intensity of 160 $\mu mol/m^2/sec$ under 24 h light and then placed in a glass green house at 23 °C. The characteristics of germinating seeds of original wild type and mutant R4-2-4 and R4-5-8 lines were analyzed.

Preparation of Crude Extracts and Soluble Fractions

Young leaves were cut out from plants grown in the field and immediately immersed in the liquid nitrogen. After chopping the leaves the leaves were well homogenized according to a method described in Haque et al. (2010) [3]. After obtaining crude fraction, the soluble fraction was obtained removing chloroplast and mitochondrial fractions.

Chloroplast and Mitochondrial Fractions

The fractionation and purification of chloroplast and mitochondria were performed according to a method described in Haque et al. (2010) [3].

Detection of Histidine Kinases in Crude Extracts

The crude extracts were prepared from the leaves of original wild type, and mutant lines, R4-2-4 and R4-5-8. Proteins of 6.25, 12.5, 25, 50 and 100 μg were mixed with 233 KBq [γ-^{32}P] ATP.

After mixing radioisotope the mixtures were kept on ice for 10 min and then SDS-sample buffer was added. The samples were loaded and subjected to SDS-PAGE.

Assay for Catalase Activity in Crude Extracts

The crude extracts were prepared from young leaves of original wild type and the mutants, R4-2-4 and R4-5-8, respectively according to the method described by Haque et al. (2010) [3].

In each of these lanes, 25 – 25 μg, 50 – 50 μg and 100 – 100 μg of crude extracts from the leaves of original wild type and the mutant R4-2-4 a), and original wild type and the mutant R4-5-8 b) were loaded and electrophoresis on native PAGE were performed. Catalase activities were assayed according to the method descrived by Haque et al. (20109 [3].

Isolation of Precursor of OsNDPK1, OsNDPK2 and OsNDPK3

The isolation methods for mRNA for precursors of OsNDPK1, OsNDPK2 and OsNDPK3 and cDNA for precursors of OsNDPK1, OsNDPK2 and OsNDPK3 were according to the methods described by Haque et al. (2010)[3].

In the text, prefix Os was omitted and described as NDPK1, NDPK2 and NDPK3

Yield Test in Kanagawa Agricultural Technology Center

The major parts of summarized report of yield test of original wild type and a mutant R4-2-4 with Kinuhikari as reference wild type line made by Ken Nomura of Kanagawa Agriculture Technology Center were reported directly with only the translation to English.

Results

Isolation of ROS Resistant Mutants of *Oryza Sativa* cv Koshihikari

By the results of two successive independent treatments of rice germinating seeds we isolated two different mutants R4-2-4 and R4-5-8. These mutants were used for the further analyses to determine the characteristics of these mutants. Starting at the middle of July 2005 we transplanted four seedlings of rice plants to plastic bathes with soil and water to check the growth pattern of original lines, and mutagenized lines, and at the end of Oct. 2005 we detected a mutant R4-2-4, when the mutant line showed apparent phenotype. As shown in Figure 1, the mutant line showed multiple tillering with very condensed stems, and the plant height is a little low compared with those of original lines. Although in the photograph the color difference is not so apparent, the mutant, R4-2-4 showed denser green color than those of original wild type lines. In the end of Autumn we started second try to isolate mutants of Koshihikari, and identified a mutant line, R4-5-8. These mutants, R4-2-4 and R4-5-8 were further grown in bathes and we identified putative homozygous mutants with very good tiller and ear numbers.

Analysis of Growth Characteristics, Tiller and Ear Numbers and Ear Weights in Original Wild Type and Mutant Lines

Seeds of original wild type and the mutant lines, R4-2-4 and R4-5-8, were sown after sterilization of seeds to vermiculite moistened with 0.2 % of Hyponex and placed in the glass green house at 23 °C in the middle of Mar. 2007. After growth of seedlings they were transplanted in bathes with soil and water, which were reused after harvesting the wild type and mutant lines. Four seedlings in original wild type and mutant R4-2-4 and R4-5-8 lines were transplanted to around 25 bathes with soil and water and placed in the field. Those plants in the bathes, which retain 4 plants, were counted for their characteristics in the numbers of tillings, ear numbers and total ear weights. As shown in Figure 2, 9 bathes showed 4 rice plants in the original wild type, 23 bathes for R4-2-4 and 6 bathes for R4-5-8.

The results of tiller numbers in the original wild type, R4-2-4 and R4-5-8 are presented in Figure 2a, and those of ear numbers and the weights were presented in Figure 2b and Figure 2c, respectively. The actual plants examined were 36, 92 and 24, respectively. We can observe that the numbers of tillers, those of ears and those of total ear weights were apparently higher in the mutants, R4-2-4 and R4-5-8, compared with those of original wild type.

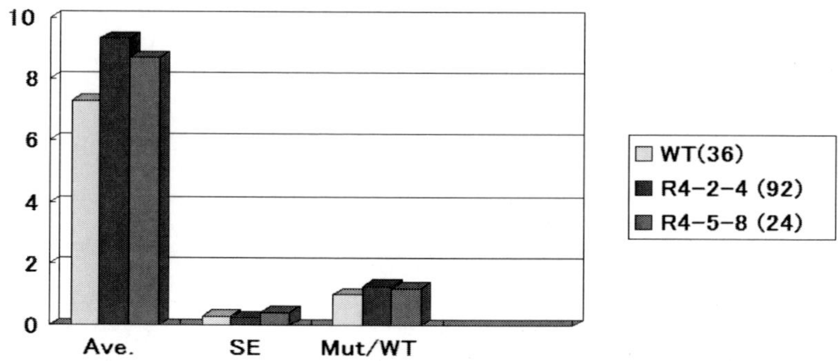

Figure 2 a. Tiller numbers of original wild type and R4-2-4 and R4-5-8 mutant lines

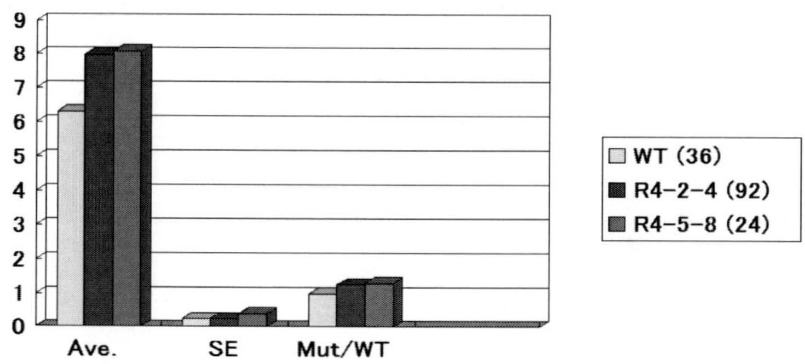

Figure 2 b Ear numbers of original wild type and R4-2-4 and R4-5-8 mutant lines

Figure 2. Continued on next page.

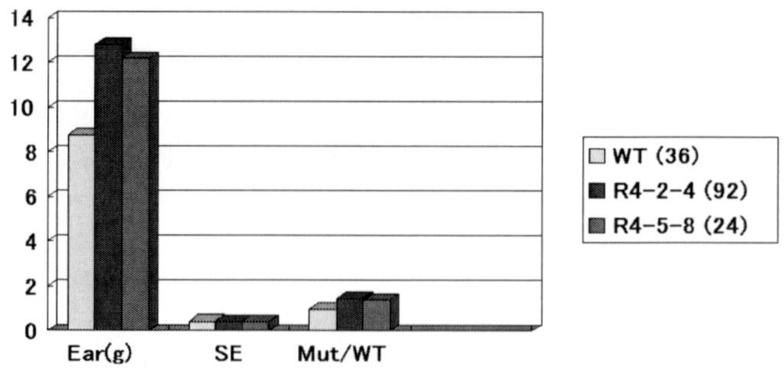

Figure 2 c Ear weights of original wild type and R4-2-4 and R4-5-8 mutant lines

Figure 2. Number of tillers, number of ears and total weight of ears in original wild type, and the R4-2-4 and R4-5-8 mutant lines in 4 planted bathes at Oct. 30, 2007. a) Numbers of tillers in original wild type (9 bathes) and the R4-2-4 (23 bathes) and R4-5-8 (6 bathes) mutant lines with SE and the ratios of mutants/original WT. b) Numbers of ears in original wild type (9 bathes) and the R4-2-4 (23 bathes) and R4-5-8 (6 bathes) mutant lines with SE and the ratios of mutants/original WT. c) Ear weights in original wild type (9 bathes) and the R4-2-4 (23 bathes) and R4-5-8 (6 bathes) mutant lines with SE and the ratios of mutants/original WT.

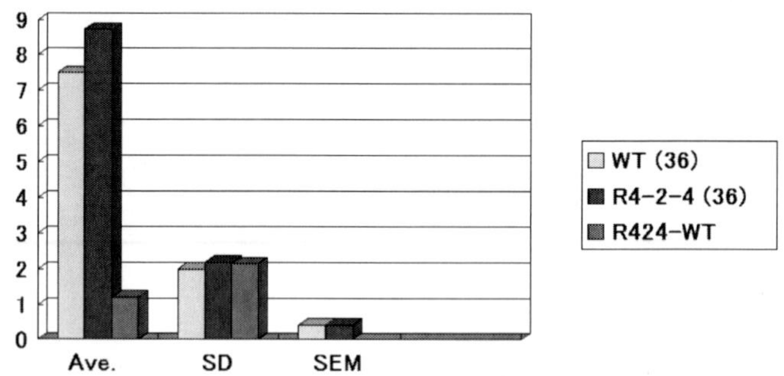

<0.5%

Figure 3 a. Tiller numbers of original wild type and R4-2-4 mutant line and the test of significance.

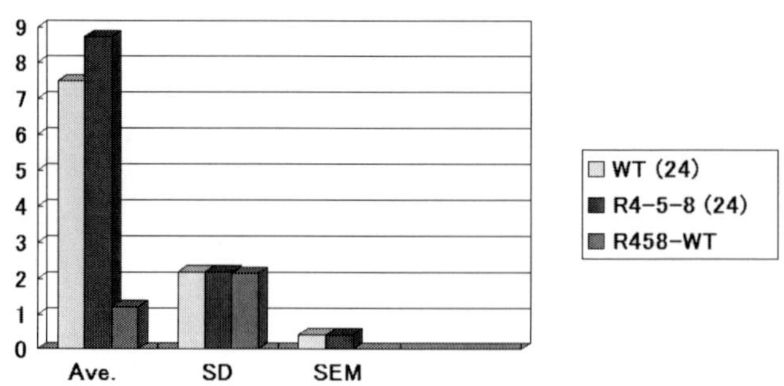

<0.5%

Figure 3 b. Tiller numbers of original wild type and mutant R4-5-8 line and the test of significance.

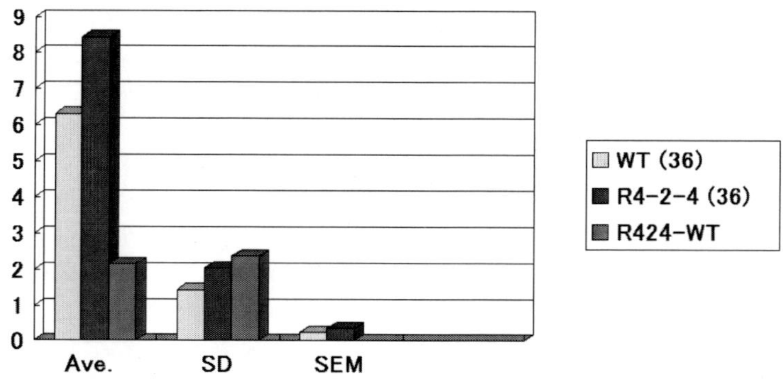

<0.005%

Figure 3 c. Ear numbers of original wild type and R4-2-4 mutant line and the test of significance.

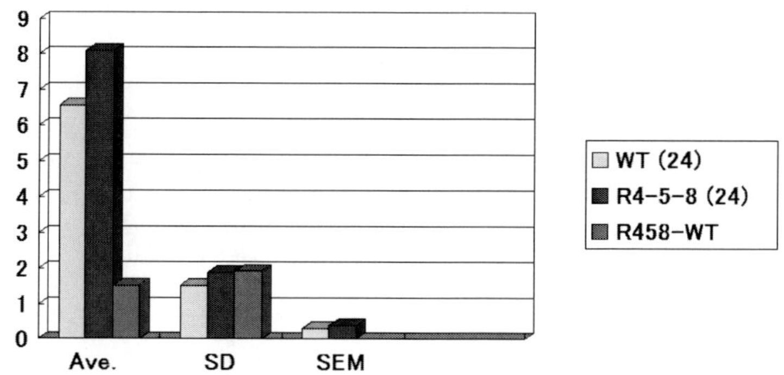

<0.05%

Figure 3 d. Ear numbers of original wild type and R4-5-8 mutant line and the test of significance.

The final yield increases corresponding to the increase in the ear weights were 1.43-fold and 1.12-fold, respectively, in R4-2-4 and R4-5-8 mutant lines compared with original wild type.

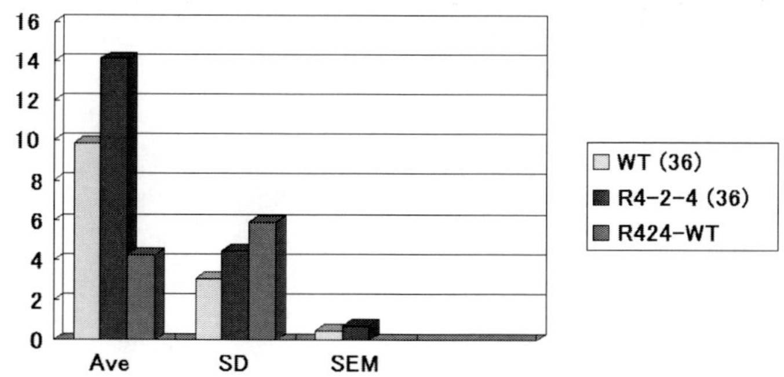

<0.005%

Figure 3 e. Ear weights of original wild type and R4-2-4 mutant line and the test of significance.

Figure 3. Continued on next page.

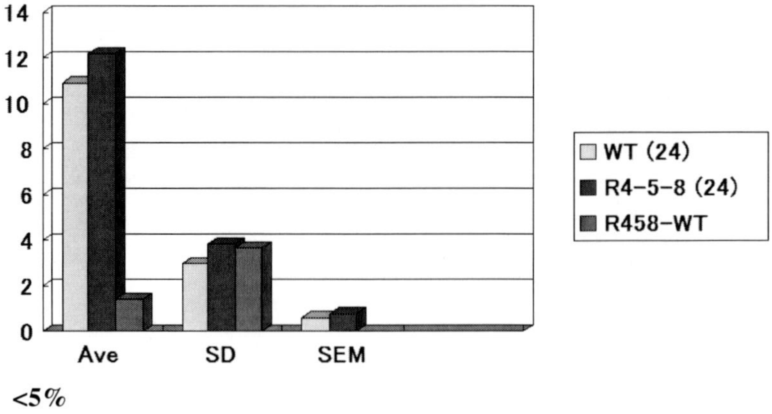

Figure 3 f. Ear weights of original wild type and R4-5-8 mutant line and the test of significance.

Figure 3. Number of tillers, number of ears and total weight of ears in original wild type and the R4-2-4 mutant line, and original wild type and the R4-5-8 mutant line in 4 planted bathes. The differences between original wild type and mutant lines were confirmed statistically and the % significance was calculated. a) Number of tillers in original wild type (9 bathes) and the R4-2-4 mutant line (9 bathes) were compared statistically, and the difference was confirmed by % significance. b) Number of tillers in original wild type (6 bathes) and the R4-5-8 mutant line (6 bathes) were compared statistically, and the difference was confirmed by % significance. c) Number of ears in original wild type (9 bathes) and the R4-2-4 mutant line (9 bathes) were compared statistically, and the difference was confirmed by % significance. a) Number of ers in original wild type (6 bathes) and the R4-5-8 mutant line (6 bathes) were compared statistically, and the difference was confirmed by % significance. b) Ear weights in original wild type (9 bathes) and the R4-2-4 mutant line (9 bathes) were compared statistically, and the difference was confirmed by % significance. f) Ear weights in original wild type (6 bathes) and the R4-5-8 mutant line (6 bathes) were compared statistically, and the difference was confirmed by % significance.

From statistical point of view the difference of tiller numbers, ear numbers and total ear weights between original wild type and R4-2-4 mutant line and between original wild type and R4-5-8 mutant lines, were confirmed by use of test of significance. Figure 3, a and b indicate that the tiller numbers of original wild type and R4-2-4 mutant line, and also original wild type and R4-5-8 showed increase in them and the differences were confirmed by test of significance. The increases in the tiller numbers in R4-2-4 and R4-5-8 are confirmed by the significance test of, less than 0.05 % and less than 0.5 %, respectively. In Figure 3, c and d, and also e and f, the increases in the ear numbers and ear weights in R4-2-4 and R4-5-8 mutant lines compared with those of original wild type were also confirmed by the test of significance, and the % significances were, less than 0.005% and 0.05%, and also less than 0.005% and 5%, respectively. The values indicate that from statistical point of view we confirmed the increases in the ear numbers and ear weights in R4-2-4 and R4-5-8 mutant lines compared with those of original wild type.

Analysis of Resistant Phenotypes to PQ in Original and Two Mutant Lines

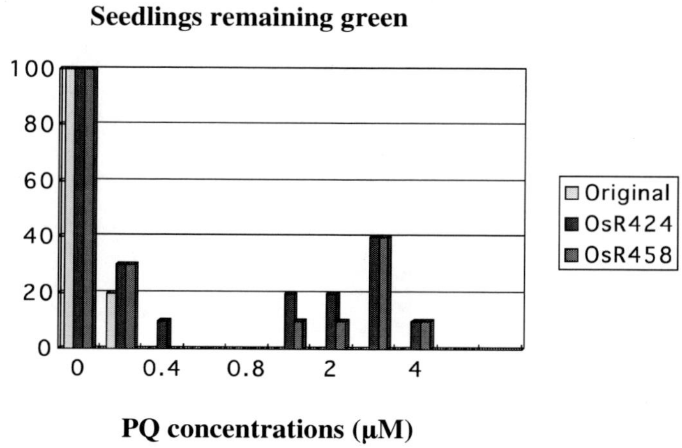

Figure 4. Resistant phenotypes of original wild type and mutant lines in the presence of various concentrations of PQ

After obtaining putative homozygous mutant lines with very good tiller phenotype, the seeds were used to compare the resistant phenotype of germinating seeds, which will remain green in the presence various concentrations of PQ. The results are presented in Figure 4.

Although more than 90% of the seeds showed germination, most of the first leaf became white in the presence of PQ in case of original line. However, mutant lines, R4-2-4 and R4-5-8 showed germination of seeds, which will remain green in the first leaves at the concentration of 1 to 4 μM showing maximum states retaining green at a concentration of 3 μM PQ. The results indicate that these two lines of mutants have the capacity to resist 1-4 μM PQ.

Analysis of Protein Distribution and the Phosphorylation Pattern of Leaf Crude Extracts from Original Wild Type and the Mutants, R4-2-4 and R4-5-8

As shown in Figure 5 a) and b), crude extracts from the leaves of original wild type and the mutant, R4-2-4, were used to identify protein distribution changing the amount of proteins from 6.25, 12.5, 25, 50 and 100 μg, and further analyzed phosphorylation pattern of the proteins mixing [γ-^{32}P]ATP and incubating the mixture on ice for 10 min. To the reaction mixtures SDS-sample buffer was added. After development of SDS-PAGE, the gels were fixed and stained by Coomasie Bririant Blue (CBB), and then the gel was exposed to isotopic image analyzer. The protein phosphorylation pattern was presented in Figure 5 b). The protein distribution and the phosphorylation patterns in original wild type and R4-5-8 mutant lines were also presented in Figure 5, c) and d), respectively. The MW marker indicates, 10, **15**, 20, **25**, 37, **50**, and **75** kD from thic band indicated by broad numerics.

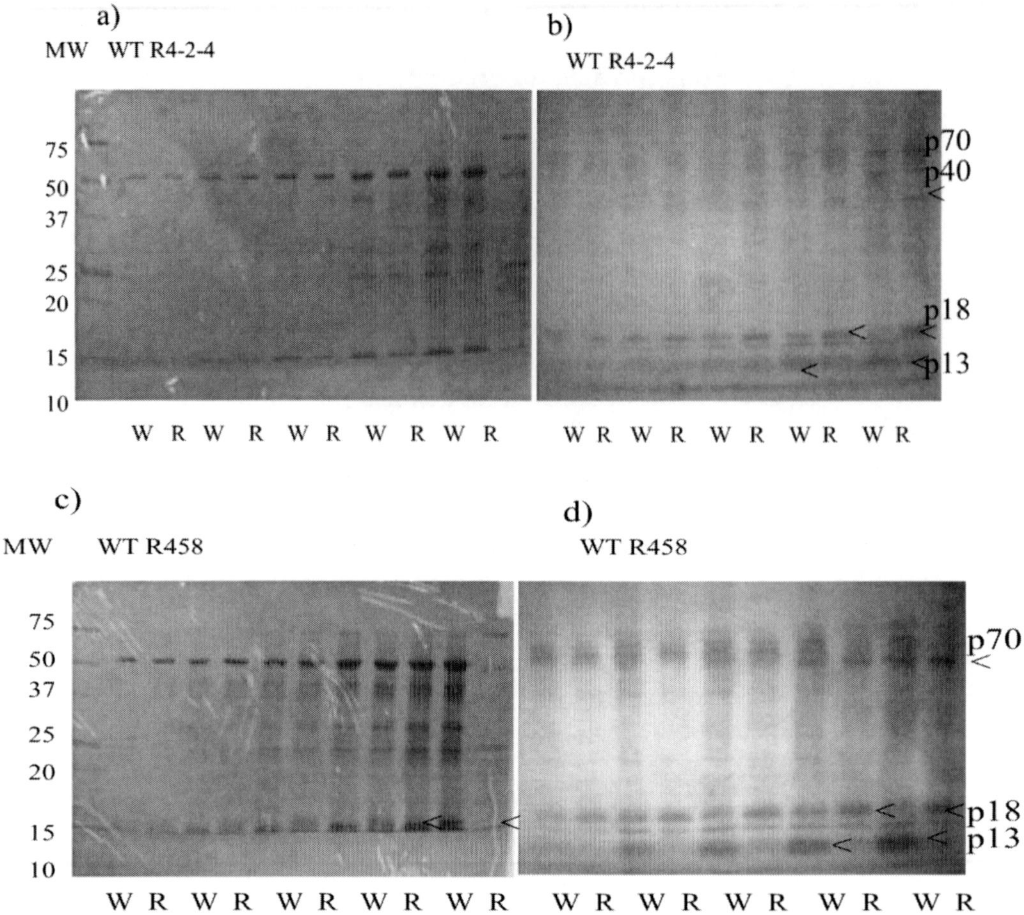

Figure 5. Analyses of proteins in the leaf crude extracts and the phosphorylation patterns of proteins changing the concentration of them. a) Comparison of proteins from leaves of original wild type and the R4-2-4 mutant after development of the proteins by SDS-PAGE followed by CBB staining. Each pair of lane included proteins of soluble fraction from original wild type and the mutant, R4-2-4 with the loading amount of 6.25, 12.5, 25, 50 and 100 μg of proteins. After addition of reaction mixture containing [γ-^{32}P] ATP on ice the reaction was stopped by the additions of SDS sample buffer. After development of proteins by SDS-PAGE, the gel was stained by CBB, and then exposed to autoradiography. b) Phosphorylation of proteins from leaves of original wild type and the mutant R4-2-4. The result of autoradiography was presented. c) Each pair of lane included proteins of soluble fraction from original wild type and the mutant, R4-5-8 with the loading amount of 6.25, 12.5, 25, 50 and 100 μg of proteins. After addition of reaction mixture including [γ-^{32}P] ATP on ice the reaction was stopped by thw additions of SDS sample buffer. After development of proteins by SDS-PAGE, the gel was stained by CBB, and then exposed to autoradiography. d) Phosphorylation of proteins from leaves of original wild type and the mutant R4-5-8. The result of autoradiography was presented.

From protein distributions between original wild type and R4-2-4, and also original wild type and R4-5-8, we could detect small difference in the low molecular weight proteins corresponding to NDPK by SDS-PAGE. From the data of phosphorylation of these proteins, we could detect significant increase in the molecules corresponding to putative precursor of NDPK in the crude extract of R4-2-4 as indicated by <p18.

There may be differences in the molecule in the amounts of proteins, and also in the capacity to be phosphorylated suggesting the presence of large amounts of precursor of NDPK in the mutant lines, R4-2-4 and also R4-5-8 compared with those of original wild type. Further analysis is required to determine this possibility. In original wild type we could detect increase in the small molecular weight radioactive protein, which may be the degradation product of NDPK as indicated by <p13.

We further could detect the difference in the amounts of phosphorylation in higher molecular weights in mutants, R4-2-4 and R4-5-8. From the condition of phosphorylation these phosphorylated proteins may be high molecular weight histidine kinases as indicated by <p40 and by <p70. The increase in the phospharylation of high molecular weight histidine kinases was suggested. On this point we need further analysis to determine it.

Analysis of Phosphorylation Patterns of proteins in Soluble Fraction, Chloroplast Fraction and Mitochondrial Fraction

From leaves of original wild type and the mutants, R4-2-4 and R4-5-8, crude extracts were prepared. From these crude extracts, soluble fractions, chloroplast fractions and mitochondrial fractions were prepared from original wild type and the mutants, R4-2-4 and R4-5-8 lines. In Figure 6 a) the protein distribution in these fractions from original wild type and the mutant R4-2-4 was compared, and large difference in the amount was not observed. However, the comparison of phosphorylated proteins informed us that in chloroplast fraction in the mutant R4-2-4 we could detect large difference in p18 and p70 as indicated in the Figure 6 a) as <p18 and <p70. From the condition of labeling, these two phosphorylated proteins were suggested to be precursor of NDPK, and high molecular weight histidine kinase, respectively.

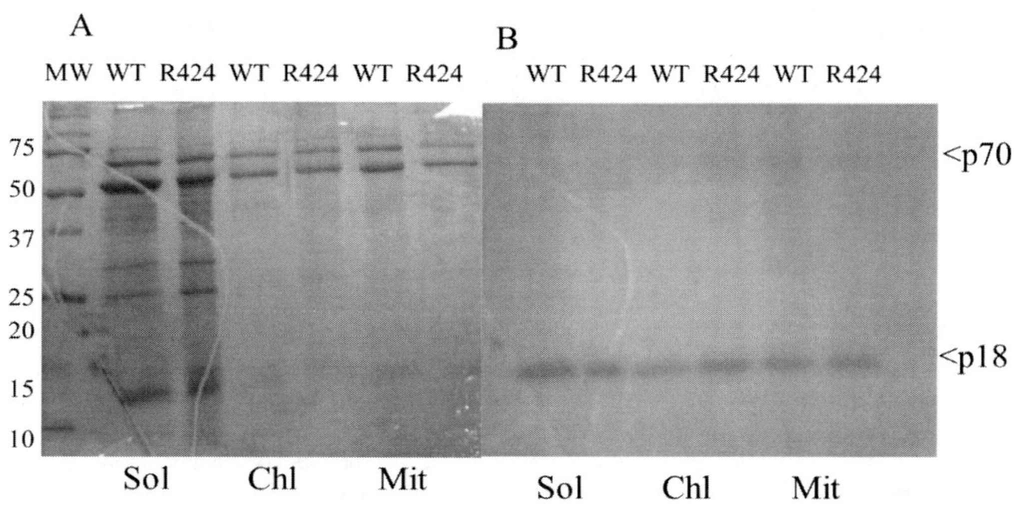

Figure 6. Continued on next page.

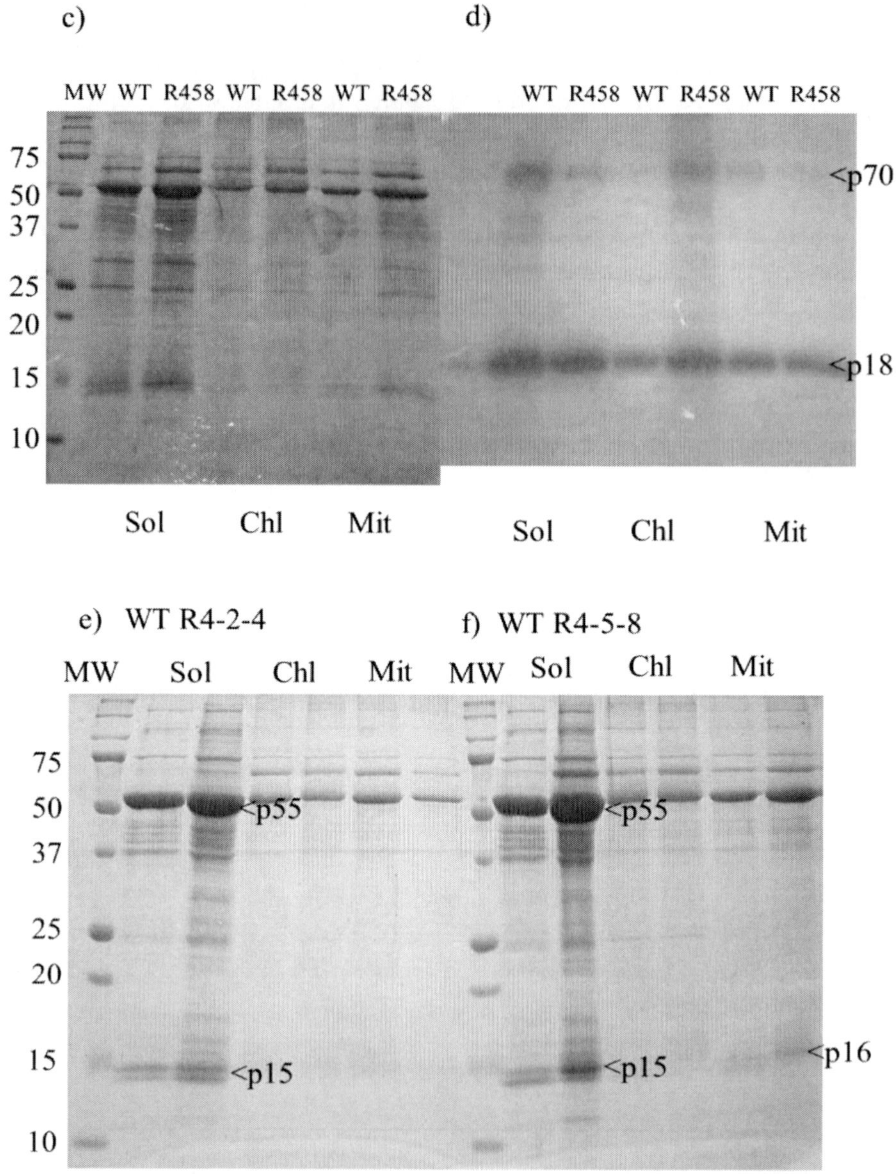

Figure 6 e and f. Protein distribution in the fractions of soluble factions (Sol), chloroplast fractions (Chl) and mitochondrial fractions (Mit) from original WT, R424 and R458

Figure 6. Protein distribution in soluble fraction, chloroplast fraction and mitochondrial fraction and the phosphorylation patterns of them in between original wild type and the mutant R4-2-4 and original wild type and R4-5-8, respectively. a) Comparison of protein distribution between original wild type and the mutant R4-2-4. b) Comparison of phosphorylation of proteins between wild type and the mutant R4-2-4. c) Comparison of protein distribution between original wild type and the mutant R4-5-8. d) Comparison of phosphorylation of proteins between wild type and the mutant R4-5-8. e) Comparison of proteins between original wild type and the mutant R4-2-4. f) Comparison of proteins between original wild type and the mutant R4-5-8.

In Figure 6 b) the comparison of protein distribution in original wild type and the mutant R4-5-8 were made in the soluble fractions, chloroplast fractions and mitochondrial fractions. In the protein distribution we could not detect large difference. However, we could also detect large difference in the phosphorylation of chloroplast fraction. The mutant R4-5-8 showed large amounts of phosphorylation in p18 and p70, which are indicated in the Figure 5 b) as <p18 and <p70.

On the protein distribution in the fractions of original wild type and the mutant lines, R4-2-4 and R4-5-8, we further confirmed it without including radioisotopic labeling in order to detect precise difference.

In the soluble fraction we could detect the increase in the amount of proteins in low molecular weight as indicated by <p15 and in high molecular weight as <p55 in the soluble fraction of both mutants, R4-2-4 and R4-5-8 were detected.

A small increase in the protein in the mitochondrial fraction of R4-5-8 was detected. The protein seems not to contribute to phosphorylation of mitochondrial protein.

Catalase Activities in the Crude Extracts Prepared from the Leaves of Original Wild Type and the Mutants R4-2-4 and R4-5-8

In Figure 7 a) and b) the catalase actibities in the crude extracts prepared from the leaves of original wild type and the mutants, R4-2-4 and R4-5-8 were presented.

In a) each of these lanes, 25 – 25 μg, 50 – 50 μg and 100 – 100 μg of crude extracts from the leaves of orignal wild type and the mutant R4-2-4, and in b) original wild type and the mutant R4-5-8 were loaded and electrophoresis on native PAGE were performed.

After electrophoresis the catalase activities in the native gels were compared. The results indicate that there was essentially no difference between original wild type and the mutants.

Original WT - R424

a) b)

W R W R W R   W R W R W R

WT - R424   WT - R458

Figure 7. The catalase activities in the crude extracts of original wild type and the mutants R4-2-4 and R4-5-8. a) Original wild type and the mutant R4-2-4. b) Original wild type and the mutant R4-5-8.

Sequence Analysis of NDPK1, 2 and 3

From original wild type and mutant mRNAs the cDNA for the precursor of NDPK1, NDPK2 and NDPK3 from original wild type and mutants, R4-2-4 and R4-5-8 were prepared. As shown in Figure 8, a), b), c), d) and e), we could not detect any difference between original wild type precursors of NDPK1, NDPK2 and NDPK3 from those of mutants, R4-2-4 and R4-5-6, respectively. However, we detected that there is a large deletion in the nucleotide sequences including transit peptide corresponding to precursor of NDPK2 in Koshihikari compared with that of Nipponbare, whose whole DNA sequences were completed. The results were shown in Figure 8, b) and c). The 78 amino acid deleted sequences really corresponded to part of transit peptide of NDPK2 in Nipponbare. Therefore, Koshihikari NDPK2 sequence lacked major part of leader sequence of original Nipponbare leader sequence of NDPK2. In ROS resistant Alaska pea the leader sequence of NDPK2 precursor included mutation of Ile12Leu by ATC to CTC exchange. This part was partly deleted in the transit peptide of NDPK2 in Koshihikari. The phophorylation analyses of histidine kinases in the former section indicate that the degradation of NDPK2 with transit peptide may be blocked in the mutant lines.

a) OsNDPK1 (Original WT, Mutant R4-2-4 and Mutant R4-5-8) (Koshihikari)

```
1   atggagcagt ccttcatcat gatcaagcct gacggcgtcc agagaggcct gattggagac
61  atcatcagca ggttcgagaa gaaaggattc ttcctgaggg ggatgaagtt catgaatgtg
121 gagaggtcct ttgcacagca gcactatgcc gacctttccg acaagccctt cttccctggg
181 ttggtggagt acatcatctc tggccccgtc gttgcgatgg tttgggaagg gaaggatgtt
241 gttgccactg gccgcaggat cattggggcc accaggccct gggaggcagc ccccggcacc
301 atccgtgctg actatgccgt ggaagttggc aggaatgtca tccatgggag cgactccgtg
361 gacaacggga agaaggagat cgctctctgg ttccctgaag gtttagctga gtggaggagc
421 aaccttcacc cttggatcta tgagtcttag agcttcgcgc ctataatcga tcacctcttg
481 gtggctagtt ctcgttaaat atttgagttt tgtttcagag ttgtcatctg gctgcctcct
541 gagtcaccat tataattctg ctgtgataca atcacagtta tatgttgcct gaacttatat
601 ggctttttaa gtttctggat tggatgtgat gggaaaataa tactagcagt cagtttttgc
661 ttataaaaaa aaaaaaaaaa aaa
```

Blue shadowed atg indicates starting codon of translation.
Brown shadowed sequences indicated translated area.

b) OsNDPK2 (Original WT, Mutant R4-2-4 and Mutant R4-5-8)(Koshihikari)

```
1   ctctctgctc ctgctcatca tcttcttctt ctccgctctc cctaatccaa aaaaaggcgg
61  ctgctttggt cgctcgctcg ctcgcgcgga cgttggagga gaagcagcta ccgggatgga
121 cgccatggcc gtgctcgcga ggacctcccg ccccgccccg accctcctcg ccgcgacctc
181 ccccgcggtg tcccggcgcc cggcggccgt ctccttcgcc gccgccgcct ccccggggtc
241 ccgcggccgc gtcgcgctga gcgccgcgtg gggcgggagg gcggccaggg ggcgcgtctc
301 cgccgccggc cgcatcgtcg cgtcgtcggt tgagcaatcc tatattatga tcaagcctga
361 tggtgttcag cgaggcctgg ttggcgagat tatttcccgt ttcgagaaga aaggctttgt
421 gttgaagggc ttaaagcttt ttcagtgccc caaggacttg gcacaggagc actacaagga
481 tttgaaggaa aaacctttct ttcccggtct gatcgagtac ataacttctg gccctgttgt
541 ttgcatggct tgggagggga atggtgttgt tgcatcagct cgcaagttaa taggagctac
```

```
601 taatcccctt caagctgaac cagggaccat acggggcgat cttgcagttc aaacaggaag
661 gaatgttgtc catggaagtg acagtcctga taacggcaag cgtgaaatcg ctctatggtt
721 caaagaaggc gaactctgcg aatgggaatc tgttctgaca ccatggcttg tagagtaatg
781 ctcccagatc agtgcaagaa ctggatgcag tgcttgttct tgatttgagg ggttatgtat
841 gagttcagat cagcagagta aagtagctta taccaaagct cggccttctc atgactgtac
901 accctgacat tcccctttt tgtaataccc aacattccct tttttgcgtt tgaccgatta
```

#Pink shadow indicates deleted part of Koshihikari compared with Nihonbare.
#Blue shadowed atg at the second line indicates the stating position of translation in Nihonbare.
#Blue shadowed atg at the 6th line indicates the starting position of translation in Koshihikari.
Gray shadowed sequences indicate translated parts of OsNDPK2

c) Nipponbare OsNDPK2 deduced aminoacid sequence

```
MIKPDGVQRGLVGEIISRFEKKGFVLKGLKLFQCPKDLAQEHYKDLKEKPFFPGLIEYITSGPVVCMAWEGNGVVAS
ARKLIGATNPLQAEPGTIRGDLAVQTGRNVVHGSDSPDNGKREIALWFKEGELCEWESVLTPWLVE
```

#Pinky shadow indicates deleted part in Koshihikari compared with Nihonbare.
d) Koshihikari OsNDPK2 deduced amino acid sequence

```
MIKPDGVQRGLVGEIISRFEKKGFVLKGLKLFQCPKDLAQEHYKDLKEKPFFPGLIEYITSGPVVCMAWEGNGVVAS
ARKLIGATNPLQAEPGTIRGDLAVQTGRNVVHGSDSPDNGKREIALWFKEGELCEWESVLTPWLVE
```

e) OsNDPK3 (Original WT, OsR424 and OsR458)(Koshihikari)

```
1    gaccgaggcg tgcgccagcc actgagaagg agaagaggtc ctccggatca aagcggcacc
61   gacgacgccg gcgaggccac tactaccata tcatagtaca tactagtaca agactggagg
121  gctagtgggg tgggggtaca atccaatcca atccaatcaa atcacatcga tcgacgccgc
181  aacccagcag cagcagcgga accctagcca agccaccatg agcaagctct gccagtcggc
241  gtgcaaggcc gccaagtctc tcctctccgc caccgccgcc gcctcctccc cgcgcacctc
301  cctcctagcc gaggggagga acgcggcgct ggccaccctc accaacctgg ggaggaagac
361  cctccccaca gcttacgcct actcctacca ccacaactcc tccgccgccg ccgccggatg
421  gctcgctgcc atccccgccg ctgtttacat gctacaggat caggaggcgc atgctgcaga
481  gatggagcgc acctttatcg ccatcaagcc tgacggcgtc caaaggggcc tgatttctga
541  gatactgtcc cgatttgaaa gaaaaggatt caagcttgtt gccatcaagc tggtggttcc
601  atccaaagaa tttgctcaga agcactatca tgatttgaaa gacagacctt tcttcaatgg
661  attgtgtgac ttccttagct ctggccctgt gcttgccatg gtttgggaag gagagggtgt
721  tatcaagtac gggagaaaac tgattggtgc tacagaccca cagaagtctg aaccaggaac
781  catcaggggt gaccttgctg ttgttgttgg caggaacatc atccatggaa gcgatggccc
841  agagactgca aaggctgaga ttggtctctg gtttgagccc agggagctgg tctcttacac
901  cagcaacgaa gagaagtgga tctacggggt caactaaccg gcgaatcatc tcccctgttt
961  gttttgtttt tttttcattt tctctcactt tacctaatta catcaccatg gaagagtgta
1021 ataaagttgc gatcgtgaca tttggggtta tgcccatgtg atctggcgca aataagaggc
1081 ctcaagaaga attgtctacc ttcgattgta gagcgtgcac tgagtaatca gaccatcacc
1141 ttcactcaaa tggtcgaacg gatggtgatg ccacttgttt tgttattgtt atataatggt
1201 actacatgtt gtc
```

Blue shadowed atg indicates starting codon.
Gray shadowed sequences indicates translated part.

Figure 8. Nucleotide sequences and amino acid sequences of precursor of NDPK1, NDPK2 and NDPK3 in Koshihikari, and those of deleted sequences of Koshihikari NDPK2 compared with those of Nipponbare. a) (described in Figure 8.) b) (described in Figure 8.) c) (described in Figure 8.) d) (described in Figure 8.) e) (described in Figure 8).

Yield Test of *Oryza Sativa* Cv Koshihinkari, Original Wild Type, 10 Lines of ROS Resistant Mutants of R4-2-4 and Kinuhikari As One of the Reference Line

The seeds of Koshihimari original wild type and the 10 mutant R4-2-4 lines grown in 2007 were used for the test of yielding in Kanagawa Agricultural Technology Center. The results of duplicate testings of yield of unhulled rice, straw, total plant weight, rice hulls, weight of 1000 seeds, water content, cleaned rice, yield of cleaned rice, ratio of cleaned rice, junk, and ratio of junk were measured. Weight of 1000 seeds, and water contents were assayed twice. Very high yielding of unhulled rice was observed in line 2, line 4 and line 8. These lines showed high yielding even in total weights of plants. The cleaned rice also showed high yielding.

These results were summarized in Table 1 and Table 2. Total weights of hulled rice in R4-2-4/ original line were 111%, 112% and 135% in line 2, line 4 and line 8, respectively. Only line 5 of R4-2-4 showed reduced weight of 94%. However, as an average of 10 lines total weights of hulled rice in the mutant lines was 108.5%, which is certainly high in the yielding values.

Even in total weights the average of yielding was 108.4%, showing essentially same value. As an average of 108.5% in the hulled rice we can expect enough yielding by use of the R4-2-4 mutant line.

DISCUSSION

The treatment of germinating seedlings in the presence of PQ induced mutations to reactive oxygen species (ROS) resistance. Among the mutants we detected mutant lines with increasing amounts of yielding, tillering and ears. Among them we identified PQ resistant mutant lines, R4-2-4 and R4-5-8, respectively. Strong sunlight includes UV C, UV B, UV A, violet, blue, green, yellow, orange, red, far-red, and thermal radiation from short wave length of UV to long wave length of thermal radiation. Short wave length light such as UV to blue could be absorbed by several photosensitizers, such as chlorophyll, FAD, FMN, riboflavin, and enhance the emission of singlet oxygen, which will be emitted by accepting energy from photosensitizers.

The lifetime of singlet oxygen is from 0.5 μsec to 4 μsec, and their flight path length is less than 500 μm. Immediately after accepting strong sunlight with shorter wavelength these photosensitizers work to emit energy, which is transferred to triplet oxygen generating singlet oxygen. Singlet oxygen generated will make flight outside of chloroplast and will be captured by catalase in the cytosol. The catalase binding singlet oxygen will interact with NDPK binding NADH, which will transfer the electron to singlet oxygen [3], generating superoxide [8]. The superoxide will be converted into hydrogen peroxide by superoxide dismutase, and then to water and triplet oxygen releasing the longer wavelength light (thermal emission) to ambient circumstances. Thus the high energy short wave length light from sunlight will inevitably be accepted by plant, and emit large amount of singlet oxygen constituting 80% of ROS.

Table 1. The results of yield test of Koshihikari mutant lines from R4-2-4 obtained in 2007. Ten lines of the mutant R4-2-4 were used and duplicate checking was performed in 2008. As a reference lines, original wild type and Kinuhikari were used

	Replic	Weight of unhulled rice (g)	Straw (g)	Total weight (g)	Rice hulls (g)	Weight of 1000 seeds (g)			Water(%)			Cleaned	Yield cleaned	Ratio cleaned	Junk	Ratio of junk
						1st	2nd	Ave	1st	2nd	Ave					
Line1	1	1223.10	1.63	2853.10	300.08	952	958	955	12.8	12.9	12.9	231.06	941.8	77.0%	12.55	5.4%
	2	1278.02	1.43	2708.02	300.08	965	976	971	13.0	13.0	13.0	226.76	965.8	75.6%	16.22	7.2%
	Ave	1250.56	1.53	2780.56	300.08	959	967	963	12.9	13.0	12.9	228.91	953.8	76.3%	14.39	6.3%
line2	1	1475.26	1.12	2595.26	300.07	938	938	938	13.1	13.0	13.1	232.27	1141.9	77.4%	10.29	4.4%
	2	1259.11	1.11	2369.11	300.08	949	951	950	13.1	13.2	13.2	233.84	981.2	77.9%	9.23	3.9%
	Ave	1367.19	1.12	2482.19	300.08	944	945	944	13.1	13.1	13.1	233.06	1061.6	77.7%	9.76	4.2%
Line3	1	1183.98	1.31	2493.98	300.06	948	951	950	13.3	13.2	13.3	285.66	1127.2	95.2%	14.26	5.0%
	2	1293.90	1.50	2793.90	300.01	960	949	955	13.2	13.2	13.2	229.09	988.0	76.4%	12.16	5.3%
	Ave	1238.94	1.41	2643.94	300.04	954	950	952	13.3	13.2	13.2	257.38	1057.6	85.8%	13.21	5.1%
Line4	1	1468.47	1.82	3288.47	300.20	942	946	944	13.4	13.0	13.2	229.30	1121.7	76.4%	11.20	4.9%
	2	1291.23	1.49	2781.23	300.10	964	951	958	13.2	13.3	13.3	229.66	988.2	76.5%	12.63	5.5%
	Ave	1379.85	1.66	3034.85	300.15	953	949	951	13.3	13.2	13.2	229.48	1054.9	76.5%	11.92	5.2%
Line5	1	1180.67	0.94	2120.67	300.05	945	956	951	13.6	13.4	13.5	284.93	1121.2	95.0%	14.36	5.0%
	2	1134.51	1.07	2204.51	300.30	961	967	964	13.3	13.4	13.4	228.32	862.6	76.0%	14.83	6.5%
	Ave	1157.59	1.01	2162.59	300.18	953	962	957	13.5	13.4	13.4	256.63	991.9	85.5%	14.60	5.8%
Line6	1	1432.32	0.98	2412.32	300.26	922	944	933	13.7	13.6	13.7	233.68	1114.7	77.8%	9.46	4.0%
	2	1147.56	0.96	2107.56	300.05	965	944	955	13.1	13.2	13.2	230.34	880.9	76.8%	11.78	5.1%
	Ave	1289.94	0.97	2259.94	300.16	944	944	944	13.4	13.4	13.4	232.01	997.8	77.3%	10.62	4.6%
Line7	1	1422.66	0.81	2232.66	300.82	983	990	987	13.0	13.1	13.1	225.29	1065.5	74.9%	17.30	7.7%
	2	1149.94	1.41	2559.94	300.09	953	958	956	13.0	12.9	13.0	230.67	883.9	76.9%	12.18	5.3%
	Ave	1286.30	1.11	2396.30	300.46	968	974	971	13.0	13.0	13.0	227.98	974.7	75.9%	14.74	6.5%
Lin8	1	1662.98	1.85	3512.98	300.32	930	931	931	13.6	13.6	13.6	231.11	1279.7	77.0%	10.81	4.7%
	2	1641.81	1.48	3121.81	300.35	929	930	930	13.4	13.5	13.5	235.52	1287.4	78.4%	9.07	3.9%
	Ave	1652.40	1.67	3317.40	300.34	930	931	930	13.5	13.6	13.5	233.32	1283.6	77.7%	9.94	4.3%
Line9	1	1510.79	1.62	3130.79	300.08	948	956	952	13.0	12.9	13.0	231.69	1166.5	77.2%	11.58	5.0%
	2	1186.05	1.04	2226.05	300.02	984	977	981	13.1	13.2	13.2	223.26	882.6	74.4%	17.65	7.9%
	Ave	1348.42	1.33	2678.42	300.05	966	967	966	13.1	13.1	13.1	227.48	1024.5	75.8%	14.62	6.5%
Line10	1	1398.18	0.91	2308.18	300.17	943	945	944	12.9	13.1	13.0	232.32	1082.1	77.4%	10.88	4.7%
	2	1307.71	1.31	2617.71	300.08	947	966	957	13.3	13.4	13.4	230.90	1006.2	76.9%	10.82	4.7%
	Ave	1352.95	1.11	2462.95	300.13	945	956	950	13.1	13.3	13.2	231.61	1044.2	77.2%	10.85	4.7%
Koshihikar original line	1	1289.56	1.30	2589.56	300.02	937	948	943	13.5	13.6	13.6	230.56	991.0	76.8%	11.82	5.1%
	2	1164.32	1.12	2284.32	300.07	934	938	936	13.2	13.2	13.2	225.66	875.6	75.2%	15.80	7.0%
	Ave	1226.94	1.21	2436.94	300.05	936	943	939	13.4	13.4	13.4	228.11	933.3	76.0%	13.81	6.1%
Kinuhikario riginal line	1	1093.72	1.12	2213.72	300.43	921	923	922	13.5	13.5	13.5	235.07	855.8	78.2%	7.05	3.0%
	2	1034.67	1.05	2084.67	300.11	953	954	954	12.9	13.1	13.0	233.43	804.8	77.8%	9.53	4.1%
	Ave	1064.20	1.09	2149.20	300.27	937	939	938	13.2	13.3	13.3	234.25	830.3	78.0%	8.29	3.5%

Table 2. The summary of yield test of Koshihikari mutant lines obtained from the mutant R4-2-4 in 2007

Total hulled rice/original line	Weight of straw (g)	Total weight (kg)	Total weight/ original line	Weight of 1000 seeds (g)	1000 seeds/ original line	Weight of cleaned rice (g)	Cleaned rice/ original line	Ratio of cleaned rice	Weight of junk rice (g)	Ratio of junk rice
102%	1.53	2.78	114%	20.8	98%	954	102%	76.3%	14.4	6.3%
111%	1.12	2.48	102%	21.2	99%	1062	114%	77.7%	9.8	4.2%
101%	1.50	2.79	115%	21.0	98%	988	106%	76.4%	12.2	5.3%
112%	1.66	3.03	125%	21.0	99%	1055	113%	76.5%	11.9	5.2%
94%	1.07	2.20	90%	20.7	97%	863	92%	76.0%	14.8	6.5%
105%	0.97	2.26	93%	21.2	100%	998	107%	77.3%	10.6	4.6%
105%	1.11	2.40	98%	20.6	97%	975	104%	75.9%	14.7	6.5%
135%	1.67	3.32	136%	21.5	101%	1284	138%	77.7%	9.9	4.3%
110%	1.33	2.68	110%	20.7	97%	1025	110%	75.8%	14.6	6.5%
110%	1.11	2.46	101%	21.0	99%	1044	112%	77.2%	10.9	4.7%
Koshihi-	1.21	2.44	-	21.3	-	933	-	76.0%	13.8	6.1%
Kinihi-	1.09	2.15	-	21.3	-	830	-	78.0%	8.3	3.5%

However, the energy of short wave length sunlight will be converted into singlet oxygen separating triplet oxygen from proton generated after photolysis of water in the chloroplast.

Therefore oxygen is functioning to reduce the energy of short wave length sunlight to longer wavelength thermal emission, thus ROS are functioning as a transformer of harmful short wavelength light to longer wavelength thermal emission working as a mediator of detoxification processes.

Short wavelength sunlight energy will make splitting of precursor of NDPK molecule directly or via singlet oxygen in original wild type of Koshihikari. However, in the mutants, R4-2-4 and R4-5-8 the precursor molecules of NDPK may be protected from splitting by either directly short wavelength light or directly by singlet oxygen. We could observe such phenomena by the analysis of phosphorylation of proteins in the crude extract prepared from original wild type and the mutants, R4-2-4 and R4-5-8. The functional precursor of NDPK will function to reduce the concentration of singlet oxygen. However, in original wild type the precursor molecules will be split either by short wavelength sunlight or singlet oxygen becoming non-functional to reduce singlet oxygen. The Koshihikari NDPK2 with transit peptide may be very fragile to strong sunlight. However, in the mutants, R4-2-4 and R4-5-8 the process of degradation of phosphorylated NDPK2 with transid peptide may be protected by some of the associating proteins as reported by Yoshida et al. (2006)[5].

The high molecular weight histidine kinase may be very close to photoreceptor or sensors of high temperature, low temperature and drought, and transfer the phosphate group to small histidine kinase such as NDPK [1].

We must consider what molecules are mutationally changed in the mutant R4-2-4 and R4-5-8. Here we have already determined that NDPK1, NDPK2 and NDPK3 precursors have no change in the primary nucleotide sequence. As another candidate we could consider that high molecular weight histidine kinase may be one of the candidate. As the other candidate the protecting protein to split NDPK into small p13 proteins may be suggested. On these points further analyses are needed to confirm these possibilities.

We know that circadian rhythm of photosynthesis in *Phaseolus vulgaris* L., was demonstrated [9]. In *Neurospora crassa* the major factor controlling circadian rhythm was demonstrated to be ROS by Yoshida et al. (2011)[10]. Histidine kinase activity was also shown to be controlled by circadian rhythm in *Arabidopsis thaliana* [11]. All of the three different factors were closely correlated with those facts that histidine kinase may control the level of ROS, which was demonstrated by Hasunuma et al. (2010) [8]. The multiple production of tillers or stems forming flower and pollen could be considered to be sexual cycles, which may be under the control of ROS. Such a situation controlling sexual cycle forming protoperithecia was demonstrated by *Neurospora crassa* [1].

Finally circadian changes of ROS in plant may control photosynthesis, which is controlled by circadian rhythm as demonstrated in *Phaseolus vulgaris* L., [9]. The fact indicate that ROS including singlet oxygen may be major factor controlling photosynthesis. The molecular mechanism of photoinhibition by strong sunlight will be solved by analyzing molecular control of histidine kinases, ROS including singlet oxygen and membrane composed of lipids accepting singlet oxygen, because the integrity of membrane is very important to keep the proton produced by photolysis to well integrated lipid thylacoid membrane.

ACKNOWLEDGMENTS

We are grateful for the support made by Nissan Motor Co., Ltd. We are also grateful to Yamada Science Foundation, Mishima Kaiun Foundation and Scientific support from YCU.

REFERENCES

[1] Hasunuma, K., Yoshida Y. and Haque, Md. E. (2012). Molecular basis of signal transduction of high intensity light via nucleoside diphosphate kinase (NDPK) in *Neurospora crassa* and *Pisum sativum* cv Alaska. in *Photoreceptors: Physiology, Types and Abnormalities*, pp 149 - 162. Eds Akutagawa, E. and Ozaki, K. *Nova Science Publishers, Inc. NY.*

[2] Haque, Md. E., Yoshida, Y. and Hasunuma, K. (2008). Paraquat-resistant mutant lines in *Pisum sativum* cv Alaska: biochemical and phenotypic characterization. *Plant Biotech. Rep.*, 2, 21 - 31.

[3] Haque, Md. E., Yoshida, Y. and Hasunuma, K. (2010). ROS play an important role in the plant growth and production in ROS-resistant *Pisum sativum* cv. Alaska. *Planta*, 232, 367-382.

[4] Errington, J. (1992). *Bacillus subtilis* sporulation: a paradigm for the spacial and temporal control of gene expression. Chapter 2, pp. 28-44. in *Development. The molecular genetic approach.* Edds. by V. E. A. Russo, S. Brody, D. Cove and S. Ottolenghi. *Springer-Verlag. Berlin.*

[5] Yoshida, Y., Ogura, Y. and Hasunuma, K. (2006). Interaction of nucleoside diphosphate kinase and catalases for stress and light responses in *Neurospora crassa. FEBS Letters* 580, 3282-3286.

[6] Fukamatsu, Y., Yabe, N. and Hasunuma, K. (2003). *Arabidopsis* NDK-1 is a component of ROS signaling by interacting with three catalases. *Plant Cell Physiol.* 44, 982-989.

[7] Wang, N., Yoshida, Y. and Hasunuma, K. (2007). Catalase-1 (CAT-1) and nucleoside diphosphate kinase-1 (NDK-1) play an important role in protecting conidial viability under light stress in *Neurospora crassa*. *Mol. Genet. Genomics,* 278, 235-242.

[8] Hasunuma, K., Yoshida, Y., Haque, M., Wang, N., Fukamatsu, Y., Miyoshi, O. and Lee, B. (2011). Global warming, plant paraquat resistance, and light signal transduction through nucleoside diphosphate kinase as a paradigm for increasing food supply. Naunyn-Schmidberg's *Arch. Pharmacol.* 384 (4-5), 391-395.

[9] Hennessey, T. L. and Field, C. B. (1991). Circaddian Rythms in Photosynthesis, Oscillation in carbon assimilation and stomatal conductance under constant conditions. *Plant Physiol.* 96, 831-836.

[10] Yoshida, Y. Iigusa, H. Wang, N. and Hasunuma, K. (2011). Cross-talk between the cellular redox state and circadian system in *Neurospora. PloS ONE* 6(11): e28227, 1-11.

[11] Mizno, T. (2004). Plant response regulators implicated in signal transduction and circadian rhythm. *Current opinion in plant biology,* 7, 1-7.
[12] Lee, B., Yoshida, Y. and Hasunuma, K. (2005). Photomorphogenetic characteristics are severely affected in nucleoside diphosphate kinase-1 (*ndk-1*)-disrupted mutants in *Neurospora crassa*. *Mol. Genet. Genomics* 275, 9-17.

INDEX

A

B

C

D

E

F

G

H

I

J

K

L

M

N

O

P

Q

R

S

T

U

V

W

X

Y

Z